AF389342

# Mathematical Modelling of Energy Systems and Fluid Machinery

# Mathematical Modelling of Energy Systems and Fluid Machinery

Editors

**Mirko Morini**
**Michele Pinelli**

MDPI • Basel • Beijing • Wuhan • Barcelona • Belgrade • Manchester • Tokyo • Cluj • Tianjin

*Editors*
Mirko Morini
University of Parma,
Department of Engineering
and Architecture
Italy

Michele Pinelli
Department of Engineering,
University of Ferrara
Italy

*Editorial Office*
MDPI
St. Alban-Anlage 66
4052 Basel, Switzerland

This is a reprint of articles from the Special Issue published online in the open access journal *Energies* (ISSN 1996-1073) (available at: https://www.mdpi.com/journal/energies/special_issues/ Mathematical_Modelling_of_Energy_Systems_and_Fluid_Machinery).

For citation purposes, cite each article independently as indicated on the article page online and as indicated below:

LastName, A.A.; LastName, B.B.; LastName, C.C. Article Title. *Journal Name* **Year**, *Volume Number*, Page Range.

ISBN 978-3-0365-0550-3 (Hbk)
ISBN 978-3-0365-0551-0 (PDF)

© 2021 by the authors. Articles in this book are Open Access and distributed under the Creative Commons Attribution (CC BY) license, which allows users to download, copy and build upon published articles, as long as the author and publisher are properly credited, which ensures maximum dissemination and a wider impact of our publications.

The book as a whole is distributed by MDPI under the terms and conditions of the Creative Commons license CC BY-NC-ND.

# Contents

# About the Editors

**Mirko Morini** graduated with a Bachelor of Science in Material Engineering in 2003, with a Master of Science in Material Engineering in 2004, and with a PhD in Industrial Engineering at the University of Ferrara in 2008. During his PhD, he had an internship at Alstom (now Ansaldo Energia) in Baden (CH). He is currently Associate Professor in Energy Systems and Fluid Machinery at the Department of Engineering and Architecture of the University of Parma. His main research activities have focused on (i) smart district energy, (ii) energy and economic analyses of energy chains based on biomass and biofuels, (iii) micro-CHP systems analysis, (iv) dynamic models for the simulation of turbomachines, (v) experimental investigation of turbomachine instabilities, (vi) analysis of the effect of blade deterioration on turbomachine performance, and (vii) analysis of start-up transients of heavy-duty gas turbines for cost-killing and reliability enhancement. This activity is documented by more than 100 scientific articles.

**Michele Pinelli** received his MSc Degree in Mechanical Engineering at the University of Bologna in 1997 and his PhD in 2001. Since 2010, he has been Associate Professor of Fluid Machinery at the Engineering Department in Ferrara. His research activity has been carried out in the field of fluid machinery and energy systems and recently has mainly dealt with the development of techniques for the numerical estimation of turbine and compressor behavior in deteriorated conditions (fouling and erosion), the development of a small-scale compressor test rig for the analysis of transient behavior, water ingestion, stall and surge conditions of small-scale compressors, and the study of innovative energy systems in micro-cogeneration applications, with particular attention to thermophotovoltaics and organic Rankine Cycle systems. His research activity is documented by more than 200 scientific articles. He received the Best Paper Award of the Oil and Gas Committee at the ASME Turbo Expo in Barcelona 2006, Montreal 2007, Orlando 2009, and Montreal 2015. He is the coordinator of a research group comprising PhD students and research fellows. He is also responsible for research collaborations with Imperial College of London (UK), St. John's College of Oxford (UK), University of Bath (UK), Institut für Luftfahrtantriebe—Universität Stuttgart (D), Southwest Research Institute (SwRI) (US), and Solar Turbines (US).

# Preface to "Mathematical Modelling of Energy Systems and Fluid Machinery"

The ongoing digitalization of the energy sector, which will make a large amount of data available, should not be viewed as a passive ICT application for energy technology or a threat to thermodynamics and fluid dynamics, in the light of the competition triggered by data mining and machine learning techniques. Digitalization creates opportunities, for example, for more sustainable energy systems through the smart management of renewable energy technologies and for more reliable fluid machines through predictive maintenance. Nevertheless, this can only be achieved if these new ICT technologies are posed on solid bases for the representation of energy systems and fluid machinery. Therefore, mathematical modelling is still relevant and its importance cannot be underestimated. The aim of this Special Issue was to collect contributions about mathematical modelling of energy systems and fluid machinery in order to build and consolidate the base of this knowledge. In this Special Issue, we collected papers dealing with many aspects of modelling techniques, from the basics of model development (e.g., problem simplification and translation, model implementation, parameter identification, and model validation) to their applications (e.g., models for optimization, 3D CFD for component design, CAE models). In this context, a relevant number of fluid machinery (e.g., axial and radial pumps, hydroturbines, turboexpanders) and energy system (e.g., thermal energy storage, refrigeration systems) typologies have been taken into consideration and studied by means of the above-mentioned models.

**Mirko Morini, Michele Pinelli**
*Editors*

*Article*

# Development and Analysis of a Multi-Node Dynamic Model for the Simulation of Stratified Thermal Energy Storage

Nora Cadau [1], Andrea De Lorenzi [1], Agostino Gambarotta [1,2], Mirko Morini [2,*] and Michele Rossi [2]

[1] CIDEA—Center for Energy and Environment, University of Parma, Parco Area delle Scienze 42/a, 43125 Parma, Italy; nora.cadau@unipr.it (N.C.); andrea.delorenzi@unipr.it (A.D.L.); agostino.gambarotta@unipr.it (A.G.)
[2] Department of Engineering and Architecture, University of Parma, Parco Area delle Scienze 181/a, 43125 Parma, Italy; mrossi@siram.it
* Correspondence: mirko.morini@unipr.it

Received: 24 September 2019; Accepted: 7 November 2019; Published: 9 November 2019

**Abstract:** To overcome non-programmability issues that limit the market penetration of renewable energies, the use of thermal energy storage has become more and more significant in several applications where there is a need for decoupling between energy supply and demand. The aim of this paper is to present a multi-node physics-based model for the simulation of stratified thermal energy storage, which allows the required level of detail in temperature vertical distribution to be varied simply by choosing the number of nodes and their relative dimensions. Thanks to the chosen causality structure, this model can be implemented into a library of components for the dynamic simulation of smart energy systems. Hence, unlike most of the solutions proposed in the literature, thermal energy storage can be considered not only as a stand-alone component, but also as an important part of a more complex system. Moreover, the model behavior has been analyzed with reference to the experimental results from the literature. The results make it possible to conclude that the model is able to accurately predict the temperature distribution within a stratified storage tank typically used in a district heating network with limitations when dealing with small storage volumes and high flow rates.

**Keywords:** thermal energy storage; stratification; dynamic simulation; heating

---

## 1. Introduction

The continuous increase in the importance of the role of energy over the last few decades, as well as the rise in fuel prices and the need to limit greenhouse gas emissions, have led to a steady growth in the use of energy saving technologies and in a more effective and extensive implementation of renewable energy sources [1]. However, despite renewable energies representing one of the best alternatives to conventional sources—such as fossil or nuclear—for energy supply in most areas of the world, renewable energies are often hampered by their discontinuous nature during the day and by the actual availability of the source during the year.

As a matter of fact, energy is not always available where, when, and how it is required, and thus storage systems (both electrical and thermal) play an important role in energy management to make it available accordingly. Therefore, to improve the availability of renewable energies in remote geographical areas and to overcome their intermittent nature, thermal energy storage (TES) represents a fundamental solution to increase their competitiveness. Thanks to its capability to allow a more sustainable use of available resources [2] and to decouple thermal energy generation and use, there is currently growing interest in thermal energy storage.

A number of applications for thermal energy storage can be seen in energy grids, since they allow (i) an effective balance in energy supply vs. demand dynamics, (ii) a decrease in heating system energy losses, reducing the number of start-up and shut-down maneuvers and the need for backup plants, (iii) an increase in deliverable capacity (heating element generation plus storage capacity), (iv) a shift in energy purchases to lower cost periods, and (v) an increase in renewable energy source exploitation.

As an example, Barbieri et al. [3] showed the storage capability to decouple electrical and thermal power production in cogeneration plants, allowing thermal energy to be stored when only the electric energy is needed and vice-versa. It is therefore apparent that thermal energy storage represents a key solution in all applications in which energy supply and demand are decoupled, giving significant advantages in terms of cost and dispatchability of the generated energy. It can also be used to support generation by conventional energy sources [4] to deal with weekly, monthly, and annual changes in energy requirements and to help peak shaving [5]. Moreover, thermal energy storage represents a fundamental technique for the optimization of overall energy conversion, transmission, and utilization processes in smart energy systems [6]. Ultimately, it has a fundamental role in energy system control strategy development [7], such as Model-in-the-Loop applications, and implementation [8,9].

Several solutions have been proposed in the literature for thermal energy storage (e.g., through sensible heat, latent heat, and thermochemical energy) [10,11] involving the most recent areas of investigation and experimentation. A detailed analysis and a performance comparison of the existing thermal energy storage systems have been performed by Sarbu and Sebarchievici [12]. Furthermore, Noro et al. [13] focused their attention on the liquid sensible and phase change material heat storage systems, pointing out their respective pros and cons. However, because of the limitations and drawbacks of latent heat and thermochemical storage solutions—e.g., large volume variations and high storage medium costs, respectively—water is still the most widely diffused energy storage material since it is easily available, cheap, non-toxic, non-flammable, and completely harmless [14]. For these reasons, this study focuses on water sensible thermal energy storage.

A multi-node model has been developed, and the effects of node number and flow rate variations have been investigated alternatively. The model has been developed with a modular approach, by considering a standardized input/output causality structure for easy implementation into a library of components for the dynamic simulation of smart energy systems [6]. Unlike other simulation tools that are currently available [15,16], the present paper shows all the fundamental equations and their implementation in an explicit way allowing easy replicability and providing high-level customization features for both discretization and numerosity of inlet or outlet ports. Then the storage model behavior has been analyzed in charge and discharge conditions, taking the experimental data from the literature as a benchmark (i.e., the experimental model developed by González-Altozano et al. [17] and the experimental study carried out by Li et al. [18], respectively). Finally, the model has been applied to a real case in an operating environment.

*Literature Review*

Accurate surveys of mathematical simulation models of TES tanks were conducted by Njoku et al. [19], and Dumont et al. [20]. It follows that these models can be divided into three main categories:

1. 0D models, comprising analytical and fully-mixed approaches;
2. 1D models, including moving boundary, plug-flow and multi-node models;
3. 2D-3D models, containing multi-zone models and CFD techniques.

Starting from the 0D models, following the analytical approach, the storage tank can be modeled as a semi-infinite body, assuming that the inlet temperature and the mass flow rates are constant and not considering mixing or ambient heat losses. It should be noted that some attempts to relax these assumptions have been successfully performed in [21,22]. The fully-mixed hypothesis is still the simplest approach, since uniform temperature distribution is considered in the whole tank volume, assuming that all the incoming water mixes perfectly with the water already in the tank. Governing equations are based on mass and energy conservation [23].

Among the 1D models, the two-zone moving boundary approach is based on the assumption that the storage volume is divided into two zones with uniform temperature separated by a thermocline to prevent the hot and cold fluid from mixing and to maintain a stable temperature gradient [24]. In order to reduce the mathematical complexity of the analytical solutions, for the sake of practicability, an integral approximate approach has been successfully applied to the 1D storage tank models by Chung and Shin [25]. In the plug-flow method the tank volume is composed of a variable number of isothermal disks moving within the tank without any mixing between them [26]. Finally, the multi-node models are developed following a 1D finite-volume method that considers a uniform horizontal temperature in each layer (called a node). Therefore, a temperature gradient in the vertical direction only (i.e., stratification) and a 1D flow inside the tank [27,28] can be modeled. Using this 1D approach, González-Altozano et al. [17] have proposed a new methodology for the estimation of numerous temperature-dependent indices employed for the characterization of thermal stratification in water storage tanks (as temperature profile and thermocline thickness) during the charging phase, while Li et al. [18] have investigated the influence of several inlet structures on the stratification effectiveness and discharging performance of a tank.

Within the 2D-3D models, the zonal approach is a 3D finite-volume method with a coarse mesh based on mass and energy balances [29], instead the CFD [30,31] is a finite element method governed by mass, energy, and momentum conservation laws.

Each approach shows advantages and disadvantages depending on its specific application: the pros and cons of each model category are briefly summarized in Table 1. It is shown that the 2D-3D models allow for a more detailed simulation of physical systems compared to the 1D models, but with a higher computational burden because of the increase in dynamic states (resulting in a larger number of differential equations to be solved). Concerning the comparison between the reliability of different models, an insightful numerical study has been proposed by Cabelli [32], which states that, under particular circumstances, the temperature profile of the stratified fluid can be reasonably predicted with a simpler 1D model instead of a 2D model. Thus, in this paper specific attention has been paid to 1D models as they represent the best compromise between accuracy (which is related to the number of differential equations representing the model) and computational time (Table 1).

It is widely known that several physical phenomena occur in a water storage system that affect its performance. Among them, stratification has particularly raised the interest of researchers [27,28,33] for its strong influence on the thermal performance of plants [34,35].

**Table 1.** Pros and cons for each thermal energy storage (TES) tank modeling approach.

| Model Dimension | Pros | Cons |
|---|---|---|
| 0D | ✓ Straightforward implementation<br>✓ Low simulation time (lower than Real Time) | ✗ Accurate only for small volume storage tanks<br>✗ Complex phenomena (e.g., convection and conduction) cannot be considered |
| 1D | ✓ Modest implementation effort<br>✓ Model simplicity<br>✓ Limited simulation time (close to Real Time)<br>✓ Reduced number of model equations | ✗ Limited accuracy for high flow rates and complex geometries<br>✗ Not able to describe flow pattern in detail especially during fast-changing operating conditions |
| 2D-3D | ✓ More detailed description of flow pattern in storage volume<br>✓ Required for detailed storage system design | ✗ Difficult implementation<br>✗ Large number of model equations<br>✗ High simulation times<br>✗ High CPU power |

For instance, Van Koppen et al. [36] and Furbo and Mikkelsen [37] found that thermal stratification in solar heating systems allows a reduction in temperature at the collector inlet (which increases its efficiency) and leads to a limited operation of the auxiliary energy supply. Accordingly, stratification improves not only the water tank efficiency, but also that of the whole extended system.

Taking into account the scope of this work, the 1D approach has been chosen by focusing the attention on multi-node architecture. Even though several approaches for multi-node modeling of thermal energy storage tanks have been proposed in the literature (as reported above), in general they do not allow variations in the number and size of nodes considered in the simulation. To allow for a more flexible customization, a new multi-node model has been developed by the authors and is presented in this paper. Both the dimensions and number of nodes of the model can be set, focusing on the desired zone of the tank, by increasing the number of nodes in that specific area.

## 2. Materials and Methods

The model of the thermal energy storage developed by the authors falls under the multi-node 1D model category. The basic approach consists of dividing the tank into a number N of zones (called "nodes") where temperature is considered uniform ($T_i$). This assumption considers the vertical gradient of the temperature in the tank, while the horizontal gradient is neglected. When numbering the nodes from the top of the tank (Figure 1), the first node is the hottest and node "N" is the last and coldest.

As the developed model is physics-based, the involved equations arise from volume (1) and energy balances (2). Thus, the continuity equation and energy conservation equation are implemented for each node in the following form:

$$\frac{dV}{dt} = \sum \dot{V}(t) = 0 \tag{1}$$

$$\frac{dE}{dt} = \sum h(t){\cdot}\dot{m}(t) + \sum \dot{Q} \tag{2}$$

The energy Equation (2) involves enthalpy flows and heat exchanges not related to mass flows. The latter are due to heat transfer inside the tank between neighboring nodes, and to ambient losses through the tank walls because of temperature differences between the storage medium, tank walls, and the surrounding environment. The continuity and energy balance equations are implemented for each node, considering enthalpy flows, heat transfers between adjacent nodes and ambient losses. A schematic representation of the generic node is shown in Figure 2. The governing dynamic equations for the ith node are ordinary differential equations representing volume (3) and energy balances (4):

$$\dot{V}_{vert_{out}} = \frac{\dot{m}_{in}}{\rho(T_{in})} + \dot{V}_{vert_{in}} - \frac{\dot{m}_{out}}{\rho(T_i)} \tag{3}$$

$$\begin{aligned} M{\cdot}c{\cdot}\frac{dT_i}{dt} = {}& \sum \dot{m}_{in}{\cdot}c{\cdot}T_{in} - \sum \dot{m}_{out}{\cdot}c{\cdot}T_i \\ & -\dot{V}_{vert_{out}}{\cdot}\rho(T_{vert_{out}}){\cdot}c{\cdot}T_{vert_{out}} + \dot{V}_{vert_{in}}{\cdot}\rho(T_{vert_{in}}){\cdot}c{\cdot}T_{vert_{in}} \\ & +k{\cdot}(T_{i-1} - T_i) - k{\cdot}(T_i - T_{i+1}) - U{\cdot}A{\cdot}(T_i - T_{amb}) \end{aligned} \tag{4}$$

The water density varies with the temperature, and it is thus assumed constant in each node. It must also be specified that $\dot{V}_{vert_{in}}$ and $\dot{V}_{vert_{out}}$ are positive when water is flowing downwards. Accordingly, the sign of $T_{vert_{in}}$ and $T_{vert_{out}}$ is defined by the following relationships:

$$T_{vert_{out}} = \begin{cases} T_i & \text{if } \dot{V}_{vert_{out}} > 0 \\ T_{i+1} & \text{if } \dot{V}_{vert_{out}} < 0 \end{cases} \tag{5}$$

$$T_{vert_{in}} = \begin{cases} T_{i-1} & \text{if } \dot{V}_{vert_{in}} > 0 \\ T_i & \text{if } \dot{V}_{vert_{in}} < 0 \end{cases} \tag{6}$$

The energy conservation equation was applied to each node in order to predict its time-temperature history taking into account the thermal losses both to the surroundings and to adjacent nodes because of the convection and conduction. The latter is represented in Equation (4) by the terms $k \cdot (T_{i-1} - T_i)$ and $k \cdot (T_i - T_{i+1})$, which can be appointed as "pseudo-conduction terms" since they are used to represent the thermal exchange that would take place between bordering nodes by convection.

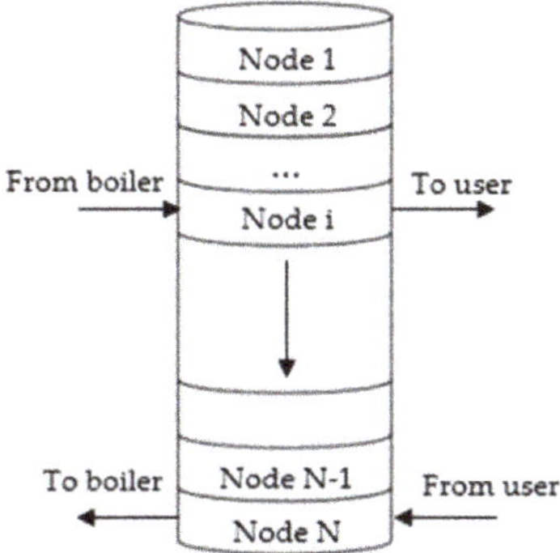



**Figure 1.** Liquid storage tank schematic representation for the multi-node modeling approach.

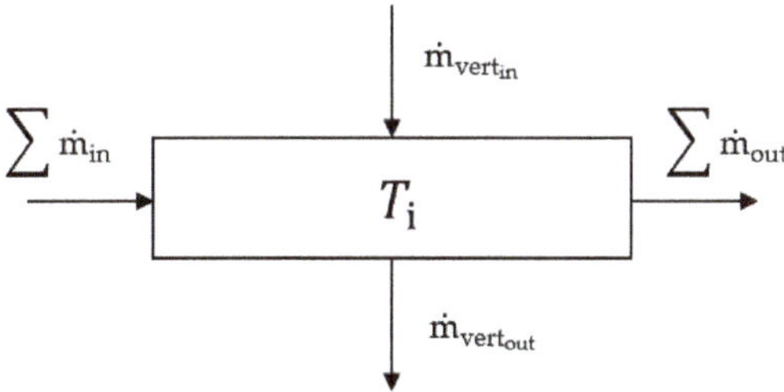



**Figure 2.** Generic node schematic representation.

This is a simplified way to consider these contributions (as suggested in [17,28]), but it has been noted by a preliminary analysis, which is not reported herein for the sake of brevity, that their influence on the global energy balance could be neglected. For this reason, this is an approach which is commonly used when dealing with multi-node models [15,16]. The terms $\dot{V}_{vert_{out}} \cdot \rho(T_{vert_{out}}) \cdot c \cdot T_{vert_{out}}$ and $\dot{V}_{vert_{in}} \cdot \rho(T_{vert_{in}}) \cdot c \cdot T_{vert_{in}}$ represent the energy associated with the incoming and outgoing flows among neighboring nodes and can be considered as "transport terms."

*Model Implementation*

The model was implemented in Matlab®/Simulink® (The MathWorks Inc., Natick, MA, USA) one of the most widely used proprietary calculation software systems, to estimate mass flow rates, temperature, and pressure in each node (respecting causal coupling with physical models of other components).

By considering the typical causality of state-determined systems, the inputs to the storage model are as follows:

- Incoming mass flow rates and their related temperatures for each node;
- Outgoing mass flow rates for each node but one (orange arrow in Figure 3) that is calculated by the model through the mass conservation equation.

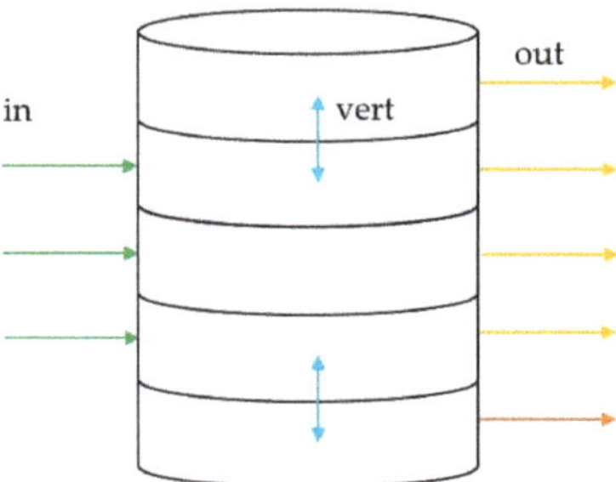



**Figure 3.** Schematic representation of nodal approach for stratified storage tank, showing incoming, outgoing, and vertical mass flows.

The entry node for each incoming flow is calculated through a dedicated algorithm which, by comparing the incoming flow temperature with the temperature of the ith node, directs the flow toward the next node with the same or a higher temperature. Therefore, the storage model should consist of at least two nodes.

A mask was created which allows the user to set the general properties of the tank such as the geometrical characteristics (height, thickness, and diameter), specifications of the insulation materials, fluid properties (i.e., thermal conductivity and initial temperature), number and dimension of nodes, and convection properties.

The model is composed of two macro sections: (i) one aimed at the calculation of thermal and geometrical characteristics (i.e., overall heat loss coefficient U, area A and conduction constant k) and (ii) one intended to solve the node balance equations.

For a proper simulation, each node sub model needs several inputs, as shown in Figure 4, including the temperature of both the previous and following nodes ($T_{i-1}$ and $T_{i+1}$). Some parameters are derived from the model dialogue mask ($U_i$, $A_i$, $k_i$), some are boundary terms equal for all nodes (i.e., $T_{amb}$), and others come from the previous and following nodes. The node outputs are the water temperature ($T_i$), the vertical mass flow rate toward the next node and where it exists, the mass flow rate which has not entered any node and is directed to the next one ($\dot{m}_{i+1}$). The model gives the temperature, the outgoing mass flow rate from the last node (bottom of the tank) and the outcoming pressure as the output.

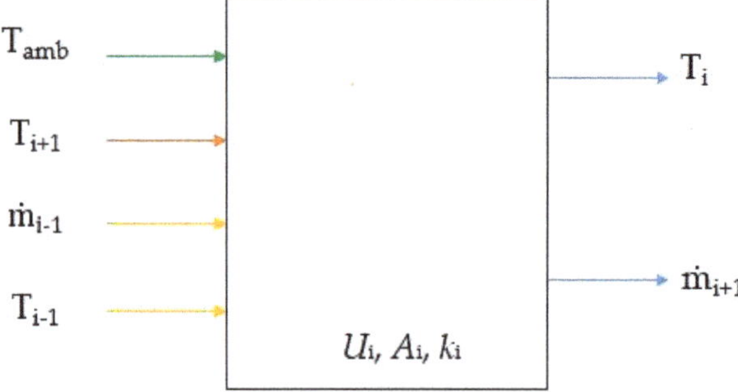



**Figure 4.** Main input and output parameters for ith node.

## 3. Results

Simulations have been conducted on a single heat storage tank both in the charge and discharge phase in order to investigate the effects of overall node number on time-temperature profiles, both during charge and discharge, and the effect of the flow rates only during discharge.

The TES tank model behavior was analyzed by comparing it with the experimental data found in the literature regarding reproducible fully-instrumented laboratory tests, i.e., given by González-Altozano et al. [17] for charging and Li et al. [18] for discharging. Then the model was applied to the simulation of a large storage tank monitored in an operating environment [38].

For all the simulation sets, a variable-step ode45 (Dormand-Prince) solver was used, since it represents a good compromise between calculation time and results accuracy.

### 3.1. Model Analysis—Charge Phase

First, the charge maneuver of a thermal energy storage tank was simulated. The tank, initially at low temperature, was heated up with a hot water flow fed at the top of the tank. A grid sensitivity analysis on node number and dimension was performed.

#### 3.1.1. Settings

According to [17], a cylindrical storage tank with a height equal to 1800 mm, an internal diameter equal to 800 mm, and a resulting volume of 0.9 m$^3$ was considered. The side walls, the top, and the bottom of the tank were insulated with a 50-mm-thick layer of fiberglass, with thermal conductivity equal to 0.043 W/m·K. The initial temperature of the water was set to 20 °C for the whole tank and a constant volume flow rate of 16 dm$^3$/min at 52 °C was introduced at the top of the tank, with a conventional inlet elbow. A summary of the measurement devices is given in Table 2.

**Table 2.** Measurement devices used for the charge experimental trial.

| Quantity | Instrument | Task |
| --- | --- | --- |
| 12 | Thermocouples (Type T—Class 1) (error < ±0.2%) | Measure the water temperature in the tank |
| 2 | Thermocouples (Type T—Class 1) (error < ±0.2%) | Measure the water inlet and outlet temperature during the charge process |
| 1 | Electromagnetic flowmeter Endress Hauser—mod. 53H08 DN8 (error < ± 0.1%) | Measure the water inlet mass flow rate |
| 1 | Data acquisition system National Instruments—mod. DAQ 9178 | Collect sensor data |
| 1 | Thermocouples input module National Instruments—mod. 9213 | TC signal conditioning and acquisition |
| 1 | Flowmeter input module National Instruments—mod. 9208 | Flow-meter signal acquisition |

González-Altozano et al. [17] placed twelve thermocouples uniformly spaced along the vertical axis of the tank, located 150 mm apart and 75 mm from both the top and the bottom of the tank, as represented in Figure 5. Therefore, N = 12 was set as the reference number for the nodes in the model.

The simulation time has been fixed at 4073 s, equal to the time declared by González-Altozano et al. [17] to replace 120% of the total storage tank volume.

#### 3.1.2. Sensitivity Analysis

Simulations have been conducted in three significant cases: for the reference number of nodes (N = 12), for twice the reference number of nodes (2·N) and for half the reference number of nodes (N/2). The trend of the temperature evolution against time during charging simulation for a storage tank model with 6 (Figure 6), 12 (Figure 7), and 24 nodes (Figure 8) has been recorded and compared to the experimental [17] time-temperature evolution.

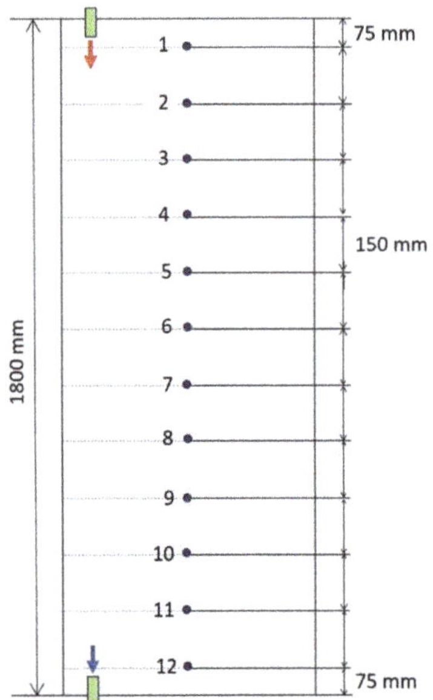


**Figure 5.** Schematic representation of the experimental tank for the charge phase [9]. The hot water inlet (red arrow) and the cold-water outlet (blue arrow) are placed at the top and the bottom of the tank, respectively.

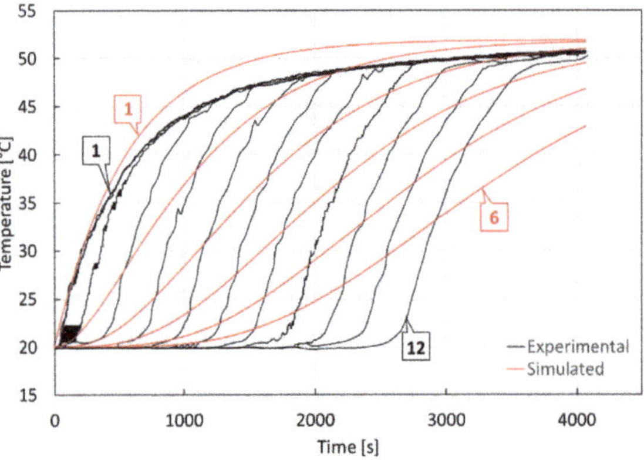


**Figure 6.** Temperature evolution during the charge phase—experimental vs. simulated (6 nodes).

As a first remark, it may be noted that simulated and experimental values of replacement time (i.e., the times required to replace the total water mass in the tank) are comparable. The results show that when the number of nodes N is increased, the thermal dynamics speeds up and the replacement time decreases toward the experimental value. Specifically, the dynamics of the upper nodes becomes faster whereas the dynamics related to the middle and the lower nodes slows down slightly in the first part—until reaching approximately 40 °C—and then increases, giving a closer match to the overall time-temperature evolution of the experiment. This latter phenomenon may be due to the thermal inertia of the upper nodes. This feature could be better appreciated in Figure 9, where for the sake of readability only the comparison of the first, middle, and last node of the three different models is shown, based on the number of nodes.

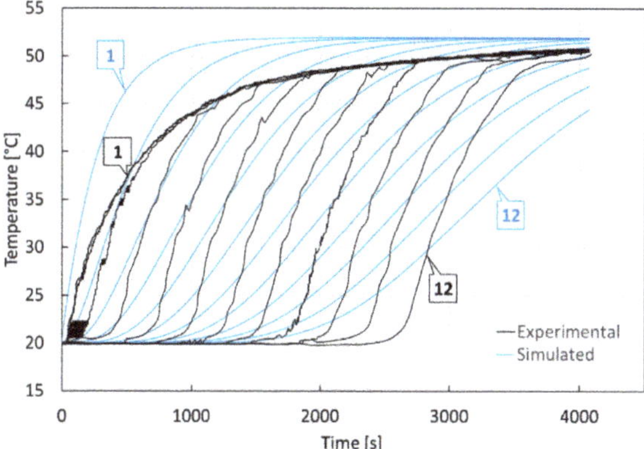

**Figure 7.** Temperature evolution during the charge phase—experimental vs. simulated (12 nodes).

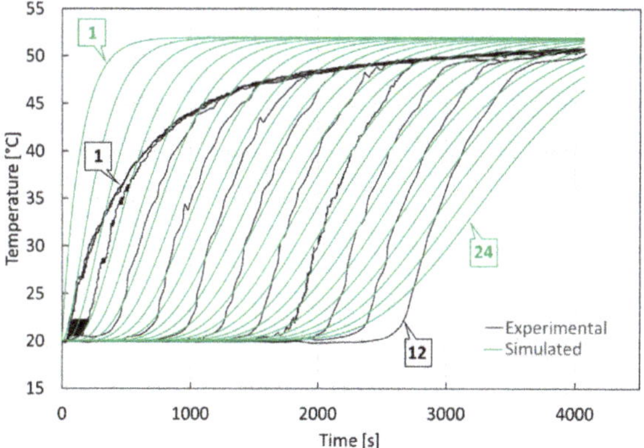

**Figure 8.** Temperature evolution during the charge phase—experimental vs. simulated (24 nodes).

Thus, on the one hand, an increase in the number of nodes leads to faster thermal dynamics and to an improvement in accuracy of the replacement time. However, on the other hand, this highlights a specific behavior of the model. In fact, the larger the volume of a node (i.e., the lower the total node number N), the larger the mass of water inside the node and the more significant the damping effect of the incoming flow temperature. This is the case of the simulation with 6 nodes where the temperature in the first node seems to better fit the experimental temperature. This may be due to the upper zone of the storage being turbulently mixed. In fact, the temperature measured by the three thermocouples at the top are almost superimposed. On the contrary, a small number of nodes would be less beneficial when large incoming flow rates (i.e., high speed of the incoming water) are involved and there are no mixing devices—i.e., diffusers—at the inlet. In this case the incoming fluid would flow very quickly through the storage directly to the outgoing zone and without exchanging thermal energy with the crossed nodes. This operating condition cannot be described by the model because of the 1D approach used.

Thus, since the number of nodes N plays a significant role, some nodes have been split to better point out the influence that node volume plays on model accuracy. Three nodes—representative of the heights where the corresponding thermocouples are located—were chosen as reference nodes (i.e., the third, sixth, and ninth) for this purpose. Each of them was split into ten nodes, as represented in Figure 10.

The temperature evolution of each reference node after splitting was recorded and compared to those related to the experimental data and simulated results from the models with six, twelve, and twenty-four nodes (Figures 11–13). The results show that by decreasing the volume of nodes and focusing the attention on a specific zone by refining the local 1D "mesh," the calculated temperature evolution matches the measurements better and better, especially for nodes belonging to the upper part of the storage tank. This is due to the fact that the lower nodes are negatively influenced by the dynamics of the upper ones.

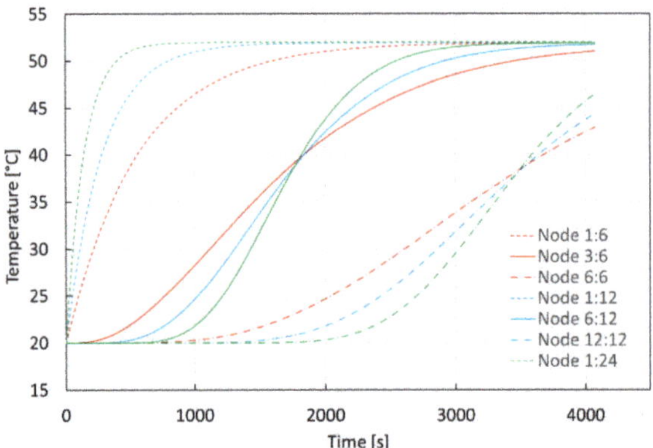


**Figure 9.** Temperature evolution during the charge phase from models with a different number of nodes.

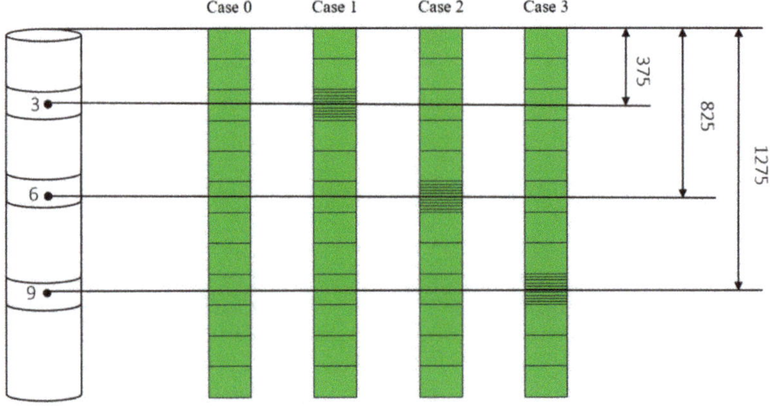


**Figure 10.** Reference node splitting for the investigation of the influence that node volume plays on model accuracy (heights in millimeters).

In conclusion, the choice of the number of nodes N determines the resolution with which the vertical temperature distribution can be modeled in the storage tank. In fact, an increase in the number of nodes will allow significant temperature gradients to be modeled more accurately (Figure 14). Thanks to versatility and accurate physical representation of stratification and heat exchanges, it seems that the model can be a useful simulation tool for the reliable prediction of temperature evolution in a stratified storage tank during charge operation.

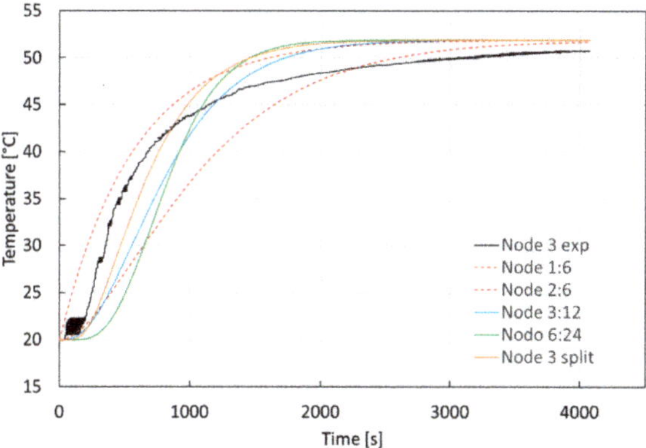

**Figure 11.** Experimental temperature evolution during the charge phase at node #3 vs. simulation results.

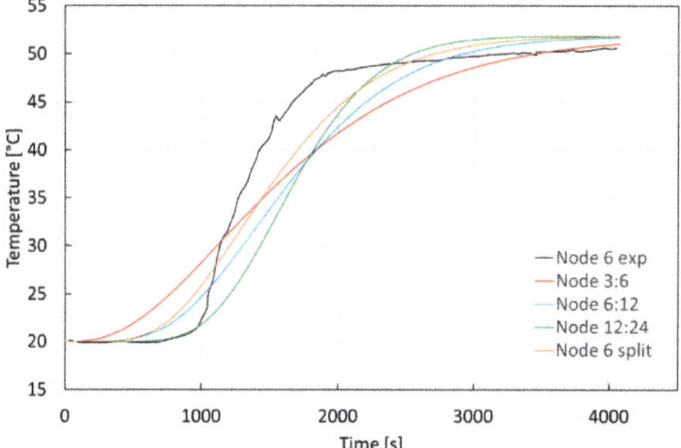

**Figure 12.** Experimental temperature evolution during the charge phase at node #6 vs. simulation results.

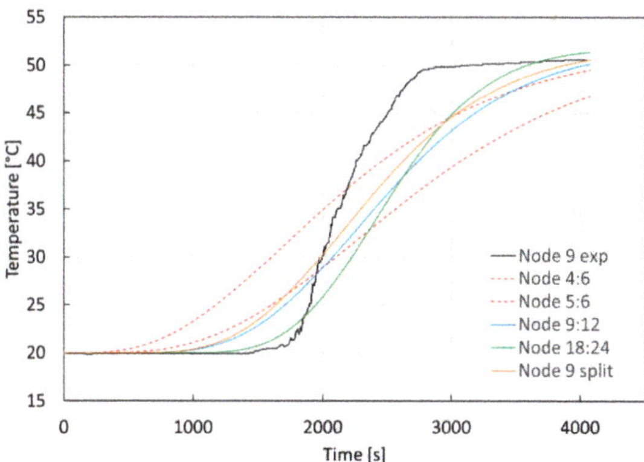

**Figure 13.** Experimental temperature evolution during the charge phase at node #9 vs. simulation results.

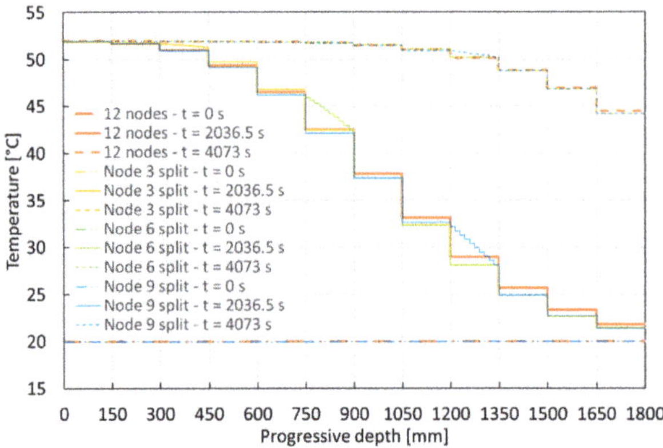



**Figure 14.** Temperature stratification (during the charge phase) inside the storage tank at three different simulation stages.

*3.2. Model Analysis—Discharge Phase*

Thereafter, the discharge maneuver of a thermal energy storage tank was simulated. The tank, initially at high temperature, was cooled down with the introduction of a cold-water flow. The effects of the node number and flow rate variations on temperature evolution were investigated.

3.2.1. Settings

According to [18], a cylindrical storage tank with a height of 800 mm and an internal diameter of 400 mm (resulting in a volume of 0.1 m$^3$) was modeled. Three different flow rates (5 dm$^3$/min, 10 dm$^3$/min, and 15 dm$^3$/min) at 15 °C were introduced—by means of a slotting-type inlet—at the bottom of the tank initially at 60 °C. The whole incoming flow rate exited the storage tank from the top. Li et al. [18] divided the water tank into eight layers of the same dimension fitted with one thermocouple each with a measurement time interval of 5 s. The first and the last thermocouples were located at 50 mm from the top and the bottom of the tank, respectively, whereas the intermediate ones are placed 100 mm apart (Figure 15). A summary of the measurement instrumentation is given in Table 3.

**Table 3.** Measurement devices used for the discharge experimental trial.

| Quantity | Instrument | Task |
| --- | --- | --- |
| 8 | Thermocouples (Type T—Class 1) | Measure the water temperature in the tank |
| 2 | Thermocouples (Type T – Class 1) (error < ±0.2 °C) | Measure the water inlet and outlet temperature during the discharge process |
| 1 | Glass rotor flowmeter | Measure the water inlet mass flow rate |
| 1 | Data acquisition system Agilent—mod. 34970A | Collect sensor data |
| 1 | Thermocouples input module National Instruments—mod. 9213 | TC signal conditioning and acquisition |

As the real tank was equipped with eight thermocouples, the model was set with a basic number of eight nodes. The simulation time was set equal to the unit replacement time, i.e., 1536 s, 768 s, and 512 s for a volume flow rate of 5 dm$^3$/min, 10 dm$^3$/min, and 15 dm$^3$/min, respectively.

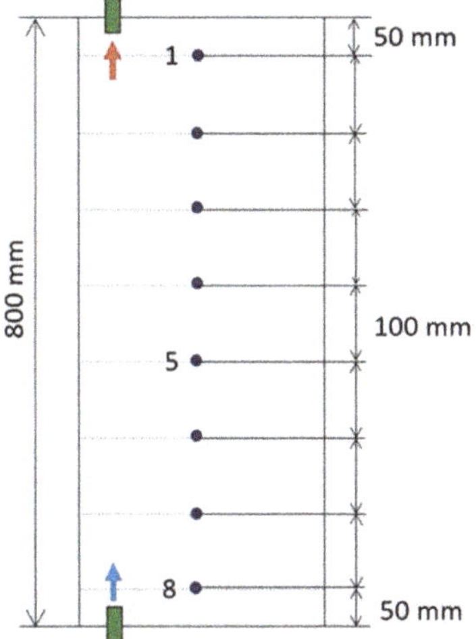

**Figure 15.** Schematic representation of the experimental tank for the discharge phase [10]. The cold-water inlet (blue arrow) and the hot water outlet (red arrow) are placed at the bottom and the top of the tank, respectively.

### 3.2.2. Sensitivity Analysis

Similarly to what was done for charging (Section 3.1), here again simulations were conducted in the three more significant cases: for the reference number of nodes (N = 8), for twice the reference number of nodes (2·N) and for half the reference number of nodes (N/2). In addition, for each specific number of nodes, the three different flow rates mentioned in the settings paragraph (Section 3.2.1) were introduced alternatively.

Finally, the fifth node—representative of the height at which the fifth thermocouple is located—was chosen as the reference node and the 100 mm above and below were split into ten nodes each (Figure 16). That choice is based on the fact that the first and the last nodes—where the first and the last thermocouples are located—are not far enough from the edges of the tank to allow the corresponding node to be split.

At the beginning, the temperature evolution trend of the first, middle, and last node was plotted and compared to the experimental time-temperature evolution [18] during discharging simulation for 5 dm³/min (Figure 17), 10 dm³/min (Figure 18), and 15 dm³/min (Figure 19) volume flow rates.

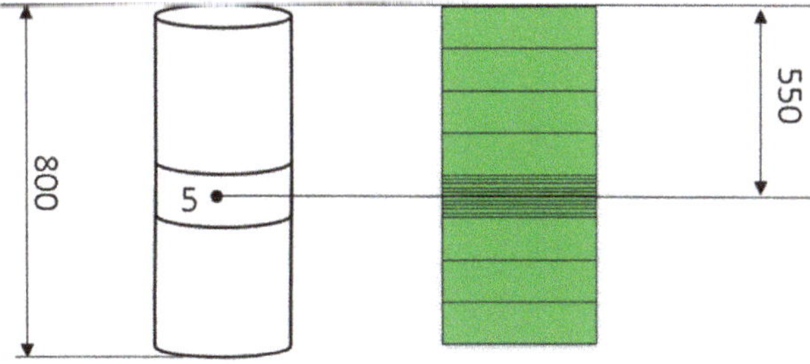

**Figure 16.** Reference node (#5) splitting for the investigation of the influence that node volume plays on model accuracy (heights in millimeters).

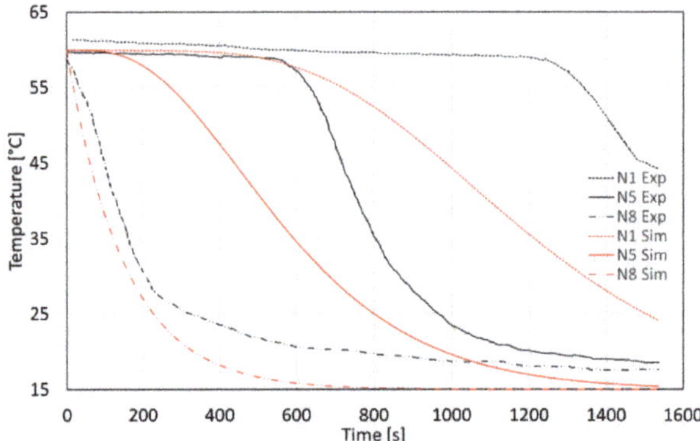

**Figure 17.** Temperature evolution during the discharge phase—experimental vs. simulated—5 dm$^3$/min volume flow rate.

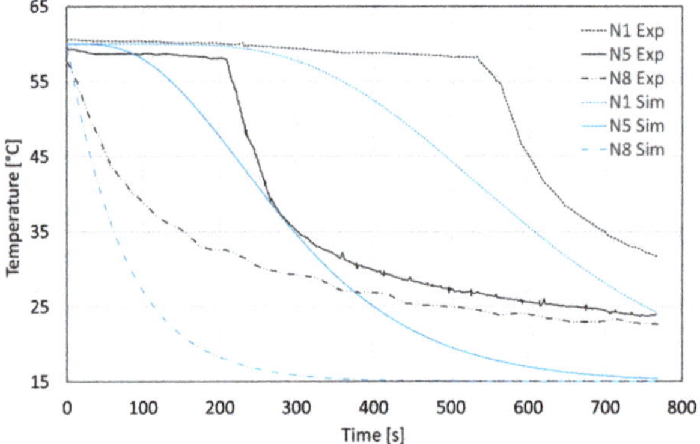

**Figure 18.** Temperature evolution during the discharge phase—experimental vs. simulated—10 dm$^3$/min volume flow rate.

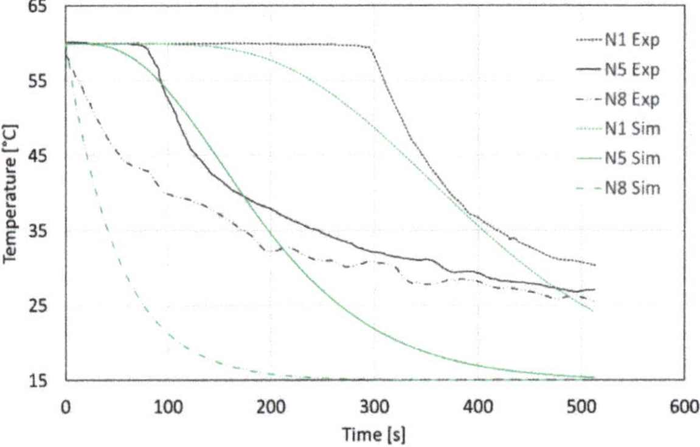

**Figure 19.** Temperature evolution during the discharge phase—experimental vs. simulated—15 dm$^3$/min volume flow rate.

Then, similarly to what has been done for charging, during discharging the calculated temperature evolution of the reference node after splitting has been plotted for the three simulations with different volume flow rates (see Section 3.2) and then compared to experimental data [18] and to the simulation with four, eight, and sixteen nodes (Figures 20–22).

A first analysis shows that unit replacement times obtained from all the simulations remain comparable to the experimental results.

It can be observed that an increase in the number of nodes brings the temperature evolution closer to the experimental data, especially for small volume flow rates. It should be noted that in this case the simulated storage is small (800 mm high and 400 mm in diameter) and node splitting does significantly improve the sixteen-node model, as the node number is already limited. This feature can be better appreciated by observing the green and the orange curves in Figures 20–22, which become closer and closer to each other as the incoming flow rate decreases.

To better investigate the effects of flow rate variations on temperature evolution, three different simulations were performed, each one keeping the number of nodes equal to that considered in the experimental model (N = 8).

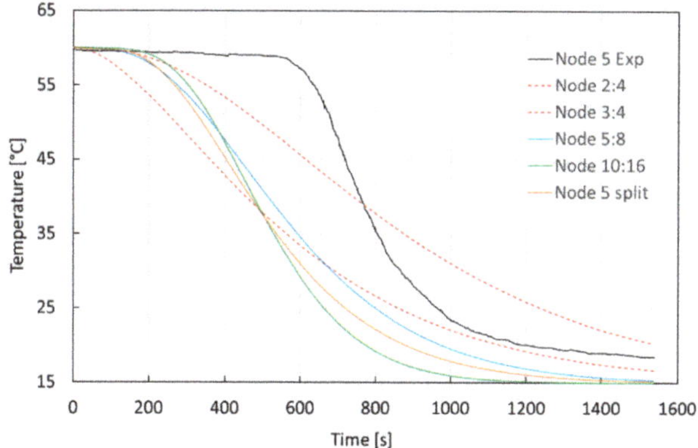

**Figure 20.** Experimental temperature evolution at node #5 vs. simulation results—5 dm$^3$/min volume flow rate—discharge phase.

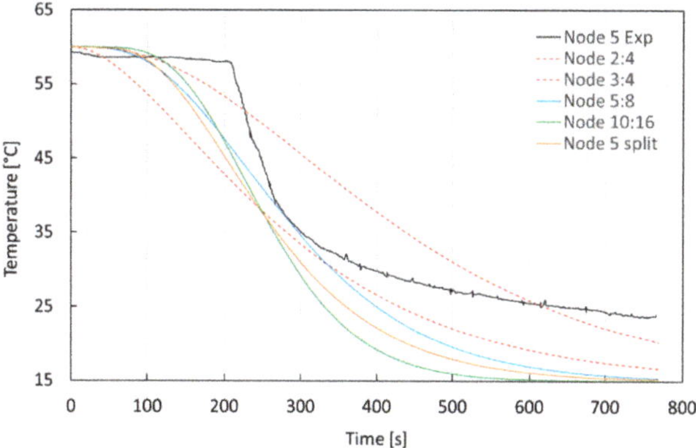

**Figure 21.** Experimental temperature evolution at node #5 vs. simulation results—10 dm$^3$/min volume flow rate—discharge phase.

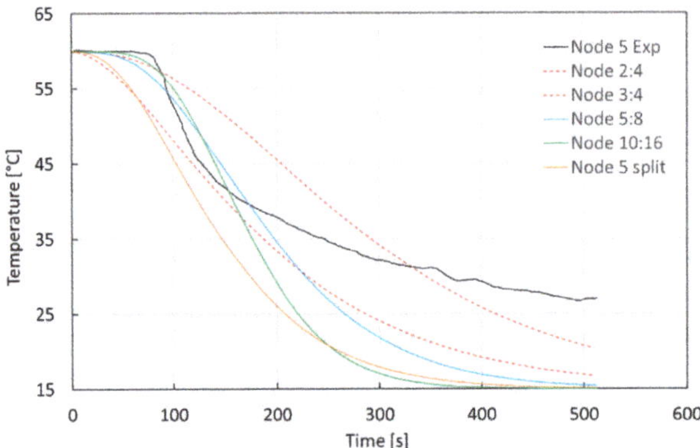

**Figure 22.** Experimental temperature evolution at node #5 vs. simulation results—15 dm$^3$/min volume flow rate—discharge phase.

When observing the temperature evolution of the tank for a flow rate of 5 dm$^3$/min (Figure 17), the time temperature evolution for node #8 is comparable to the experimental data. The mismatch slightly increases as it continues toward the upper nodes. It is clear that node #1, being the furthest from the fluid inlet, has a slower response to heat exchange because it is the last one to interact with the cold inlet (then its dynamics is affected by all the other nodes with which the heat exchange takes place beforehand). That trend is reversed with an increase in the inlet flow rate (Figures 18 and 19) in the sense that the higher the flow rate the better the curve representing node #1 fits the experimental data and vice versa for those representing the lower nodes (nodes #5 and #8). As a matter of fact, by observing Figures 18 and 19 compared to the above-mentioned Figure 17, it can be noted that the dynamic behavior becomes faster and faster (as a greater mass of fluid exchanges heat with the water mass in the tank). The temperature at node #8 takes around 300 s for a 15 dm$^3$/min volume flow rate, about 400 s and 800 s for a 10 dm$^3$/min and 5 dm$^3$/min volume flow rate, respectively, to reach the injected flow temperature (i.e., 15 °C).

Looking at the experimental curves, even after the unit replacement time (i.e., the time required to replace the whole water mass in the tank) has passed, the water temperature does not reach the temperature of the incoming flow (i.e., 15 °C). This behavior might be due to the fact that—given the experimental storage size—for high flow rates, part of the incoming fluid can be directed straight to the exit duct, and it has no time to exchange heat with the water in the tank. Thus, the water remains at a higher temperature (around 25 °C). In other words, a fraction of the inlet flow rate is bypassed, and its thermal energy is not stored in the tank.

The above-mentioned temperature behavior is not present in the results given by the simulations, as the approach followed to build up the model does not allow the flow bypass event to be considered (i.e., the entire incoming flow passes through all nodes without exchanging heat).

From these simulations it is apparent that the model results in terms of temperature changes inside the tank show deviations from the experimental data when the 3D effects become more and more significant (i.e., when the size of the tank becomes smaller and fluid velocity becomes higher). However, it should be noted that the proposed model can satisfactorily describe the physical behavior of the component within a complex energy grid, without unacceptable increases in the calculation burden.

This work points out further topics for the future development of the model. For high volume flow rates (compared to the size of the tank), the nodes furthest from the inlet section are less affected by the dynamics of the intermediate nodes. Moreover, the global heat exchange is faster, as a greater mass of fluid exchanges energy with the water in the tank.

Finally, referring to the number of nodes, it can be concluded that for small-size storage tanks the node-splitting technique does not significantly improve the accuracy of the temperature evolution inside the tank.

### 3.3. Model Application

In the last case study, the real daily operation of the upper region of a large atmospheric two-zone heat storage tank was investigated [38]. The aim of this last trial is to examine how the model behaves when representing the real operation of a large Thermal Energy Storage tank.

### 3.3.1. Settings

Unlike the previous cases, the tank in question is divided into two regions by an insulated intermediate floor. A vertical open-ended pipe serves as a hydraulic connection between the upper and the lower zone: it compensates the water density changes in the lower region and it prevents the whole tank from under-pressure or over-pressure. Furthermore, the intermediate floor—acting like an obstacle to natural convection—allows hot water in the lower zone to be stored at higher pressure if compared to an ordinary TES tank.

Since the simulation of a two-zone heat storage tank falls outside the scope of this work, only the upper region of the tank has been modeled: according to [38], it is a cylindrical tank with a height of 30 m and an internal diameter of 20 m (resulting in a volume of 9420 m$^3$). A constant volume flow of 50 m$^3$/h at 95 °C was introduced throughout the day by means of a radial diffuser placed at the top of the tank. The same inverse mass flow exited the storage tank from the bottom outlet. In order to evaluate the thermal stratification, the tank was equipped with a distributed temperature sensing (DTS) measurement system (resolution of about 0.1 °C) and 28 vertically aligned PT100 sensors.

### 3.3.2. Simulation Results

The tank model is made up of 25 evenly spaced nodes. Initial conditions were set on the basis of the information given in [38]. For ease of comparison, the water temperature daily evolution was recorded for seven specific tank heights, at four-hour intervals. The simulation was performed by means of the Matlab$^®$ ode45 variable-step solver, which took 0.71 s to simulate the daily operation of the storage tank, with a standard laptop.

The simulation result and absolute error are shown in Table 4, together with the experimental temperatures mentioned in the reference article.

The simulated temperature profile is in good accordance with the experimental data both for the top and the bottom of the considered region; the average absolute error is equal to 1 °C. It should be observed that a slight increase in the absolute error occurs over time at the half-height of the tank region; it is probably due to the mixing of chilled water—entering the upper zone at the temperature of about 62 °C—from the compensation pipe, as reported in [38]. Since the interaction between the two zones of the storage tank is not considered in the proposed model, the simulated temperature gradients seem to be smaller than the experimental one.

Even considering this issue, the model has proved to be performant in the representation of the real operation of a large heat storage tank: the maximum absolute error does not exceed 6 °C.

**Table 4.** Comparative table of experimental and simulated daily temperature evolution inside the TES tank.

| | Temperature (°C) | | | | | | | | | | | | | | | | | | | | |
|---|---|---|---|---|---|---|---|---|---|---|---|---|---|---|---|---|---|---|---|---|---|
| Time (h) | Exp. | Sim. | Abs. | Exp. | Sim. | Abs. | Exp. | Sim. | Abs. | Exp. | Sim. | Abs. | Exp. | Sim. | Abs. | Exp. | Sim. | Abs. | Exp. | Sim. | Abs. |
| 0 | 52 | 52 | 0 | 52 | 52 | 0 | 53 | 53 | 0 | 54 | 54 | 0 | 75 | 71 | 4 | 91 | 89 | 2 | 99 | 99 | 0 |
| 4 | 52 | 52 | 0 | 52 | 52 | 0 | 53 | 53 | 0 | 54 | 54 | 0 | 82 | 77 | 5 | 91 | 90 | 1 | 99 | 99 | 0 |
| 8 | 52 | 51 | 1 | 52 | 52 | 0 | 54 | 53 | 0 | 54 | 54 | 0 | 86 | 81 | 5 | 91 | 91 | 0 | 99 | 98 | 1 |
| 12 | 52 | 51 | 1 | 52 | 52 | 0 | 54 | 53 | 0 | 54 | 55 | 1 | 87 | 83 | 4 | 93 | 91 | 2 | 99 | 98 | 1 |
| 16 | 53 | 51 | 1 | 53 | 53 | 0 | 54 | 54 | 0 | 55 | 57 | 2 | 87 | 85 | 2 | 93 | 92 | 1 | 99 | 98 | 1 |
| 20 | 53 | 52 | 1 | 53 | 53 | 0 | 54 | 54 | 0 | 55 | 59 | 4 | 87 | 86 | 1 | 93 | 92 | 1 | 99 | 98 | 1 |
| 24 | 53 | 52 | 1 | 53 | 53 | 0 | 54 | 54 | 0 | 56 | 62 | 6 | 87 | 88 | 1 | 93 | 93 | 0 | 99 | 97 | 2 |
| | 0 | | | 5 | | | 10 | | | 15 | | | 20 | | | 25 | | | 30 | | |
| | Height (m) | | | | | | | | | | | | | | | | | | | | |

Note: Exp. = experimental; Sim. = simulated; Abs. = absolute error.

## 4. Discussion

In this paper, a new Matlab®/Simulink® model for the simulation of stratified sensible heat storage systems was presented. The 1D model was built using the multi-node approach, solving the volume and energy balance equations for each node. Because of its innovative structure, the model is highly customizable in node number and dimension, enabling detailed investigation of the thermal stratification phenomenon.

Three experimental datasets from the literature were used as references for the model analysis and validation, by considering the charge phase, the discharge phase, and the nominal operation of heat storage tanks with different sizes.

In the case of the charge phase, the tank—initially at a low temperature—was heated up with the injection of hot-water at the top. This simulation was performed three times, by varying the number of nodes while maintaining all the other parameter values (such as tank dimension, inlet and outlet flow rates, and initial temperature of the stored water).

In the case of the discharge phase, the tank—initially at a warm temperature—was cooled down with the injection of cold-water at the bottom. In this case, both the number of nodes and the inlet flow rates were varied throughout the simulations, while maintaining the other parameter values. For a better comparison of the results, an appropriate point was chosen as reference; it corresponds to the location of the fifth thermocouple in the experimental system, it is far enough from the edges of the tank and it was matched with a specific node in every simulation.

In the real operation case, the tank was fed with hot water from the top radial diffuser. The daily temperature evolution was simulated by means of a model made up of 25 evenly spaced nodes. The simulated and experimental data were compared in order to evaluate the model performance when dealing with large storage tanks, commonly used in district heating applications.

The comparison between simulated and experimental data confirms that the choice of the number of nodes plays a significant role in the representation accuracy of thermal stratification inside the storage tank: an increase in the number of nodes—which, for a given tank capacity, corresponds to a decrease in node volume—improves the simulation results (i.e., temperature variations in the tank) allowing more accurate temperature gradients.

The model proposed in this paper proved to be able to give an accurate physical representation of stratification and heat exchange phenomena in sensible heat storage systems and can be a useful tool to reliably simulate temperature changes in stratified storage tanks. Its innovative features are flexibility and adaptability, which make it possible to choose the number and dimensions of each node in the model, allowing the user to focus the simulation on a specific zone of interest.

However, the model shows limitations for some specific storage configurations. If the ratio between the inlet flow rate and the node volume is too small, the thermal stratification dynamics slows down and the simulation becomes inaccurate; the larger the node volume, the greater the mass of water contained and the higher the mixing effect that dampens the incoming flow temperature. Furthermore, the simplified representation of convection can lead to discrepancies between the real and the simulated temperature evolutions, as is the case for the models reported in the literature. Another limitation was detected in small storage tank simulations; for high flow rates and large temperature differences between the incoming and the stored fluid, a fraction of the incoming fluid may flow directly to the exit duct without exchanging heat with the stored water. In other words, part of the fluid is bypassed, and its thermal energy is not stored. The developed model is not able to reproduce this phenomenon as—due to the 1D approach followed—all the incoming flows pass through all the nodes.

It should be recalled that the stratified storage model was developed to become part of a library of physics-based components for the dynamic simulation of district heating networks. In these applications the involved storage tanks are large and the temperature differences between the incoming and stored water temperature are usually fairly low, and therefore the above-mentioned drawbacks of the proposed 1D model are acceptable, compared to the advantage of keeping low calculation times.

**Author Contributions:** Conceptualization, A.G., M.M. and M.R.; funding acquisition and supervision, A.G. and M.M.; investigation, validation and writing, N.C. and A.D.L.

**Funding:** This work was supported by the "Efficity—Efficient Energy Systems for Smart Urban Districts" project (CUP E38I16000130007) co-funded by the Emilia-Romagna Region through the European Regional Development fund 2014–2020.

**Conflicts of Interest:** The authors declare no conflict of interest.

## Nomenclature

| | | |
|---|---|---|
| A | $m^2$ | area |
| E | J | energy |
| M | Kg | mass |
| $\dot{Q}$ | W | heat flow |
| T | °C | temperature |
| U | $W/(m^2 \cdot K)$ | overall heat exchange coefficient |
| $\dot{V}$ | $m^3/s$ | volume flow |
| c | $J/(kg \cdot K)$ | specific heat |
| h | $J/(kg \cdot K)$ | specific enthalpy |
| k | W/K | conduction constant |
| $\dot{m}$ | kg/s | mass flow rate |
| t | s | time |

## Subscripts and Superscripts

| | |
|---|---|
| amb. | ambient |
| i | index |
| in | incoming |
| out | outgoing |
| vert | vertical |

## References

1. Sharma, A.; Tyagi, V.V.; Chen, C.R.; Buddhi, D. Review on thermal energy storage with phase change materials and applications. *Renew. Sustain. Energy Rev.* **2009**, *13*, 318–345. [CrossRef]
2. Nkwetta, D.N.; Haghighat, F. Thermal energy storage with phase change material—A state-of-the art review. *Sustain. Cities Soc.* **2014**, *10*, 87–100. [CrossRef]
3. Barbieri, E.S.; Melino, F.; Morini, M. Influence of the thermal energy storage on the profitability of micro-CHP systems for residential building applications. *Appl. Energy* **2012**, *97*, 714–722. [CrossRef]
4. Ibrahim, H.; Ilinca, A.; Perron, J. Energy storage systems-Characteristics and comparisons. *Renew. Sustain. Energy Rev.* **2008**, *12*, 1221–1250. [CrossRef]
5. Dincer, I.; Rosen, M.A. *Thermal Energy Storage: Systems and Applications*, 2nd ed.; John Wiley & Sons: Oshawa, ON, Canada, 2011.
6. Gambarotta, A.; Morini, M.; Rossi, M.; Stonfer, M. A Library for the Simulation of Smart Energy Systems: The Case of the Campus of the University of Parma. *Energy Procedia* **2017**, *105*, 1776–1781. [CrossRef]
7. Cadau, N.; De Lorenzi, A.; Gambarotta, A.; Morini, M.; Saletti, C. A Model-in-the-Loop application of a Predictive Controller to a District Heating system. *Energy Procedia* **2018**, *148*, 352–359. [CrossRef]
8. Dainese, C.; Faè, M.; Gambarotta, A.; Morini, M.; Premoli, M.; Randazzo, G.; Rossi, M.; Rovati, M.; Saletti, C. Development and application of a Predictive Controller to a mini district heating network fed by a biomass boiler. *Energy Procedia* **2019**, *159*, 48–53. [CrossRef]
9. Gambarotta, A.; Morini, M.; Saletti, C. Development of a Model-based Predictive Controller for a heat distribution network. *Energy Procedia* **2018**, 2896–2901. [CrossRef]
10. Kalaiselvam, S.; Parameshwaran, R. *Thermal Energy Storage Technologies for Sustainability*, 1st ed.; Academic Press: Cambridge, MA, USA, 2014.
11. Ould Amrouche, S.; Rekioua, D.; Rekioua, T.; Bacha, S. Overview of energy storage in renewable energy systems. *Int. J. Hydrog. Energy* **2016**, *41*, 20914–20927. [CrossRef]

12. Sarbu, I.; Sebarchievici, C. A Comprehensive Review of Thermal Energy Storage. *Sustainability* **2018**, *10*, 191. [CrossRef]
13. Noro, M.; Lazzarin, R.M.; Busato, F. Solar cooling and heating plants: An energy and economic analysis of liquid sensible vs phase change material (PCM) heat storage. *Int. J. Refrig.* **2014**, *39*, 104–116. [CrossRef]
14. Alva, G.; Lin, Y.; Fang, G. An overview of thermal energy storage systems. *Energy* **2018**, *144*, 341–378. [CrossRef]
15. Drück, H.; Pauschinger, T. *Multiport Store—Model for TRNSYS—Type 340*; Institut für Thermodynamik und Wärmetechnik (ITW) Universität Stuttgart: Stuttgart, Germany, 2006.
16. Wemhöner, C.; Hafner, B.; Schwarzer, K. Simulation of Solar Thermal Systems with Carnot Blockset in the Environment Matlab® Simulink®. In Proceedings of the Eurosun Conference 2000, Copenhagen, Denmark, 19–23 June 2000.
17. González-Altozano, P.; Gasque, M.; Ibáñez, F.; Gutiérrez-Colomer, R.P. New methodology for the characterisation of thermal performance in a hot water storage tank during charging. *Appl. Therm. Eng.* **2015**, *84*, 196–205. [CrossRef]
18. Li, S.H.; Zhang, Y.X.; Li, Y.; Zhang, X.S. Experimental study of inlet structure on the discharging performance of a solar water storage tank. *Energy Build.* **2014**, *70*, 490–496. [CrossRef]
19. Njoku, H.O.; Ekechukwu, O.V.; Onyegegbu, S.O. Analysis of stratified thermal storage systems: An overview. *Heat Mass Transf.* **2014**, *50*, 1017–1030. [CrossRef]
20. Dumont, O.; Carmo, C.; Dickes, R.; Georges, E.; Quoilin, S.; Lemort, V. Hot water tanks: How to select the optimal modelling approach? In Proceedings of the 12th REHVA World Congress, Aalborg, Denmark, 22–25 May 2016.
21. Yoo, H.; Kim, C.J.; Kim, C.W. Approximate analytical solutions for stratified thermal storage under variable inlet temperature. *Sol. Energy* **1999**, *66*, 47–56. [CrossRef]
22. Yoo, H.; Pak, E.T. Analytical solutions to a one-dimensional finite-domain model for stratified thermal storage tanks. *Sol. Energy* **1996**, *56*, 315–322. [CrossRef]
23. Campos Celador, A.; Odriozola, M.; Sala, J.M. Implications of the modelling of stratified hot water storage tanks in the simulation of CHP plants. *Energy Convers. Manag.* **2011**, *52*, 3018–3026. [CrossRef]
24. Dickes, R.; Desideri, A.; Lemort, V.; Quoilin, S. Model reduction for simulating the dynamic behavior of parabolic troughs and a thermocline energy storage in a micro-solar power unit. In Proceedings of the ECOS Conference 2015, Pau, France, 29 June–3 July 2015.
25. Chung, J.D.; Shin, Y. Integral approximate solution for the charging process in stratified thermal storage tanks. *Sol. Energy* **2011**, *85*, 3010–3016. [CrossRef]
26. Duffie, J.A.; Beckman, W.A. *Solar Engineering of Thermal Processes*, 4th ed.; John Wiley & Sons: Oshawa, ON, Canada, 2013.
27. Nash, A.; Jain, N. Dynamic Modeling and Performance Analysis of Sensible Thermal Energy Storage Systems. In Proceedings of the 4th International High-Performance Buildings Conference, Purdue, Indiana, 11–14 July 2016.
28. Chacker, A.; Bouchetiba, H. Thermal behaviour of a storage tank of solar water heater in cases of charge and discharge. *Int'l J. Comput. Commun. Instrum. Engg* **2017**, *4*, 73–77. [CrossRef]
29. Franke, R. Object-oriented modeling of solar heating systems. *Sol. Energy* **1997**, *60*, 171–180. [CrossRef]
30. De Césaro Oliveski, R.; Krenzinger, A.; Vielmo, H.A. Comparison between models for the simulation of hot water storage tanks. *Sol. Energy* **2003**, *75*, 121–134. [CrossRef]
31. Bouhal, T.; Fertahi, S.; Agrouaz, Y.; El Rhafiki, T.; Kousksou, T.; Jamil, A. Numerical modeling and optimization of thermal stratification in solar hot water storage tanks for domestic applications: CFD study. *Sol. Energy* **2017**, *157*, 441–455. [CrossRef]
32. Cabelli, A. Storage tanks—A numerical experiment. *Sol. Energy* **1977**, *19*, 45–54. [CrossRef]
33. Badescu, V. Optimal operation of thermal energy storage units based on stratified and fully mixed water tanks. *Appl. Therm. Eng.* **2004**, *24*, 2101–2116. [CrossRef]
34. Aisa, A.; Iqbal, T. Modelling and simulation of a solar water heating system with thermal storage. In Proceedings of the 7th IEEE Annual Information Technology, Electronics and Mobile Communication Conference, Vancouver, BC, Canada, 13–15 October 2016. [CrossRef]
35. Cruickshank, C. Evaluation of a Stratified Multi-Tank Thermal Storage for Solar Heating Applications. Ph.D. Thesis, Queen's University, Kingston, ON, Canada, June 2009.

36. van Koppen, C.W.J.; Thomas, P.S.; Veltkamp, W.B. Actual benefits of thermally stratified storage in a small and a medium size solar system. In Proceedings of the International Solar Energy Society Silver Jubilee Congress, Atlanta, GA, USA, 28 May–1 June 1979.
37. Furbo, S.; Mikkelsen, S.E. Is low flow operation an advantage for solar heating systems? In Proceedings of the Biennial Congress of the International Solar Energy Society, Hamburg, Germany, 13–18 September 1987; pp. 962–966. [CrossRef]
38. Herwig, A.; Umbreit, L.; Rühling, K. Measurement-based modelling of large atmospheric heat storage tanks. In Proceedings of the 16th International Symposium on District Heating and Cooling, Hamburg, Germany, 9–12 September 2018. [CrossRef]

 © 2019 by the authors. Licensee MDPI, Basel, Switzerland. This article is an open access article distributed under the terms and conditions of the Creative Commons Attribution (CC BY) license (http://creativecommons.org/licenses/by/4.0/).

Article

# Derivation and Uncertainty Quantification of a Data-Driven Subcooled Boiling Model

Jerol Soibam [1,*], Achref Rabhi [1], Ioanna Aslanidou [1], Konstantinos Kyprianidis [1] and Rebei Bel Fdhila [1,2]

1  School of Business, Society and Engineering, Mälardalen University, 72123 Västerås, Sweden; achref.rabhi@mdh.se (A.R.); ioanna.aslanidou@mdh.se (I.A.); konstantinos.kyprianidis@mdh.se (K.K.); rebei.bel_fdhila@hitachi-powergrids.com (R.B.F.)
2  Hitachi ABB Power Grids, 72226 Västerås, Sweden
*  Correspondence: jerol.soibam@mdh.se

Received: 12 October 2020; Accepted: 12 November 2020; Published: 16 November 2020

**Abstract:** Subcooled flow boiling occurs in many industrial applications where enormous heat transfer is needed. Boiling is a complex physical process that involves phase change, two-phase flow, and interactions between heated surfaces and fluids. In general, boiling heat transfer is usually predicted by empirical or semiempirical models, which are horizontal to uncertainty. In this work, a data-driven method based on artificial neural networks has been implemented to study the heat transfer behavior of a subcooled boiling model. The proposed method considers the near local flow behavior to predict wall temperature and void fraction of a subcooled minichannel. The input of the network consists of pressure gradients, momentum convection, energy convection, turbulent viscosity, liquid and gas velocities, and surface information. The outputs of the models are based on the quantities of interest in a boiling system wall temperature and void fraction. To train the network, high-fidelity simulations based on the Eulerian two-fluid approach are carried out for varying heat flux and inlet velocity in the minichannel. Two classes of the deep learning model have been investigated for this work. The first one focuses on predicting the deterministic value of the quantities of interest. The second one focuses on predicting the uncertainty present in the deep learning model while estimating the quantities of interest. Deep ensemble and Monte Carlo Dropout methods are close representatives of maximum likelihood and Bayesian inference approach respectively, and they are used to derive the uncertainty present in the model. The results of this study prove that the models used here are capable of predicting the quantities of interest accurately and are capable of estimating the uncertainty present. The models are capable of accurately reproducing the physics on unseen data and show the degree of uncertainty when there is a shift of physics in the boiling regime.

**Keywords:** computational fluid dynamics (CFD); artificial neural network (ANN); subcooled boiling flows; uncertainty quantification (UQ); Monte Carlo dropout; deep ensemble; deep neural network (DNN)

## 1. Introduction

Engineering applications with high heat flux generally involve boiling heat transfer. The high heat transfer coefficient reached by boiling flows makes boiling heat transfer relevant for research where thermal performance enhancement is needed. Although boiling heat transfer can improve the cooling performance of a system, the underlying physics are not fully understood yet. Therefore, it remains a major challenge to model the boiling heat transfer behavior for a boiling system.

Among other methods, the two-fluid model-based computational fluid dynamics (CFD) has shown a good capability of dealing with boiling flow heat transfer problems. In such an approach, information of the interface between vapor and liquid is taken as an average using closure equations, resulting in lower computational power requirements. Two-fluid models represent a promising solution to develop high fidelity representations, where all the concerned fields can be predicted with good accuracy. One of the most known two-fluid approach based CFD models to simulate boiling flows is the Rensselaer Polytechnic Institute (RPI) model, proposed by Kurul et al. [1]. The RPI model proposes to decompose the total applied heat flux on the wall on three components to take into account the evaporation, forced convection, and the quenching. Many authors adopted this approach to model the boiling flows in different geometries and for multiple operating conditions [2–5]. When comparing their results to the available experimental data, fair agreements were obtained only for few fields, while bad estimations were obtained for many others. The fields of interest here are the vapor void fraction, the different phases velocities and temperatures, and the heated surface temperature. These failed estimations are related principally to the use of several closure models representing phenomenons occurring at different length scales, as the momentum and mass transfer at the interface, bubbles interaction in the bulk flow, and more importantly mechanical and thermal interactions on the heated surface, leading to nucleation, evaporation, bubble growth and departure, etc. These closure equations are developed based on empirical or mechanistic treatments. The first is based on experimental data that are valid only for reduced ranges of operating conditions, and when they are extrapolated, accurate results are no longer guaranteed. Mechanistic treatments are based on several assumptions that generally neglect the complicated interactions between bubbles, interactions between the heated surface and fluids, and they represent a hard linking between the occurring phenomenons time and space scales.

The field of fluid dynamics is closely linked to massive amounts of data from experiments and high-fidelity simulations. Big data in fluid mechanics has become a reality [6] due to advancements in experimental techniques and high-performance computing (HPC). Machine learning (ML) algorithms are rapidly advancing within the fluid mechanics domain, and ML algorithms can be an additional tool for shaping, speeding up, and understanding complex problems that are yet to be fully solved. They provide a flexible framework that can be tailored for a particular problem, such as reduced-order modeling (ROM), optimization, turbulence closure modeling, heat transfer, or experimental data processing. For example, proper orthogonal decomposition (POD) has been successfully implemented to obtain a low-dimensional set of ordinary differential equations, from the Navier–Stokes equation, via Galerkin projection [7,8]. POD technique has been used to investigate the flow structures in the near-wall region based on measurements and information from the upper buffer layer at different Reynolds number for a turbulent channel [9]. However, it has been reported that the POD method lacks in handling huge amount of data when the size of the computational domain increases [10]. On the other hand, artificial neural networks (ANN) are capable of handling huge data size and nonlinear problems of near-wall turbulent flow [11]. ANNs have been used to learn from large eddy simulation (LES) channel flow data, and it is capable of identifying and reproducing highly nonlinear behavior of the turbulent flows [12]. In the last few years, there have been multiple studies related to the use of ANNs for estimating the subgrid-scale in turbulence modeling [13–15]. Among other neural networks, convolution neural networks (CNNs) have been widely used for image processing due to their unique feature of capturing the spatial structure of input data [16]. This feature is an advantage when dealing with fluid mechanics problems since CNN allows us to capture spatial and temporal information of the flows. Multiple CNN structure [17] has been proposed to predict the lift coefficient of airfoils with different shapes at different Mach numbers, Reynolds number, and angle of attack. A combination of CNN with multi-layer perceptron (MLP) [18] has been used to generate a time-dependent turbulent inflow generator for a fully developed turbulent channel flow. The data used for that work have been obtained from direct numerical simulation (DNS), and the model was able to predict the turbulent flow long enough to accumulate turbulent statistics.

Machine learning techniques are rapidly making inroads within heat transfer and multiphase problems too. These ML methods consist of a series of data-driven algorithms that can detect a pattern in large datasets and build predictive models. They are specially designed to deal with large, high-dimensional datasets, and have the potential to transform, quantify, and address the uncertainties of the model. ANN-based on the back-propagation model has been used to predict convective heat transfer coefficients in a tube with good accuracy [19]. Experimental data from impingement jet with varying nozzle diameter and Reynolds number have been used to build an ANN model. This model was then used to predict the heat transfer (Nusselt number) with an error bellow 4.5% [20]. More recently, researchers have used ANN for modeling boiling flow heat transfer of pure fluids [21]. Neural networks have been used to fit DNS data to develop closure model relation for the average two-fluid equations of vertical bubble channel [22]. The model trained on DNS data was then used for different initial velocities and void fraction, to predict the main aspects of DNS results. Different ML algorithms [23] have been examined to study pool boiling heat transfer coefficient of aluminum water-based nanofluids, and their result showed that the MLP network gave the best result. In the past, CNNs model has been used even as an image processing technique for two-phase bubbly flow, and it has been shown that they can determine the overlapping of bubbles: blurred and nonspherical [24].

Although there is research related to boiling heat transfer and machine learning algorithms, according to the authors' knowledge all the above-mentioned models focus on deterministic ANN models. When working with a reduced order model or black-box model such as deep learning models for physical applications, it becomes of prime importance to know the model confidence and capability. If these models are to be used for a new set of operating conditions, the first thing to consider is how reliable this model is and how accurate the predicted value is. Therefore, it is crucial to know the uncertainty present in the predictions made.

The aforementioned ML techniques applied to heat transfer and fluid mechanics are mostly deterministic approaches that lack in providing predictive uncertainty information. At the same time, deterministic ANN models tend to produce overconfident predictions, which may lead to unpredictable situations when used in real-life industrial applications. Therefore, it is essential to quantify the uncertainty present in the model for any practical applications. Uncertainty models provide the user with the confidence level, and they allow them to make better decisions based on engineering judgment.

In Bayesian learning, a priori distribution is defined upon the parameters of a NNs, then given the training data, the posterior distribution over the parameter is computed, which is used to estimate the predictive uncertainty [25]. The fundamental concern in Bayesian learning while analyzing data or making decisions is to be able to tell whether the model is certain about its output. Bayesian-based models offer a mathematically grounded framework to reason about model uncertainty, but the computational cost rapidly increases as the data size increases. Due to computational constraints in performing exact Bayesian inference, several approximation methods have been proposed such as; Markov chain Monte Carlo (MCMC) [26] method, variational Bayesian methods [27,28], Laplace approximation [29], probabilistic backpropagation (PBP) [30]. The nature of predictive quality using Bayesian NNs depends on correct priori distribution and degree of approximation of Bayesian inference due to high computational cost [31]. However, Bayesian models are harder to implement, they make it harder to define the correct priori properties, and they are computationally slower compared to traditional neural networks. Another approach to quantify the uncertainty of an ANN model is the Deep Ensemble method [32], which is inspired by the bootstrap technique [33]. In this method, it is assumed that the data has a parametrized distribution where the parameters depend on the input. Finding these parameters is the aim of the training process, i.e., the prediction network will not output a single value but instead will output the distributional parameters of the given input.

It is worth mentioning that this work presents one of the first attempts to implement deep learning techniques and quantify the model uncertainty of a data-driven subcooled boiling model. In this work, three data-driven models using deep learning techniques have been investigated to study the heat transfer behavior of a subcooled boiling in a minichannel with varying inflow velocity and heat fluxes. The first model focuses on the prediction of the deterministic value of the void fraction and wall temperature of the minichannel which are quantities of interest (QoIs). The second and third model focuses on probabilistic deep learning techniques to derive the uncertainty in the models when predicting the QoIs. The two methods used are Deep Ensemble, which is representative of the Maximum Likelihood approach, and Monte Carlo Dropout, which is representative of Bayesian inference. These two models are capable of capturing the nonlinear behavior that exists in the subcooled boiling data and is capable of reproducing the physics of unseen data for both interpolation and extrapolation datasets.

## 2. Methodology

### 2.1. CFD Modeling

There are two methods when it comes to modeling of boiling flows. The first approach is Volume of Fluid (VoF), which employs interface tracking to give a better understanding of the bubbles nucleation and ebullition cycle [34,35]. However, an extremely large grid needs to be employed to perform this kind of simulation and it is computationally expensive. On the other hand, the Eulerian approach is based on averaging the conservation equations with selected interfacial terms. This makes it more appropriate for this work, since the goal is to model boiling in a full channel scale. Therefore, in this work CFD simulations were carried out based on an Eulerian two-fluid approach for modeling the subcooled nucleate boiling flows. The conservation equations for each phase, i.e., liquid and vapor phases, are solved numerically based on a Finite-Volume discretization implemented in the open-source platform OpenFOAM. The presented number of model details, correlations, and assumptions have been evaluated and used in [5].

The mass conservation equation for each phase, liquid continuous phase or vapor dispersed phase, can be written as:

$$\frac{\partial \alpha_k \rho_k}{\partial t} + \nabla \cdot (\alpha_k \rho_k \mathbf{U}_k) = \Gamma_{ki} - \Gamma_{ik} \tag{1}$$

where the subscript $k$ denotes the phase, $\alpha$ is the void fraction, $\rho$ is the phase density, $\mathbf{U}$ is the phase velocity and $\Gamma$ is the mass transfer rate per unit volume that denotes evaporation or condensation, calculated based on the boiling equations that will be presented later.

For each phase $k$, the following momentum conservation is solved based on the following equation:

$$\frac{\partial \alpha_k \rho_k \mathbf{U}_k}{\partial t} + \nabla \cdot (\alpha_k \rho_k \mathbf{U}_k \mathbf{U}_k) = -\alpha_k \nabla p + \mathbf{R}_k + \mathbf{M}_k + \alpha_k \rho_k \mathbf{g} + (\Gamma_{ki} \mathbf{U}_i - \Gamma_{ki} \mathbf{U}_k) \tag{2}$$

where $\nabla p$ is the pressure gradient, $\mathbf{R}$ is the combined turbulent and laminar stress term, calculated based on the Reynolds analogy, $\mathbf{g}$ is the gravitational acceleration, and $\mathbf{M}$ is the interfacial momentum transfer, accounted for the drag forces calculated based on Schiller and Naumann [36] drag coefficient, the virtual mass forces calculated based on a constant coefficient equal to 0.5, and the turbulent dispersion forces calculated based on the model of de Bertodano [37].

The energy transport equation is written in terms of specific enthalpy $h$ for each phase $k$ as follows:

$$\frac{\partial \alpha_k \rho_k h_k}{\partial t} + \nabla \cdot (\alpha_k \rho_k \mathbf{U}_k h_k) = \alpha_k \frac{Dp}{Dt} + \nabla \left( \alpha_k D_{t,k}^{eff} \nabla h_k \right) + \Gamma_{ki} h_i - \Gamma_{ik} h_k + Q_{wall,k} \tag{3}$$

where $D_{t,k}^{eff}$ is the effective thermal diffusivity, and $Q_{wall,k}$ is the product of the applied heat flux on the wall with the contact area with the wall per unit volume.

In order to account for the turbulent behavior of the dispersed phase flow, the $k - \varepsilon$ turbulence model was used for the vapor phase. However, the bubbles-induced turbulence needs to be accounted also in the turbulent flow behavior of the continuous phase. Hence, the Lahey $k - \varepsilon$ [38] turbulence model is adopted for the liquid phase.

In order to calculate the mass transfer rates, a boiling model is needed. The approach implemented follows the very well known RPI model after Kurul and Podowski [1], where the total applied heat flux $q_w''$ on the wall is divided into three components, evaporation heat flux $q_{w,e}''$, forced convection heat flux $q_{w,c}''$, and quenching heat flux $q_{w,q}''$. The total applied heat flux can be written as follows:

$$q_w'' = q_{w,e}'' + q_{w,c}'' + q_{w,q}'' \tag{4}$$

To be able to calculate each heat flux contribution, the closure boiling equation applied at the heated surface needs to be specified, as the active nucleation site density $N_a$, the bubble departure diameter $d_{dep}$, and the bubble departure frequency $f_{dep}$. The mathematical expressions of each heat flux contribution can be found in [1,39].

The Active Nucleation Site Density (ANSD) that represents the cavities from where bubbles are nucleated, is calculated based on the correlation of Benjamin and Balakrishnan [40] given by:

$$N_a = 218\mathrm{Pr}_l^{1.63}\Delta T_{sup}^3 \gamma^{-1}\theta^{-0.4} \tag{5}$$

where $\mathrm{Pr}_l$ is the liquid Prandtl number, $\Delta T_{sup}$ is the wall superheat, $\gamma$ is a coefficient taking into account the liquid and heated surface thermophysical properties and $\theta$ is a coefficient taking into account the heated surface roughness and the system pressure.

The Bubble Departure Diameter (BDD) is a parameter that represents the nucleating bubble critical diameter, beyond which the bubble leaves its nucleation site. In this work, this parameter is calculated based on the semiempirical model of Ünal [41], given as follows:

$$d_{dep} = \frac{2.42\ 10^{-5}p^{0.709}a}{\sqrt{b\phi}} \tag{6}$$

where $a$ and $b$ are the model coefficients, taking into account the working fluid and the heated surface thermophysical properties, and $\phi$ is a parameter controlled by the local flow velocity.

The last closure equation for the boiling model is the Bubble Departure Frequency (BDF), representing the number of bubbles leaving a nucleation site per unit time. It is calculated according to the mechanistic model of Brooks and Hibiki [42] as:

$$f_{dep} = \frac{C_{fd}\mathrm{Ja}_w^{0.82}N_T^{-1.46}\rho^{*-0.93}\mathrm{Pr}_{sat}^{2.36}}{d_{dep}^2} \tag{7}$$

where $C_{fd}$ is the model coefficient depending on the size of the channel where boiling occurs, $N_T$ is a dimensionless temperature, $\rho^*$ is a dimensionless density ratio, $\mathrm{Ja}_w$ is a modified Jacob number, and $\mathrm{Pr}_{sat}$ is the liquid Prandtl number evaluated at the corresponding saturation temperature.

Now, the mass transfer rates can be calculated as follows:

$$\Gamma_{evap} = \frac{A_{w,b,e}^f}{6}\rho_v d_{dep}f_{dep} \tag{8}$$

$$\Gamma_{cond} = \frac{h_c\left(T_{sat}(p) - T_l\right)}{h_{lg}}A_s \tag{9}$$

where $A_{w,b,e}^f$ is an area fraction of the heated surface not affected by bubbles, $h_c$ is a condensation heat transfer coefficient calculated based on the correlation of Ranz et al. [43], and $h_{lg}$ is the latent heat of vaporization.

In the previously presented equation system, the selected models and correlations are developed and used in [5]. This numerical model is also used in this work to provide complete datasets of results needed for the present investigation.

### 2.2. CFD Simulation Data

Data used in this work are obtained from 2D CFD simulations based on the Eulerian two-fluid approach. The computational domain is illustrated in Figure 1. It consists of a narrow rectangular upward channel (0.003 × 0.4 m), heated from one side by a constant uniform heat flux. Water as working fluid is flowing upward the channel at atmospheric pressure. At the channel inlet, the liquid velocity and temperature are set. At the channel walls, a nonslip boundary condition is set for both phases, liquid, and vapor. Since the channel is heated only from one side, the boiling closure equations, i.e., active nucleation site density, bubble departure diameter, and frequency models, are applied on this particular wall. The thermophysical properties of the liquid are calculated based on the inlet temperature, while the vapor and heated surface thermophysical properties are evaluated at the saturation temperature. However, the current numerical code allows the calculation of the local saturation temperature based on the computed local pressure. Multiple heat fluxes and inlet velocities were used to conduct the simulations. In total, 102 simulations have been performed for inlet velocity ranging from 0.05 ms$^{-1}$ to 0.2 ms$^{-1}$ and heat flux ranging from 1000 Wm$^{-2}$ to 40,000 Wm$^{-2}$. The developed CFD model was validated in a previous work presented in [5] based on experimental result of [44,45]. The heated surface temperature measurements were compared against the CFD predictions and good agreements were obtained.

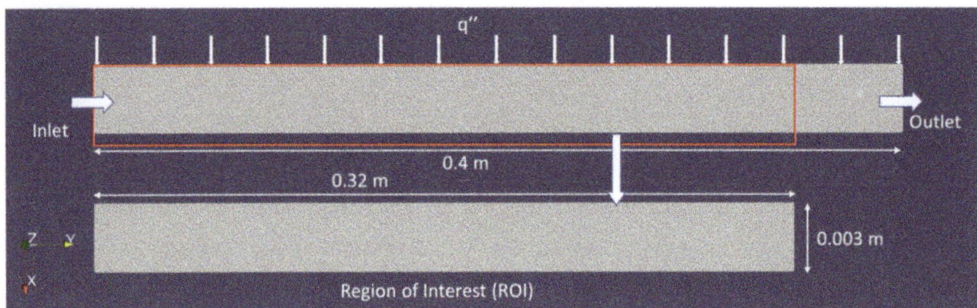


**Figure 1.** CFD simulation domain and selected region of interest for data extraction.

### 2.3. CFD Validation

To validate the CFD results used in this work predictions of the onset of nucleate boiling are compared to the experimental measurements of Kromer et al. [45], as shown in Table 1. The simulated heated surface temperature, is compared to the measurements of Al-Maeeni [44] and is presented in Figure 2. The experimental results of Kromer et al. [45] and Al-Maeeni [44] were based on upward flow boiling in a narrow aluminum rectangular channel (3 × 10 × 400 mm). They used water as working fluid medium, and the channel was heated from one side with a constant heat flux. Kromer et al. [45] used a mass flux of 58.1 kgm$^{-2}$s$^{-1}$ with two different heat fluxes, $q'' = 20,000$ Wm$^{-2}$ and $q'' = 30,000$ Wm$^{-2}$, whereas Al-Maeeni used a mass flux of 134.2 kgm$^{-2}$s$^{-1}$ with a heat flux of $q'' = 50,000$ Wm$^{-2}$. It is to be noted that both the experiments were conducted under the same inlet subcooling $\Delta T_{sub,in} = 10$ K. From Table 1 it can be seen that a reasonable agreement is achieved for the onset nucleate boiling for both tested heat fluxes, with a maximum error less than 25%. However, a much better agreement is achieved for the heated surface temperature, with a maximum error of 3% between the measurements and the CFD predictions as shown in Figure 2. The predictions associated with the boiling closure equations used in this work are the ANSD model of Benjamin and Balakrishanan [40], the BDD model of Ünal [41], and the BDF model of Brooks and Hibiki [42] (Current model). The CFD results are further compared with the predictions associated with the boiling closure equations given by Benjamin and Balakrishnan [40] for the ANSD, the BDD of Tolubinski and Kostanchuk [46], and the BDF of Cole [47] (Previous model). The comparison of the experiments and the models predictions are shown in Figure 2. It can be noted from the plot that the current model used in this work shows better prediction of the heat surface temperature.

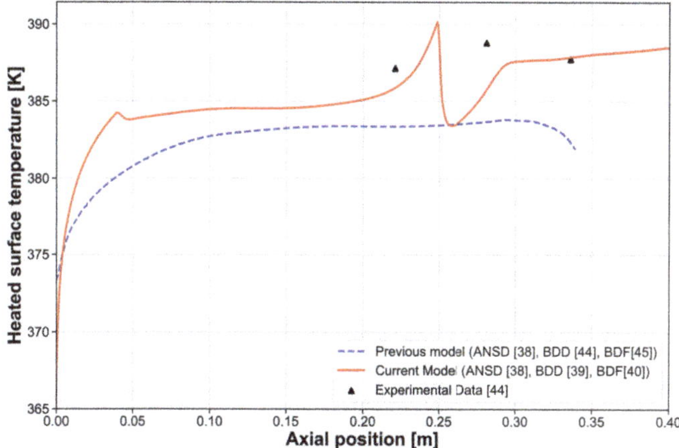

**Figure 2.** CFD validation for the heated surface temperature with the experimental data provided by [44].

**Table 1.** CFD validation for the void fraction with the experimental data provided by [45].

| Heat Flux $q''$ $Wm^{-2}$ | $Z_{ONB}$ [Kromer et al. [45]] (mm) | $Z_{ONB,CFD}$ (mm) | Error % |
|---|---|---|---|
| 30,000 | 188.9 ± 0.5 | 172 | 9 |
| 20,000 | 192.3 ± 0.5 | 240 | 25 |

*2.4. Data Handling*

To train the deep neural network models, data were extracted from the CFD simulations in the region 0 to 0.32 m, which is the selected region of interest (ROI), this is done to avoid the influence of boundary near the outlet. The ROI for extracting the data is shown in Figure 1. The number of cells present in the ROI are 321 × 26 resulting in 8346 data points for each case. The domain axes (x and y) are further converted into nondimensional numbers by diving with the maximum length value of the minichannel. This is done so that the model is not constrained to learn based on the height of the channel and applicable for other channel lengths. Out of the 102 simulated cases, 96 cases are used for training and validating the models, while the remaining 6 cases are used for further analyzing the model performance. These 96 cases are split into 80% (Training Dataset) for training the model and 20% (Validation Dataset) for validation. The validation and test dataset are used for evaluating the models. The validation dataset is predominately used to describe the evaluation of models when tuning hyperparameters and data preparation, and the test dataset is predominately used to describe the evaluation of a final model when comparing it to other final models. Hence, the remaining 6 test cases are used to provide an unbiased evaluation of the final model fit on the training dataset.

The selected feature input signals obtained from CFD simulations are shown in Table 2. These inputs are chosen based on their influence on the quantities of interest. The expected outputs (void fraction and wall temperature) from the DNN models are presented in Table 3. Before feeding the training data into the network, the input and output features are normalized between 0 and 1. This way the ML algorithms can learn better since the scale of data used in this study is very sparse.

**Table 2.** Input features used for training the network.

| Input Features | Feature Expressions |
| --- | --- |
| Pressure gradient | $\dfrac{\partial\langle p\rangle}{\partial x}$ <br> $\dfrac{\partial\langle p\rangle}{\partial y}$ |
| Momentum convection | $\dfrac{\partial\langle p\rangle\langle u\rangle\langle u\rangle}{\partial x}$ <br> $\dfrac{\partial\langle p\rangle\langle u\rangle\langle v\rangle}{\partial x}$ <br> $\dfrac{\partial\langle p\rangle\langle u\rangle\langle v\rangle}{\partial y}$ <br> $\dfrac{\partial\langle p\rangle\langle v\rangle\langle v\rangle}{\partial y}$ |
| Energy convection | $\dfrac{\partial\langle p\rangle\langle T\rangle\langle u\rangle}{\partial x}$ <br> $\dfrac{\partial\langle p\rangle\langle T\rangle\langle v\rangle}{\partial y}$ |
| Total heat flux | $q''$ |
| Inlet velocity | $u_{inlet}$ |
| Inlet pressure | $p_{inlet}$ |
| Temperature inlet | $T_{inlet}$ |
| Ambient pressure | $p_{amb}$ |
| Fluid and gas viscosity | $\mu_l \; \mu_g$ |
| Nondimensional x and y axis | $\dfrac{x}{x_{max}}, \; \dfrac{y}{y_{max}}$ |
| Nondimensional arc length | $\dfrac{l}{l_{max}}$ |

**Table 3.** Output features (quantities of interest).

| Output Features | Feature Expressions |
| --- | --- |
| Wall Temperature | $T_{wall}$ |
| Void Fraction | $\alpha$ |

*2.5. Deep Neural Networks Architectures*

Artificial neural networks are computational models that are inspired by the way neurons in brains work. They have the ability to acquire and retain information and generally comprise an input layer, hidden layer, and output layer. These layers are sets of processing units, represented by so-called artificial neurons, interlinked by multiple interconnections, implemented with vectors and matrices of synaptic weights. An ANN with multiple hidden layers is generally known as deep neural network (DNNs) or multi-layer perceptron (MLP).

Two stages are involved while training the MLP network with the back-propagation technique also known as the generalized Delta rule. These stages are illustrated in Figure 3, which shows a MLP with 5 hidden layers, 18 signals on its input layer, $n_1 \ldots n_5$ neurons in the hidden layers, and finally 2 output signals. In the first stage, the signals $\{x_1, x_2, \ldots \ldots, x_n\}$ of the training sets are inserted into the network inputs and are propagated layer-by-layer until the production of the corresponding outputs. Thus, this stage propagates only in the forward direction to obtained the responses (outputs) from the network, hence, it is called the feed-forward phase. The network undergoes series of nonlinear transformation controlled by parameters like weights **W** and biases **b**, followed by a nonlinear activation function ($g\,(x)$). There is a wide number of activation functions that can be used depending on classification or regression problems. In this work, a nonlinear activation function called rectified linear units (ReLU) is used. The main advantage of ReLU function is that it does not activate all the neurons at the same time. It only activates the neuron if the linear transformation is above zero value and it is computationally efficient. The MLP in Figure 3 can be interpreted as:

$$h_1 = g(W_1^T x + b_1)$$
$$..$$
$$h_5 = g(W_5^T h_4 + b_5)$$
$$\hat{y} = g(W_6^T h_5 + b_6)$$

(10)

$$g(x) = \begin{cases} 0 & \text{for } x < 0 \\ x & \text{for } x \geq 0 \end{cases}$$

(11)

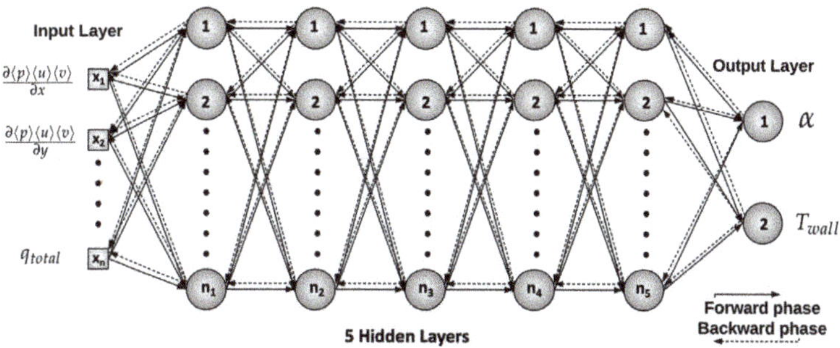


**Figure 3.** Back propagation deep neural network (DNN) architecture used in this work.

A cost or loss function $\mathcal{L}$ is defined while training the network to measure the error between the predicted value $\hat{y}$ and the target value $y$. The type of loss function used while training a neural network is problem specific and depending on whether it is a classification or a regression problem. For this work, mean squared error (MSE) loss is used for computing the loss function. Once the loss is computed, the error gradient concerning weights and biases in all the layers can be computed through a backward phase. The main objective of the backward phase is to estimate the gradient of the loss function for the different weights by using the chain rule of differential calculus. These computed gradients are then used to update the weights and biases of all the hidden layers. In this work, the Adaptive Moment Estimation (Adam) [48] optimization technique is used to update the weights and biases of the network with a learning rate of Lr = $1e^{-4}$. The Adam optimization technique has been chosen due to its capability of handling large datasets, high-dimensional parameters, and sparse gradients. Since these gradients are estimated in the backward direction, starting from the output node, this learning process is referred to as the backward phase.

$$\mathcal{L} = MSE(y, \hat{y})$$

(12)

While training a deep neural network, it is a common issue that the model becomes overfitted. Overfitting occurs when an algorithm is tailored to a particular dataset and is not generalized to deal with other datasets. To avoid overfitting of the DNN model, a regularization term is generally introduced in the loss function $\mathcal{L}$. The two most common regularization approaches are $\mathcal{L}1$ norm (*Lasso Regression*) and $\mathcal{L}2$ norm (*Ridge Regression*) and they are expressed:

$$\mathcal{L} = MSE(y, \hat{y}) + \lambda \sum_{i=1}^{N} |w_i|$$

$$L_1 \, norm$$

(13)

$$\mathcal{L} = MSE(y, \hat{y}) + \lambda \sum_{i=1}^{N} w_i^2$$

$$L_2 \, norm$$

where $\lambda$ is a positive hyperparameter that influences the regularization term, with a larger value of $\lambda$ indicating strong regularization.

These regularization terms regularize or shrink the coefficient estimates towards zero, and it has been shown that shrinking the coefficient estimates can significantly reduce the variance [49]. $\mathcal{L}1$ regularization forces the weight parameters to become zero, and $\mathcal{L}2$ regularization forces the weight parameters towards zero but never exactly zero. When regularization term is applied to DNN it results in smaller weight parameters by making some neurons neglectable. This makes the network less complex and avoids overfitting. For this work, $\mathcal{L}2$ norm (*Ridge Regression*) regularization is applied in all the hidden layers while training the deep neural network.

### 2.6. Uncertainty of Deep Learning

Deep learning techniques have attracted considerable attention in fields such as physics, fluid mechanics, and manufacturing [50–52]. In these fields estimating model uncertainty is of crucial importance since it is vital to understanding the interpolation and extrapolation ability of the model. Deep learning algorithms are able to learn powerful representations that can map high dimensional data to an array of outputs. These mapping functions are often taken blindly and assumed to be accurate, which may not be always true. Hence, it is of paramount importance to be able to quantify the uncertainty present and justify the behavior of these models. Therefore, in this work, two uncertainty quantification (UQ) models have been investigated to justify the nature of the model and they are described in detail.

### 2.6.1. Monte Carlo (MC) Dropout method

The deep learning models can be cast as Bayesian models, without changing either the model or the optimization process. This can be done by an approach called dropout during training and prediction, and it has been proven that dropout in ANNs can be interpreted as a Bayesian approximation of a well known probabilistic model, the Gaussian process (GP) [53]. Dropout has been used in many deep learning models as a way to avoid overfitting [54] like regularization technique. Dropout technique in a neural network with some modifications can be used for estimating the uncertainty, as described by Gal et al. [55]. Their method implies that as long as the neural network is trained with few dropout layers, it can be used to estimate the uncertainty of the model during the time of prediction. Unlike traditional Dropout networks, Monte Carlo Dropout (MC Dropout) networks apply dropout both at the training and testing phase.

When a network with input feature $X^*$ is trained with dropout, the model is expected to give an output with predictive mean $\mathbb{E}(y^*)$ and the predictive variance $Var(y^*)$. To approximate the model as a Gaussian Process, a prior length scale $l$ is defined and captures the belief over the function frequency. A short length-scale $l$ corresponds to high frequency data, and a long length-scale $l$ corresponds to low frequency data. Mathematically the Gaussian process precision ($\tau$) is given as:

$$\tau = \frac{l^2 p}{2N\lambda_w} \tag{14}$$

where $p$ is the probability of the units (artificial neurons) not dropped during the process of training, $\lambda_w$ is the weight decay, and $N$ is the size of the dataset. Similarly, dropout is activated during the prediction phase of a new set of data (validation or test data) $x^*$, i.e., randomly units are dropped during the prediction phase. The prediction step is repeated several times ($T$) with different units dropped every time, and results are collected $\{\hat{y}_t^*(x^*)\}$. The empirical estimator of the predictive mean of the approximated posterior and the predictive variance (uncertainty) of the new test data is given by the following equations:

$$\mathbb{E}(y^*) \approx \frac{1}{T} \sum_{t=1}^{T} \hat{y}_t^*(x^*) \tag{15}$$

$$Var(y^*) \approx \tau^{-1}I_D + \frac{1}{T}\sum_{t=1}^{T}\hat{y}_t^*(x^*)^T\hat{y}_t^*(x^*) - \mathbb{E}(y^*)^T\mathbb{E}(y^*) \tag{16}$$

### 2.6.2. Deep Ensemble

Deep Ensemble [32] is a non-Bayesian method for uncertainty quantification in machine learning models. Deep ensemble learning is a learning paradigm where ensembles of several neural networks show improved generalization capabilities that outperform those of a single network. It has been shown that an ensemble model has good predictive quality and can produce good estimates of uncertainty [32]. In general, while training a neural network for a regression problem, the goal is to minimize the error between the target value $\mathbf{y}$ and the predicted value $\hat{y}$ using mean squared error (mse) loss. However, to obtain the uncertainty estimates, the model has to be expressed in the form of a probabilistic model. Hence, this approach assumes that given an input $X$, the target $y$ has a normal distribution with a mean and a variance depending on the values of $X$. This results in modification of the loss function; instead of minimizing the difference of target and predictive value, the goal is to minimize the predictive distribution to the target distribution using the Negative Log-Likelihood (NLL) loss. It is important to use the correct scoring rule while determining the predictive uncertainty of a model, and NLL has been proven to be a proper scoring rule for evaluating predictive uncertainty [56].

$$\mathcal{L}_{loss} = -logp_\theta\frac{y_n}{X_n} = \frac{log\sigma^2(X)}{2} + \frac{(y - \mu_\theta(X))^2}{2\sigma^2(X)} + c \tag{17}$$

where, $\mu_\theta(X)$ and $\sigma^2(X)$ are the predictive mean and the variance, $c$ is a constant. Intuitively, the goal is to minimize the difference between the predictive distribution to the target distribution using the negative log-likelihood loss.

In a Deep Ensemble model, $M$ networks are trained with different random initialization. It is to be noted that as the number of networks increases, the computational cost also increases during the time of training. Therefore, the number of networks to be considered is a trade-off between computational speed and accuracy of the prediction. For this study, 5 networks with random initialization were trained to create the deep ensemble model. Ensemble results are then treated as a uniformly-weighted mixture model, although for simplicity the ensemble prediction is approximated to be a Gaussian distribution whose mean and variance are respectively the mean and variance of the mixture. The mean and variance mixture is given by:

$$\mu_*(X) = \frac{1}{M}\sum_m \mu_{\theta m}(X) \tag{18}$$

$$\sigma_*^2(X) = \frac{1}{M}\sum_m (\sigma_{\theta m}^2(X) + \mu_{\theta m}^2(X)) - \mu_*^2(X) \tag{19}$$

where $\mu_{\theta m}(X)$ and $\sigma_{\theta m}^2(X)$ are the mean and variance of individual networks, $\mu_*(X)$ and $\sigma_*^2(X)$ are the mean and variance of the ensemble model respectively. The detailed explanation and benchmark of the deep ensemble model and equations can be found in the research conducted by Lakshminarayanan et al. [32]. To implement the above method in a standard deep learning architecture the following steps were taken: first, a custom loss function with NLL is defined, then another custom layer has been defined to extract the mean and variance as an output of the network.

The summary of all the models investigated in this work is shown in Table 4. The Monte Carlo Dropout model has the possibility to predict the quantities of interest with uncertainty (MC Dropout) and without uncertainty (MC no-Dropout) for the same trained model. To avoid overfitting of all the models used in this work, the following measures have been taken into consideration while training the network:

- $L_2$ regularization term is introduced in the loss function.

- Callback functions are defined to save only the best weights of the network, with early stopping of the training if the validation loss does not improve in the next epoch.
- Several batch size were tested and for this study batch size of 256 gave the best results. Batch size in machine learning is the number of training data used in one iteration.
- The best weight saved during the training phase by the callbacks is loaded before the prediction phase.

This way it ensures that the model is not an overfitted model and it uses the best weights of the model while predicting a new unseen case.

**Table 4.** Specification of the models, MSE: Mean Squared Error, NLL: Negative Log-Likelihood, Adam: Adaptive Moment Estimation, Lr: Learning rate, MLP: Multi Layer Perceptron, MC: Monte Carlo.

| Models | Input Signals | Hidden Layers | Output Signals | Batch Size | Uncertainty Quantification | Cost/Loss Function | Optimizer | Lr |
|---|---|---|---|---|---|---|---|---|
| MLP | 18 | 5 | 2 | 256 | No | MSE | Adam | $1e^{-4}$ |
| MC no-Dropout | 18 | 5 | 2 | 256 | No | MSE | Adam | $1e^{-4}$ |
| MC Dropout | 18 | 5 | 2 | 256 | Yes | MSE | Adam | $1e^{-4}$ |
| Deep Ensemble | 18 | 5 | 2 | 256 | Yes | NLL | Adam | $1e^{-4}$ |

## 3. Results and Discussion

In this work, the open-source deep-learning library Tensorflow 1.14 along with python 3.5 are used to build the architecture of the deep learning models (MLP, MC Dropout, and Deep Ensemble). In total there are 102 CFD datasets for varying heat flux and velocities, out of which 96 cases are split into training data (80%) and validation data (20%). Each case consists of 8346 data points. The remaining 6 cases are purely used for further testing of the model, of which 3 cases for interpolation, and the remaining 3 cases for extrapolation. The model performance is first evaluated using the validation dataset first, then tested on the interpolation dataset, and finally on the extrapolation dataset. The MLP model is a deterministic model whereas the MC Dropout and Deep Ensemble models provide the deterministic values as well as the expected variance of the predicted value. The validation and testing performance of the MLP model is relatively lower when compared to MC Dropout and Deep Ensemble models and the statistic performance can be seen in Table 5. From the table it can be noted that the DE model shows the best performance. The models used in this study are capable of predicting the void fraction field and temperature field of the minichannel domain. However, the discussion presented in the Sections 3.2 and 3.3 will be focused on the near-wall region of the minichannel. The near-wall region is of particular interest since there is a sudden shift in physics/behaviors and the motivation was to demonstrate the performance and robustness of the DNN models in this region. For the interested reader, the prediction of the full field flow of the void fraction and temperature in the minichannel is demonstrated in Appendix A through Figures A5–A8.

**Table 5.** Performance of the models on validation dataset of the computational domain, VF: Void Fraction, Temp: Temperature.

| Case Dataset | Models | RMSEP | | $R^2$ | |
|---|---|---|---|---|---|
| | | VF | Temp | VF | Temp |
| Validation | MLP | 0.008 | 0.152 | 0.991 | 0.987 |
| | MC No-Dropout | 0.013 | 0.265 | 0.995 | 0.989 |
| | MC Dropout | 0.006 | 0.125 | 0.9991 | 0.997 |
| | Deep Ensemble | 0.002 | 0.081 | 0.9998 | 0.998 |

*3.1. Validation Case Studies*

The performance of the models is demonstrated using scatter plots, where the models predicted values are plotted against the CFD values for the void fraction as shown in Figure 4. The scatter points in the scatter plots are the full computational domain predicted values using the DNN models. The predicted void fraction using a standard MLP/DNN is plotted against the CFD values, as shown in Figure 4a. The predicted values are mostly concentrated near the x = y line, this line is the true line where the predicted values would perfectly match the CFD results. However, the MLP model has trouble predicting accurately near the void fraction 0 to 0.001, where the nucleation starts in the channel. It can be noted that the MLP model shows nonlinearity around void fraction 0.7 to 0.8, which could be related to change in the flow regime induced by massive generation of bubbles and the MLP model fails to capture the physics in this region. Furthermore, it can be observed that the MLP model shows a void fraction value above unity, which is nonphysical.

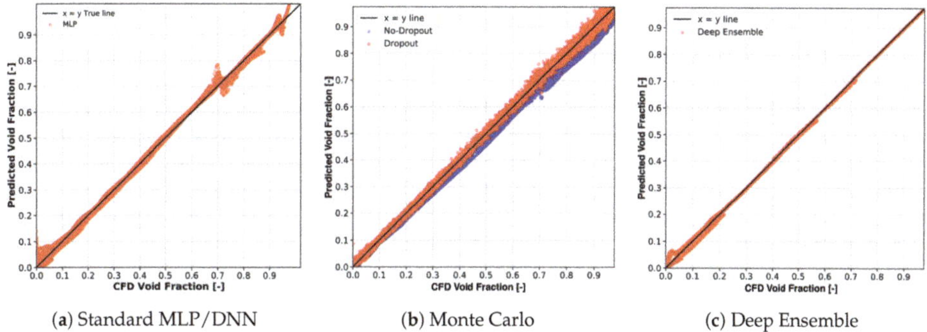

(a) Standard MLP/DNN        (b) Monte Carlo        (c) Deep Ensemble

**Figure 4.** Validation Void Fraction regression chart of the computational domain.

The void fraction values predicted using the Monte Carlo Dropout method are presented in Figure 4b. It can be observed that there are two scatter set of results "red": *dropout prediction* and "blue": no-dropout prediction. During the dropout prediction phase, 20% of the neurons in each hidden layer are randomly deactivated every time for 1000 iteration, resulting in 1000 values for each void fraction, the mean value of each void fraction is then plotted. This nature of dropout prediction allows us to quantify the variance of the predicted value, hence it can indicate the degree of uncertainty present in this model. Whereas the blue scatter plot is the predictive value obtained from the model with no-dropout during the prediction phase. It can be seen from the plot that the performance of the dropout prediction outperforms the no-dropout prediction and fits the x = y line better. It can be further observed from the plot that the predicted values are concentrated near the x = y line at low values of void fraction. Then the predicted values start to get sparse as the void fraction value increases and become more sparse around 0.8. This sparsity in fitting the data can be related to change in flow and boiling regime where the underlying physical phenomenon becomes very complex and includes a strong instability in the flow. From the validation dataset, it is evident that the Monte Carlo dropout model captures well when void fraction is low and its performance starts to deteriorate as the void fraction value increases.

The predictive performance of the Deep Ensemble (DE) model on the validation dataset is shown in Figure 4c, where the predicted values are plotted against the CFD value. Unlike the MLP and MC Dropout model, the void fraction value predicted by the DE model does not suffer from sparsity, and the predicted values are concentrated near the x = y line, meaning this model is capable of reproducing the CFD values accurately. It can be further noticed that DE models perform well for void fraction ranging from 0.2 and above, which implies that the DE model seems to capture the physics that exists in these regimes from the data used in training. However, there are some uncaptured nonlinearities present in the model when predicting void fraction values between 0 to 0.05, and this is the region where subcooled boiling starts in the channel. In the region of subcooled boiling, small bubbles start to appear, which changes the dynamics of the flow in the channel. The DE model marginally fails to capture the sudden shift in physics for some of the data points; nevertheless, its performance improves when predicting void fraction for the rest of the subcooled boiling regime. From this, it can be concluded that the DE model captures the main flow features very well for the validation dataset for most of the boiling regimes with some small deviation near the onset of nucleate boiling.

The predicted temperature values are plotted against the CFD values for the computational domain and are shown in Figure 5. The predictive capability of the standard MLP/DNN model is demonstrated in Figure 5a. From the plot, it is evident that the MLP model performs poorly when predicting the temperature. The MLP model shows high nonlinearity between the predicted and CFD values starting from 372 K to 379 K; however, it is still worth mentioning that the maximum relative error percentage is under 0.6%. Figure 5b shows the scatter plot of the CFD value and the MC model predicted values. The scatter plot with blue represents the prediction with no-dropout and the scatter plot with red represents the prediction with dropout activated. It is clear from the plot that the prediction with no-dropout deviates away from the x = y line and it under predicts the temperature. However, when dropout is activated during the prediction phase, the predictions are more concentrated around the x = y line. It can be further noted that the sparsity of the predicted values increases in the range of 373.15 K to 375 K, because for most of the case data subcooled boiling starts around this temperature. The maximum relative error between the CFD and the predicted value for the dropout prediction is below 0.3%. In contrast, the predictive nature of the DE model outperforms the MLP and MC dropout model as shown in Figure 5c. It is evident from the figure that the DE model is capable of perfectly capturing the temperature for all the boiling regimes. After closer inspection, it can be noted that the DE model slightly suffers while predicting the temperature around 374 K to 379 K. The DE model predicts the CFD values very well, with a maximum relative error below 0.05%.

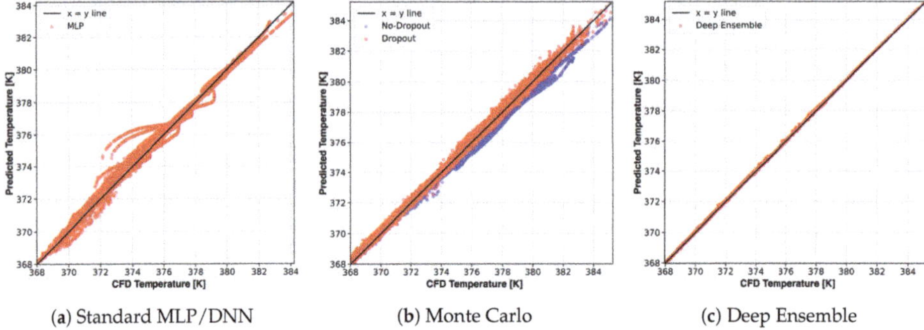

| (a) Standard MLP/DNN | (b) Monte Carlo | (c) Deep Ensemble |
|---|---|---|

**Figure 5.** Validation Temperature regression chart of the computational domain.

Overall, it can be concluded that the MLP model has lower performance when compared to the MC Dropout and the Deep Ensemble model. Therefore, in the following sections the results obtained from the MLP model for interpolation and extrapolation predictions will not be included in the main part of the discussions. For the interested readers the results are presented in the Appendix A through Figures A1–A4.

### 3.2. Interpolation Case Studies

To further demonstrate the predictive performance of the models, further unused test datasets are used to independently evaluate the model behavior. Here the interpolation dataset refers to data that were not used during training or the validation as presented in Section 3.1, with heat flux and inlet velocity values within the range of the training data. The statistics of the predictive performance of all the models tested on 3 unseen interpolation cases are given in Table 6. It can be seen that the DE model has the lowest RMSEP for the predicted void fraction and wall temperature.

**Table 6.** Performance of the models on interpolation test dataset, VF: Void Fraction, Temp: Temperature. Interpolation * is the tested data presented in the results.

| Case Datasets | U ms$^{-1}$ | $q''$ Wm$^{-2}$ | Models | RMSEP | | $R^2$ | |
|---|---|---|---|---|---|---|---|
| | | | | VF | Temp | VF | Temp |
| Interpolation * | 0.05 | 14,000 | MLP | 0.0050 | 0.162 | 0.998 | 0.953 |
| | | | MC No-Dropout | 0.0310 | 0.417 | 0.992 | 0.695 |
| | | | MC Dropout | 0.0030 | 0.146 | 0.999 | 0.962 |
| | | | Deep Ensemble | 0.0005 | 0.011 | 0.999 | 0.999 |
| Interpolation | 0.075 | 20,000 | MLP | 0.006 | 0.154 | 0.996 | 0.972 |
| | | | MC No-Dropout | 0.024 | 0.381 | 0.994 | 0.832 |
| | | | MC Dropout | 0.004 | 0.117 | 0.999 | 0.984 |
| | | | Deep Ensemble | 0.001 | 0.009 | 0.999 | 0.999 |
| Interpolation | 0.15 | 21,000 | MLP | 0.0040 | 0.052 | 0.992 | 0.998 |
| | | | MC No-Dropout | 0.0040 | 0.185 | 0.993 | 0.976 |
| | | | MC Dropout | 0.0030 | 0.037 | 0.997 | 0.999 |
| | | | Deep Ensemble | 0.0006 | 0.006 | 0.999 | 0.999 |

The results presented here are for a heat flux of $q'' = 14{,}000$ Wm$^{-2}$ and an inlet velocity of $u = 0.05$ ms$^{-1}$. The scatter plots of the predicted void fraction by MC model and DE model are plotted against the CFD value and are shown in Figure 6a,b. From Figure 6a, it is evident that the prediction with dropout activated outperforms the prediction with no dropout. In the case of no-dropout prediction, the MC model suffers as the void fraction increases and starts to deviate from the x = y line. The deviation in the predicted value is due to change in the boiling regime, and the no-dropout prediction fails to capture the change of regime. However, when dropout is activated in the MC model the predicted values fit well to the x = y line. Still, there is a slight predictive performance deterioration near the zero value, where the bubble starts to form in the channel. Aside from that, the dropout model is capable of accurately replicating the CFD values. The DE model has superior prediction performance for void fraction as shown in Figure 6b.

The predicted wall temperature values against the CFD values are shown in the Figure 6c,d. The values predicted using the MC model are presented in the Figure 6c, and it can be noticed that there is poor performance for wall temperatures above 375 K. The MC Dropout model slightly suffers to predict accurately the trend of wall temperature. Nonetheless, it is worth mentioning that the regression plot shows a maximum relative error between the CFD value and the MC dropout predicted value below 0.2%. The wall temperature values predicted by the DE model are shown in Figure 6d. From the scatter plot it can be inferred that the predicted values fit well with the x = y line. Unlike the MC dropout model, the DE model is capable of predicting all the boiling flow regimes with excellent accuracy.

The predicted profile of the void fraction and the uncertainty of the predicted values by the DNN probabilistic models along the nondimensional arc length are presented in Figure 7. Nondimensional arc length refers to the height of the minichannel near the wall. The prediction obtained from MC Dropout is illustrated in Figure 7a. It can be observed that the prediction with no-dropout under performs starting from the void fraction of 0.55 to 0.6, while the prediction with dropout keeps up with the CFD trend. The uncertainty of the predictive value is presented in terms of standard deviation $\sigma$; the filled region represents $\pm 3\,\sigma$. The filled region indicates the confidence level of the model when used for predicting unseen cases. It can be further noted that the model confidence level varies along the nondimensional arc length depending on the regime of subcooled boiling. It can be seen that there is a sudden increase in uncertainty just before a void fraction value of 0.1. This is due to nucleation occurring near the wall, causing phase change. The uncertainty of the model expands as the void fraction approaches unity; this could be related to very complex phenomena and strong instability in the flow. Whereas in Figure 7b, it can be seen that the variation in uncertainty is considerably lower compared to that of MC Dropout predicted value, which makes the DE model more robust. The DE model shows very little uncertainty near the onset of nucleate boiling and almost zero uncertainty until the void fraction approaches unity. From this, it can be noted that the DE model is capable of accurately reproducing the physical phenomena on an unseen interpolation case study.

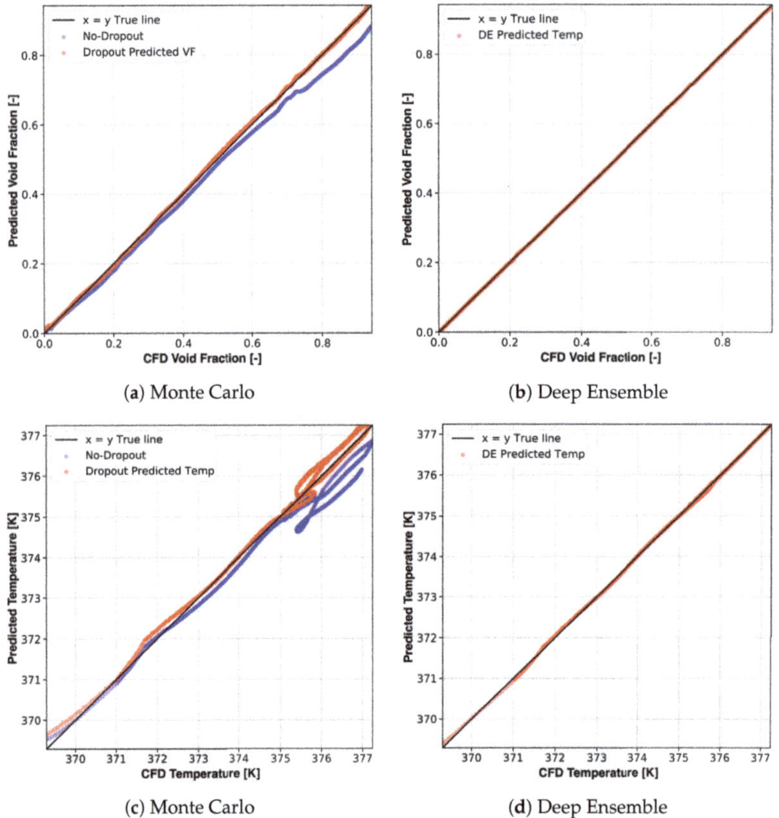

(a) Monte Carlo

(b) Deep Ensemble

(c) Monte Carlo

(d) Deep Ensemble

**Figure 6.** Regression chart for interpolation dataset using Monte Carlo and Deep Ensemble for void fraction and wall temperature at $q'' =$ 14,000 Wm$^{-2}$ and u = 0.05 ms$^{-1}$.

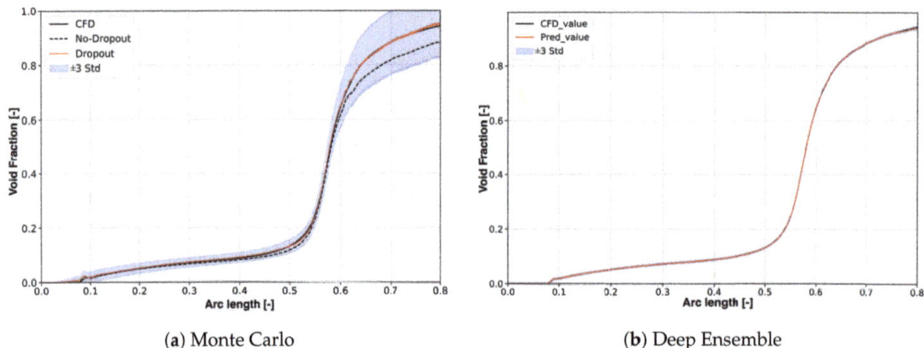

(a) Monte Carlo

(b) Deep Ensemble

**Figure 7.** Void Fraction profile along arc length for interpolation dataset using Monte Carlo and Deep Ensemble models at $q'' =$ 14,000 Wm$^{-2}$ and u = 0.05 ms$^{-1}$.

The predicted wall temperature profile and the uncertainty present along the arc length near the wall of the minichannel is shown in Figure 8. The comparison of CFD vs no-dropout, and dropout temperature profile is presented in Figure 8a. It can be noted that the value predicted with no-dropout is far away from the CFD value, especially after the critical heat flux point (377.2 K). On the other hand, the value predicted with dropout is in close range to that of CFD value, but it slightly over predicts, i.e., between an arc length of 0.6 to 0.8. The filled region in the plot presents the confidence level of the model and it is represented as $\pm 3\,\sigma$. It is evident from the plot that the uncertainty of the MC model is fairly constant until the nondimensional arc length of 0.6. Then the uncertainty of the model starts to peak as the arc length increases. To sum up, the MC model is capable of indicating where the model performance is likely to be good and deteriorate depending on the subcooled flow boiling regimes. In contrast, the $\pm 3\,\sigma$ for the DE model is relatively small compared to the MC dropout model and it is shown in Figure 8b. The DE model predicted wall temperature over lapse with the CFD wall temperature, which signifies that the model is capable of closely replicating the CFD data. It can be further noted that the DE model accurately predicts the temperature profile near with low uncertainty.

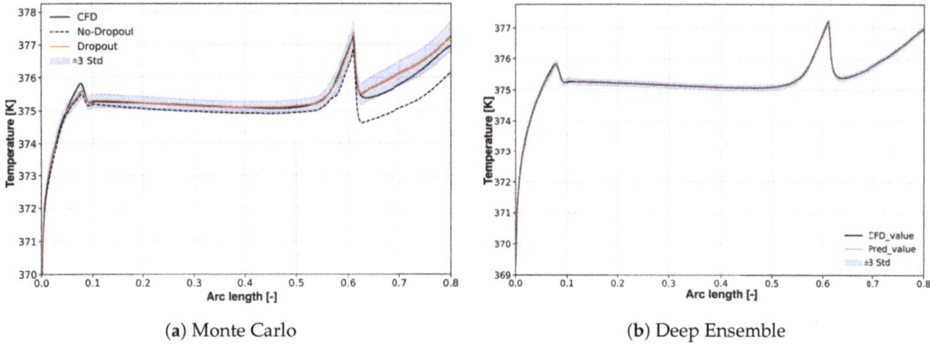

(a) Monte Carlo    (b) Deep Ensemble

**Figure 8.** Wall temperature profile along arc length for interpolation dataset using Monte Carlo and Deep Ensemble models at $q'' = 14{,}000\ \mathrm{Wm^{-2}}$ and $u = 0.05\ \mathrm{ms^{-1}}$.

From the results seen above and from Table 6 for both the models tested on the interpolation dataset, it can be concluded that the DE model has a better performance in predicting wall temperature and void fraction. The DE model also showed less uncertainty variation compared to the MC Dropout model. Therefore, the DE model is more robust for this specific problem and datatype. For the interested readers, the correlation and sensitivity that exist between the wall temperature and void fraction are shown in Appendix A through Figure A9. A detailed flow field prediction of void fraction and temperature of the minichannel by the MLP model and the DE model is shown in the Appendix A through Figures A5 and A6.

### 3.3. Extrapolation Case Studies

As expected, the models performed well for the interpolation data. This leads naturally to the next step or evaluating the models' prediction performance on an extrapolation dataset. To evaluate model capability, tests are performed on three cases where heat flux values are not within the range of the original training datasets. The statistical performance of all the DNN models when tested on 3 unseen extrapolation cases are listed in Table 7. Once again, it can be noted from the table that the DE model outperforms other models.

The results presented here for the extreme extrapolation dataset have a heat flux of $q'' = 40{,}000\ \mathrm{Wm^{-2}}$ and an inlet velocity of $0.2\ \mathrm{ms^{-1}}$. It is worth mentioning that the highest heat flux value present in the training data was $q'' = 29{,}000\ \mathrm{W/m^2}$, which implies that there is a huge gap in heat flux between the training data and the tested extrapolation dataset. The main motivation was to see if the data-driven models are capable of capturing the physics from the data used for training.

Interestingly, it will be shown that these models are capable of accurately replicating the quantities of interest.

**Table 7.** Performance of the models tested on extrapolation dataset, VF: Void Fraction, Temp: Temperature. Extreme Extrapolation * is the tested data presented in the results.

| Case Datasets | U $\mathrm{ms}^{-1}$ | $q''$ $\mathrm{Wm}^{-2}$ | Models | RMSEP | | $R^2$ | |
|---|---|---|---|---|---|---|---|
| | | | | VF | Temp | VF | Temp |
| Extrapolation | 0.125 | 25,000 | MLP | 0.0040 | 0.082 | 0.998 | 0.993 |
| | | | MC No-Dropout | 0.0130 | 0.285 | 0.987 | 0.917 |
| | | | MC Dropout | 0.0020 | 0.055 | 0.999 | 0.996 |
| | | | Deep Ensemble | 0.0004 | 0.010 | 0.999 | 0.999 |
| Extrapolation | 0.1 | 30,000 | MLP | 0.0110 | 0.180 | 0.990 | 0.986 |
| | | | MC No-Dropout | 0.0250 | 0.581 | 0.994 | 0.866 |
| | | | MC Dropout | 0.0090 | 0.101 | 0.999 | 0.995 |
| | | | Deep Ensemble | 0.0006 | 0.015 | 0.999 | 0.999 |
| Extreme Extrapolation * | 0.2 | 40,000 | MLP | 0.016 | 0.298 | 0.989 | 0.962 |
| | | | MC No-Dropout | 0.021 | 0.577 | 0.982 | 0.861 |
| | | | MC Dropout | 0.006 | 0.092 | 0.998 | 0.996 |
| | | | Deep Ensemble | 0.005 | 0.064 | 0.999 | 0.998 |

The scatter plot in Figure 9 shows the predicted void fraction and the wall temperature by the DNN probabilistic models against the CFD values. From Figure 9a, it can be noted that the prediction with no-dropout deviates from the x = y line, especially above a void fraction value of 0.2. It is clear from the regression plot that this model under predicts. Both dropout and no-dropout predictions suffer near the start of the subcooled boiling regime where nucleation is initiated near the wall. However, the dropout model prediction recovers as the void fraction increases. From the scatter plot it can be observed that the MC dropout model prediction slightly suffers when the void fraction value is around 0.2 to 0.3. In contrast, the DE model shows good predictive quality within the start phase of subcooled boiling, as seen in Figure 9b. Although the model has no problem to accurately predict the beginning of bubble formation, there is mild under prediction for the void fraction range of 0.15 to 0.25, where the number of bubbles generated grows in the channel. Nonetheless, the DE model accurately predicts for void fraction above 0.25 and the rest of the boiling regime.

The predicted wall temperature with MC dropout and MC no-dropout is presented in Figure 9c. From the plot, it can be noted that there is poor performance near 374 K for both the predictions. This is maybe related to the variation of the heat transfer coefficient near the inlet of the channel. Similar to the void fraction, the prediction obtained from the no-dropout model also deviates substantially from the x = y line. However, when dropout is activated during the prediction phase, the MC model shows better performance in predicting the wall temperature. From the plot, it can be noted that there is another slight shift near 377 K, which is likely caused by the massive generation of bubbles in the channel. The deflection of the MC dropout prediction from the CFD value has a maximum relative error of 0.2%. On the other hand, the DE model once again shows a good predictive value of the wall temperature and fits well to the x = y line as shown in Figure 9d. From the plot, it can be seen that the DE model slightly over predicts near 374 K, similar to that of the MC Dropout model. For the remaining predicted wall temperature, the DE model coincides with the x = y line.

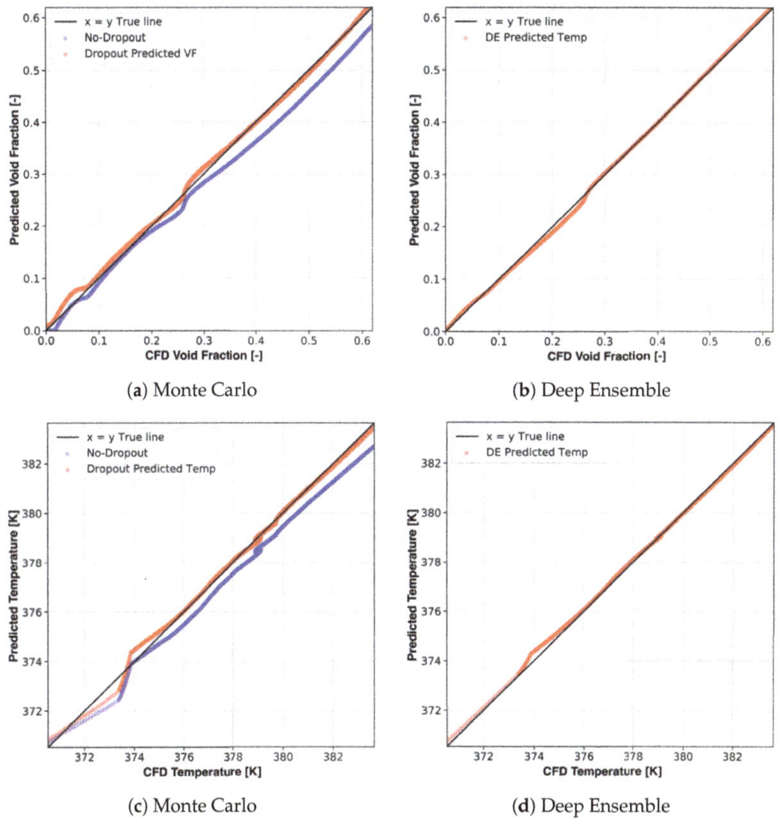



(**a**) Monte Carlo

(**b**) Deep Ensemble

(**c**) Monte Carlo

(**d**) Deep Ensemble

**Figure 9.** Regression chart for interpolation dataset using Monte Carlo and Deep Ensemble for void fraction and wall temperature at $q'' = 40{,}000\ \mathrm{Wm}^{-2}$ and $u = 0.2\ \mathrm{ms}^{-1}$.

The comparison of CFD and predicted void fraction along the nondimensional arc length of the minichannel is shown in Figure 10. The uncertainty of the models are shown as the standard deviation ($\sigma$) from the mean value, and the area filled with blue is $\pm 3\,\sigma$. The filled region indicates the model confidence when predicting an unseen dataset, and it also indicates which region is likely to be more uncertain of its predicted values. In this case, the onset nucleate boiling starts around an arc length of 0.2 though the heat flux is high when compared to the interpolation dataset presented above. The delay in the onset of nucleate boiling is because of the difference in velocity of the flow in the channel, for the interpolation data it had an inlet velocity of $u = 0.05\ \mathrm{ms}^{-1}$, and for this case, it has an inlet velocity of $u = 0.2\ \mathrm{ms}^{-1}$. This indicates that the saturated temperature is reached slower when the inlet velocity is higher, therefore resulting in a delay of the bubble formation in the channel. The void fraction prediction obtained from the MC Dropout model is illustrated in Figure 10a, where the black dash line is the no-dropout prediction and the one in red is the dropout prediction. Comparing both results to the CFD void fraction it is evident that the dropout prediction performs better in following the trend of the CFD values. The uncertainty for this model starts to peak around 0.1 arc length and gradually grows until 0.2 arc length, then there is a sudden jump in the variation of $\pm 3\sigma$ which is related to change in phase from the liquid to bubble formation. Though the mean value represented by the dropout curve is close to the CFD curve, the $\pm 3\sigma$ variation for the rest of the subcooled regime remains constant and starts to narrow down as the arc length increases. In conclusion, the MC Dropout model features considerable uncertainty near the nucleate boiling regime, which is maybe due to the continuous generation of bubbles.

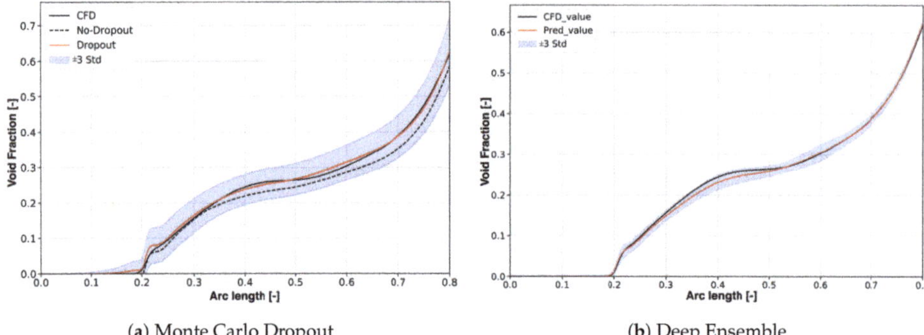

(a) Monte Carlo Dropout            (b) Deep Ensemble

**Figure 10.** Void Fraction profile along arc length for extreme extrapolation dataset using Monte Carlo and Deep Ensemble models at $q'' = 40{,}000$ Wm$^{-2}$ and u = 0.2 ms$^{-1}$.

In contrast, the DE model has a lower $\pm 3\sigma$ variation throughout the prediction of void fraction along the nondimensional arc length as shown in Figure 10b. From the plot, it can be seen that there is very little uncertainty present before the start of subcooled boiling, and the model accurately predicts the onset of nucleate boiling. However, as the void fraction increases, the uncertainty of the prediction grows until an arc length of 0.7. This increase in uncertainty is most likely due to the coalescence of tiny bubbles to larger bubbles. Finally, the degree of uncertainty diminishes as the arc length increases and the DE predicted void fraction overlaps with the CFD values. A possible explanation for this behavior of the DE model is that it can identify when there is a change in physics from liquid to vapor or when there are lots of bubbles, and it shows higher uncertainty in such a region.

The predicted wall temperature profile and the uncertainty present in the models are shown in Figure 11. Once again for the MC model, prediction obtained with no-dropout showed lower performance and its value is far away from the CFD values as illustrated by the black dash line in Figure 11a. However, when dropout is activated the predicted values showed good agreement to the CFD results, except near the region of 0.2 arc length where subcooled boiling begins. The degree of uncertainty increases as it approaches the arc length of 0.2, then it approximately remains constant for the rest of the subcooled boiling. The uncertainty trend present in the predicted wall temperature is similar to the one in the predicted void fraction as presented earlier. This implies that there is a correlation between them which is shown in the Appendix A through Figure A10. Interestingly, the uncertainty present in the DE model is relatively smaller compared to the MC Dropout model, as shown in Figure 11b. The predicted wall temperature values are very close and overlay with most of the CFD results, indicating that the model is capable of capturing the physics in the boiling regime from the training data and reproducing on unseen extrapolation dataset. It can be further noted that there is higher uncertainty near the arc length of 0.1, and this is due to unstable forced convective heat transfer between the wall and the fluid in the channel, just before the formation of bubbles.

In conclusion, both models have shown promising results on the extrapolation dataset. The models used are capable of indicating the regions of higher uncertainty while predicting the void fraction and wall temperature. From the results presented above, it can be noted that the DE model has exceptional predictive performance with lower uncertainty and is overall very robust. For the interested readers, the $\sigma$ variation between the wall temperature and void fraction for the extrapolation case are shown in the Appendix A through Figure A10. A detailed flow field prediction of the extrapolation dataset for both void fraction and temperature field is presented in Appendix A through Figures A7 and A8.

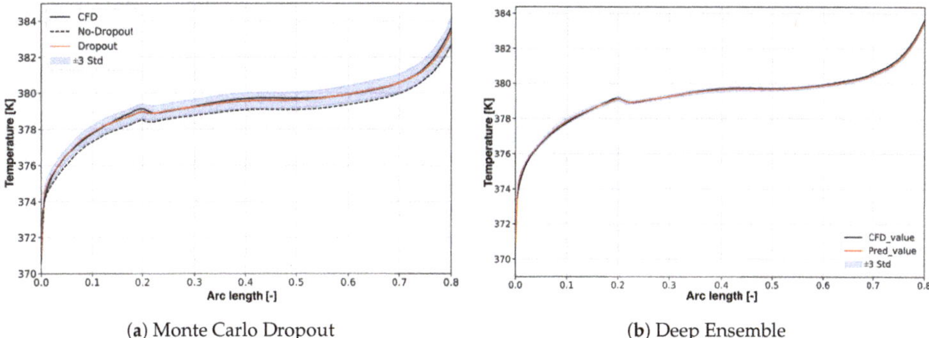

(**a**) Monte Carlo Dropout  (**b**) Deep Ensemble

**Figure 11.** Wall temperature profile along arc length for extreme extrapolation dataset using Monte Carlo and Deep Ensemble models at $q'' = 40{,}000\ \mathrm{Wm}^{-2}$ and u = 0.2 $\mathrm{ms}^{-1}$.

## 4. Conclusions

The objective of this study is twofold: firstly, to measure the accuracy of the predictions of the deep learning models compared to the CFD results and secondly to quantify the confidence level of the predictions. In this work, three supervised deep learning models have been investigated to study the subcooled boiling heat transfer in a vertical minichannel. The first method focuses on the deterministic approach, whereas the second and the third focus on the probabilistic approach to quantify the uncertainty present in the model while predicting the outputs (QoIs). The training data are obtained from CFD simulations based on the Eulerian two-fluid approach, for varying heat fluxes and inlet velocities. In total 102 cases were simulated, out of which 96 cases were used for training (80%) and validation (20%), and the remaining 6 cases were purely used for in-depth evaluation of the model's interpolation and extrapolation performance.

The models presented in this study showed a good level of accuracy while predicting the void fraction and the wall temperature. However, it has been observed that the deterministic model (standard DNN/ MLP) showed lower performance when predicting the wall temperature and void fraction. It is crucial to be able to justify the predictive nature and the uncertainty present in the model. Therefore, the probabilistic models' Monte Carlo Dropout and Deep Ensemble methods were investigated to quantify the predictive uncertainty and the confidence level of these DNNs. The output obtained from these probabilistic models is presented in the form of normal distribution rather than a deterministic value, from which the mean value and the variance of the predicted values are calculated.

According to the results presented, it can be stated that both the MC Dropout and Deep Ensemble models were able to capture the physics well from the given training data. Furthermore, they were able to reproduce these physics on unseen interpolation and extrapolation dataset. The predicted mean values, i.e., the void fraction and the wall temperature were very close to the CFD results and both performed better than the deterministic MLP model. In particular, the DE model showed exceptional predictive performance with low uncertainty. It is worth highlighting that both models were able to capture the change in boiling regimes accurately and showed higher uncertainty when there is a sudden shift in physics, for example when the nucleation starts in the minichannel. Moreover, the probabilistic models were able to reproduce the physics with good accuracy on an extreme extrapolation dataset at a heat flux of $q'' = 40{,}000\ \mathrm{Wm}^{-2}$, even though the maximum heat flux used while training the models was 29,000 $\mathrm{Wm}^{-2}$. The uncertainty quantification of the models further explains the steep change in void fraction and wall temperature when heat flux and inflow velocity are varied. On average, all the models had a $RMSEP$ error under 5% for the wall temperature and $RMSEP$ error under 2% for the void fraction with coefficient of determination $R^2 : 0.998$ and $R^2 : 0.999$, respectively. This shows that the current study can capture the underlying physics that exist in the boiling data and serves as an independent method to predict the QoIs for a new case study.

The only shortcoming of the uncertainty models compared to the standard MLP model is the computational speed, the predictive time of the uncertainty models are one order of magnitude slower compared to the deterministic MLP model. Nevertheless, the predictive time for the uncertainty models are still two orders of magnitude faster than CFD simulation. The predictive speed of uncertainty models is acceptable considering it provides better performance with the confidence levels and is reasonable for the system-level design process. Therefore, the DNN with uncertainty models can be used as a promising tool to speed up the design phase/initial guesses in the thermal management of a subcooled boiling system.

**Author Contributions:** J.S., K.K. and R.B.F. conceptualized; A.R. performed numerical simulations and wrote the CFD modeling section; J.S. outlined the Deep Learning methodology and performed the simulations, analyzed the results, and wrote the paper; I.A., K.K. and R.B.F. reviewed the paper and supervised the work. All authors have read and agree to the published version of the manuscript.

**Funding:** This research was funded by the Swedish Research Foundation under the national project Digi-Boil.

**Acknowledgments:** The authors gratefully acknowledge ABB AB, Westinghouse Electric Sweden AB, HITACHI ABB Power Grids and the Swedish Knowledge Foundation (KKS) for their support and would like to particularly thank ABB AB for providing an HPC platform.

**Conflicts of Interest:** The authors declare no conflict of interest.

## Abbreviations

| | |
|---|---|
| Adam | Adaptive Moment Estimation |
| ANN | Artificial Neural Network |
| ANSD | Active Nucleation Site Density |
| BDD | Bubble Departure Diameter |
| BDF | Bubble Departure Frequency |
| CFD | Computational Fluid Dynamics |
| CNN | Convolution Neural Networks |
| DE | Deep Ensemble |
| DNN | Deep Neural Networks |
| DNS | Direct Numerical Simulation |
| HPC | High-Performance Computing |
| LES | Large Eddy Simulation |
| MC | Monte Carlo |
| MCMC | Markov Chain Monte Carlo |
| ML | Machine Learning |
| MLP | Multi-Layer Perceptron |
| MSE | Mean Square Error |
| NLL | Negative Log-Likelihood |
| PBP | Probabilistic Backpropagation |
| POD | Proper Orthogonal Decomposition |
| QoIs | quantities of Interest |
| ROI | Region of Interest |
| ROM | Reduce-Order Modeling |
| RPI | Rensselaer Polytechnic Institute |
| UQ | Uncertainty Quantification |
| VoF | Volume of Fluid |

## Nomenclature

### Physics Constants

| | | |
|---|---|---|
| $k$ | Phase of the fluid | - |
| $\alpha$ | Void Fraction | - |
| $\rho$ | Density | $\mathrm{kgm^{-3}}$ |
| $\mathbf{U}$ | Velocity | $\mathrm{ms^{-1}}$ |
| $\Gamma$ | Rate of mass transfer per unit volume | $\mathrm{kgm^{-3}s^{-1}}$ |
| $\nabla p$ | Pressure gradient | $\mathrm{Pam^{-1}}$ |
| $\mathbf{R}$ | Combined turbulent and laminar stress, calculated based on the Reynolds analogy | $\mathrm{Nm^{-2}}$ |
| $\mathbf{g}$ | Gravitational acceleration | $\mathrm{ms^{-2}}$ |
| $\mathbf{M}$ | Interfatial momentum transfer | $\mathrm{kgm^{-2}s^{-2}}$ |
| $D_{t,k}^{eff}$ | Effective thermal diffusivity | $\mathrm{m^2s^{-1}}$ |
| $Q_{wall,k}$ | Heat flux | $\mathrm{Wm^{-2}}$ |
| $q_w''$ | Total heat flux | $\mathrm{Wm^{-2}}$ |
| $q_{w,c}''$ | Forced convection heat flux | $\mathrm{Wm^{-2}}$ |
| $q_{w,q}''$ | Quenching heat flux | $\mathrm{Wm^{-2}}$ |
| $d_{dep}$ | Bubble departure diameter | $\mathrm{m}$ |
| $f_{dep}$ | Bubble departure frequency | $\mathrm{s^{-1}}$ |
| $\mathrm{Pr}_l$ | Liquid Prandtl number | - |
| $\Delta T_{sup}$ | Wall superheat Temperature | $\mathrm{K}$ |
| $N_T$ | Dimensionless temperature | - |
| $rho^*$ | Dimensionless density ratio | - |
| $\mathrm{Ja}_w$ | Modified Jacob number | - |
| $\mathrm{Pr}_{sat}$ | liquid Prandtl number for saturated temperature | - |
| $A_{w,b,e}^{f}$ | Area fraction | $\mathrm{m^2}$ |
| $h_c$ | Condensation heat transfer coefficient | $\mathrm{Wm^{-2}K^{-1}}$ |

### Machine Learning Constants

| | | |
|---|---|---|
| $\mathbf{W}$ | Weights | - |
| $\mathbf{b}$ | Biases | - |
| $g(x)$ | Activation function | - |
| $\mathcal{L}$ | Loss function | - |
| $\hat{y}$ | Predicted value | - |
| $y$ | Target value | - |
| $\mathcal{L}1$ | Lasso Regression | - |
| $\mathcal{L}2$ | Ridge Regression | - |
| $\lambda$ | Hyperparameter | - |
| $X^*$ | Input features | - |
| $\mathbb{E}(y^*)$ | Predictive mean | - |
| $Var(y^*)$ | Predictive variance | - |
| $l$ | Prior length | - |
| $(\tau)$ | Gaussian process precision | - |
| $p$ | Probability of the neurons | - |
| $\lambda_w$ | Weight decay | - |
| $N$ | Size of the dataset | - |

## Appendix A

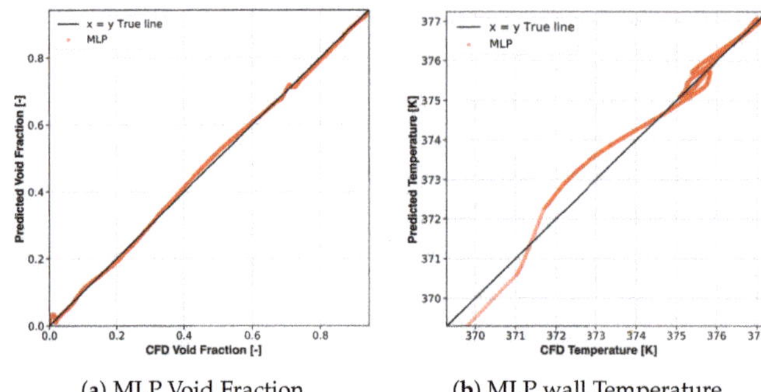

(**a**) MLP Void Fraction

(**b**) MLP wall Temperature

**Figure A1.** Predicted void fraction and wall temperature using DNN / MLP model for an interpolation dataset at $q'' = 14{,}000\ \mathrm{W/m^2}$, $u = 0.05\ \mathrm{ms^{-1}}$, from the plot it can be noted it has high nonlinearity while predicting the wall temperature.

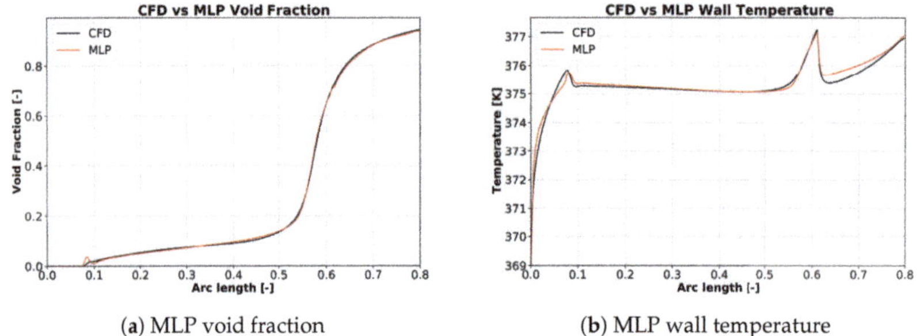

(**a**) MLP void fraction

(**b**) MLP wall temperature

**Figure A2.** Comparison between CFD and predicted void fraction and wall temperature along the arc length for an interpolation dataset at $q'' = 14{,}000\ \mathrm{W/m^2}$, $u = 0.05\ \mathrm{ms^{-1}}$. From the plot it can be noted that the MLP model has some artefacts while predicting the void fraction. It showed a sharp jump in void fraction around 0.8 arc length where the nucleation starts in the minichannel.

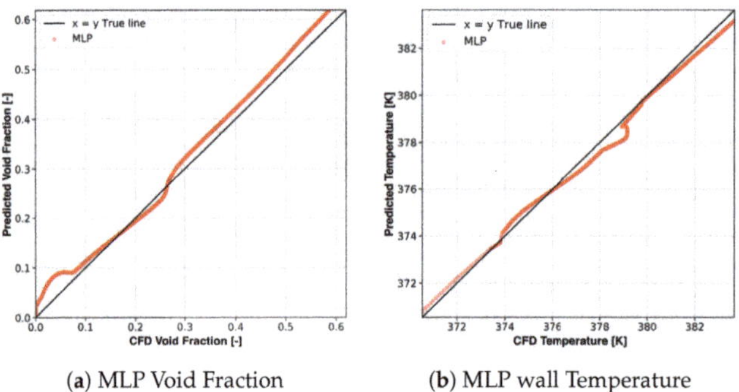

(**a**) MLP Void Fraction

(**b**) MLP wall Temperature

**Figure A3.** Regression chart of CFD vs DNN/MLP predicted void fraction and wall temperature for an extrapolation dataset at $q'' = 40{,}000\ \mathrm{W/m^2}$, $u = 0.2\ \mathrm{ms^{-1}}$. From both the plot it is evident that the MLP model lacks in reproducing the physics on an extrapolated dataset.

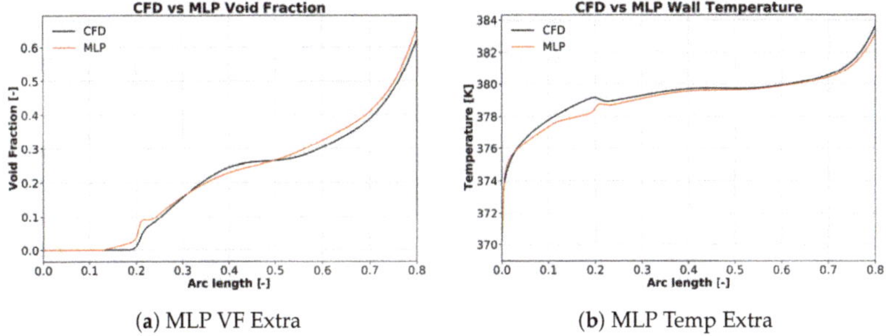

(**a**) MLP VF Extra          (**b**) MLP Temp Extra

**Figure A4.** Comparison between CFD and DNN/MLP model predicted void fraction and wall temperature along the arc for an interpolation dataset at $q'' = 40{,}000$ W/m$^2$, u = 0.2 ms$^{-1}$. From the plot it can be noted that the MLP model shows overconfident values for the void fraction and under predicted values for the wall temperature. Although the MLP model fails to capture the physics accurately it still showed good trend to that of CFD data.

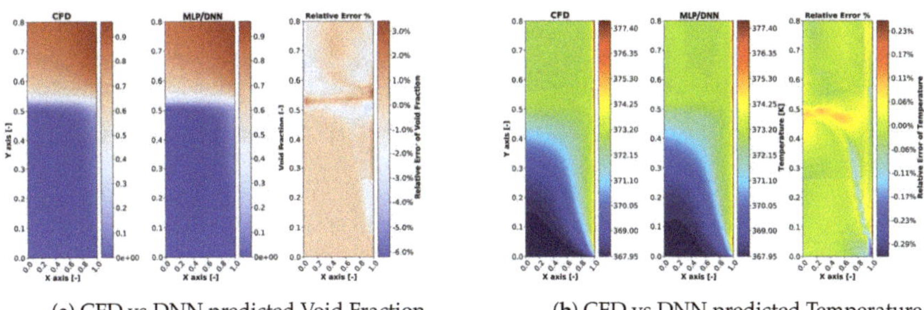

(**a**) CFD vs DNN predicted Void Fraction        (**b**) CFD vs DNN predicted Temperature

**Figure A5.** Interpolation dataset: Void fraction and temperature field of CFD and predicted by the DNN/MLP model and the relative error for an interpolation dataset at $q'' = 14{,}000$ W/m$^2$, u = 0.05 ms$^{-1}$ is presented. The plot above presents the predictive nature of the DNN model to predict the full flow field of the mininchannel. It can be seen that the DNN model is capable of reproducing the void fraction field with a maximum relative error of $-6\%$. It can be further noted that the error increases as the void fraction increases in the mininchannel. Temperature field using the DNN model is presented in Figure A5b. Compared to the void fraction prediction the DNN model has better performance when predicting the temperature field with a maximum relative error of 0.3%.

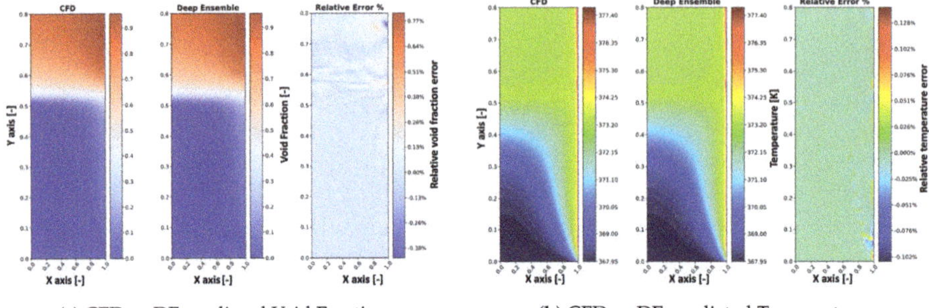

(**a**) CFD vs DE predicted Void Fraction  (**b**) CFD vs DE predicted Temperature

**Figure A6.** Interpolation dataset: Comparison of CFD and DE model prediction for $q'' = 14,000$ W/m$^2$, u = 0.05 ms$^{-1}$. It can be noted from Figure A6a that the DE model shows good performance when predicting the void fraction field with a maximum relative error of 0.77%. Similarly, the DE model shows an exceptional predicting capability for the temperature field with a maximum relative error of 0.13%. From this, it can be concluded the DE model shows almost an order of better accuracy when compared to the DNN model for the interpolation datasets.

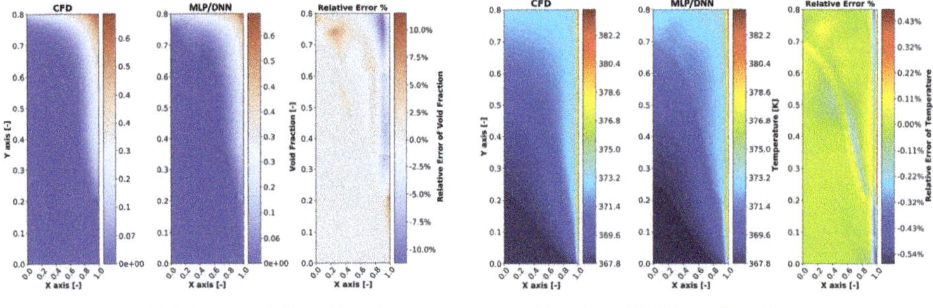

(**a**) CFD vs DNN predicted Void Fraction  (**b**) CFD vs DNN predicted Temperature

**Figure A7.** Extrapolation dataset: The CFD, DNN prediction and the relative error for $q'' = 40,000$ W/m$^2$, u = 0.2 ms$^{-1}$. It can be depicted from Figure A7a that the DNN models fail to accurately predict the void fraction field and have a maximum relative error of 10.5%. However, the DNN model shows an acceptable performance when predicting the temperature field with a maximum relative error of 0.55% as shown in Figure A7b.

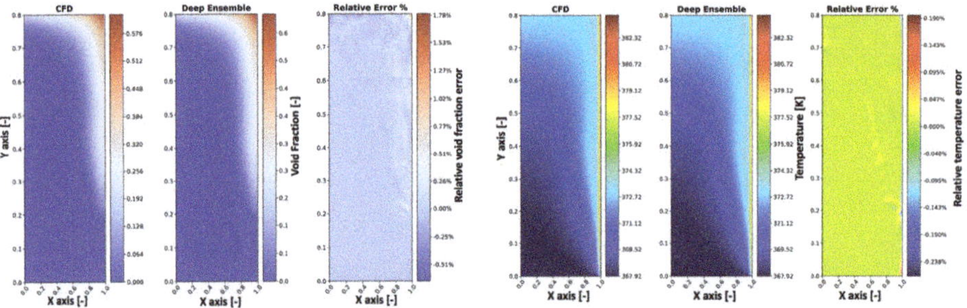

(**a**) CFD vs DE predicted Void Fraction  (**b**) CFD vs DE predicted Temperature

**Figure A8.** Extrapolation dataset: Comparison of CFD and DE model prediction for $q'' = 40,000$ W/m$^2$, u = 0.2 ms$^{-1}$. It can be again noted that the DE model outperforms the DNN model when predicting both void fraction and temperature field. The DE model has maximum relative error of 1.78% for void fraction and 0.28% for temperature field as shown in Figure A8a,b.

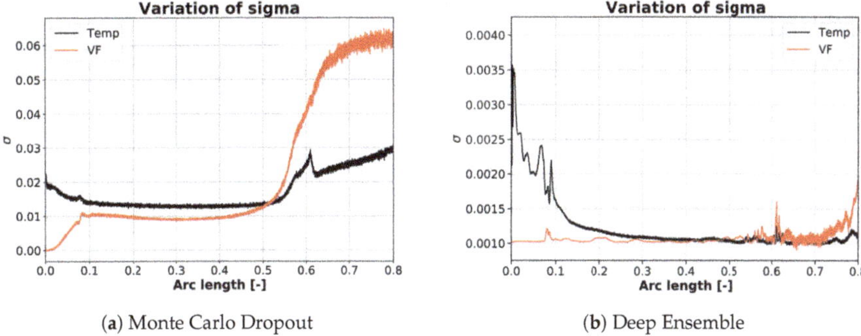

(**a**) Monte Carlo Dropout

(**b**) Deep Ensemble

**Figure A9.** Interpolation dataset: $q'' = 14{,}000\ \text{W/m}^2$, u =0.05 ms$^{-1}$. In the Figure the standard deviation of the wall temperature and void fraction along the arc length for both the models. When comparing the $\sigma$ variation between MC Dropout and DE models, it is clear that the $\sigma$ variation of DE is smaller by approximately one order of magnitude. The correlation and sensitivity that exist between the wall temperature and void fraction is shown. From the plot is evident that slight change in $\sigma$ for the void fraction influences the $\sigma$ of wall temperature. There is a sharp increase in $\sigma$ for the void fraction in MC dropout model and this is due to transition of regime from saturated boiling to film boiling.

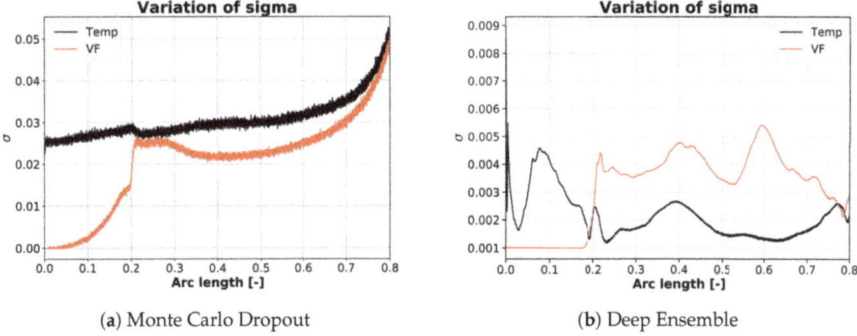

(**a**) Monte Carlo Dropout

(**b**) Deep Ensemble

**Figure A10.** Extrapolation dataset: $q'' = 14{,}000\ \text{W/m}^2$, u = 0.05 ms$^{-1}$. It is again seen that the $\sigma$ of the DE model is one order of magnitude lesser compared to that of the MC Dropout model. This implies that the DE model is less uncertain about its predicted value and is more robust in nature.

## References


1. Kurul, N.; Podowski, M. On the modeling of multidimensional effects in boiling channels. In Proceedings of the 27th National Heat Transfer Conference, Minneapolis, MN, USA, 28–31 July 1991.
2. Lai, J.; Farouk, B. Numerical simulation of subcooled boiling and heat transfer in vertical ducts. *Int. J. Heat Mass Transf.* **1993**, *36*, 1541–1551. [CrossRef]
3. Anglart, H.; Nylund, O. CFD application to prediction of void distribution in two-phase bubbly flows in rod bundles. *Nucl. Eng. Des.* **1996**, *163*, 81–98. [CrossRef]
4. Končar, B.; Kljenak, I.; Mavko, B. Modelling of local two-phase flow parameters in upward subcooled flow boiling at low pressure. *Int. J. Heat Mass Transf.* **2004**, *47*, 1499–1513. [CrossRef]
5. Rabhi, A.; Bel Fdhila, R. Evaluation and Analysis of Active Nucleation Site density Models in Boiling. In Proceedings of the Second Pacific Rim Thermal Engineering Conference, Maui, HI, USA, 13–17 December 2019.
6. Pollard, A.; Castillo, L.; Danaila, L.; Glauser, M. *Whither Turbulence and Big Data in the 21st Century?* Springer: Cham, Switzerland, 2016. [CrossRef]
7. Aubry, N.; Holmes, P.; Lumley, J.L.; Stone, E. The dynamics of coherent structures in the wall region of a turbulent boundary layer. *J. Fluid Mech.* **1988**, *192*, 115–173. [CrossRef]

8. Berkooz, G.; Holmes, P.; Lumley, J.L. The proper orthogonal decomposition in the analysis of turbulent flows. *Annu. Rev. Fluid Mech.* **1993**, *25*, 539–575. [CrossRef]

9. Podvin, B.; Fraigneau, Y.; Jouanguy, J.; Laval, J.P. On self-similarity in the inner wall layer of a turbulent channel flow. *J. Fluids Eng.* **2010**, *132*. [CrossRef]

10. Chambers, D.; Adrian, R.; Moin, P.; Stewart, D.; Sung, H.J. Karhunen–Loéve expansion of Burgers' model of turbulence. *Phys. Fluids* **1988**, *31*, 2573–2582. [CrossRef]

11. Milano, M.; Koumoutsakos, P. Neural network modeling for near wall turbulent flow. *J. Comput. Phys.* **2002**, *182*, 1–26. [CrossRef]

12. Sarghini, F.; De Felice, G.; Santini, S. Neural networks based subgrid scale modeling in large eddy simulations. *Comput. Fluids* **2003**, *32*, 97–108. [CrossRef]

13. Ling, J.; Kurzawski, A.; Templeton, J. Reynolds averaged turbulence modelling using deep neural networks with embedded invariance. *J. Fluid Mech.* **2016**, *807*, 155–166. [CrossRef]

14. Maulik, R.; San, O.; Rasheed, A.; Vedula, P. Subgrid modelling for two-dimensional turbulence using neural networks. *J. Fluid Mech.* **2019**, *858*, 122–144. [CrossRef]

15. Gamahara, M.; Hattori, Y. Searching for turbulence models by artificial neural network. *Phys. Rev. Fluids* **2017**, *2*, 054604. [CrossRef]

16. Lcun, Y.; Bottou, L.; Bengio, Y.; Haffner, P. Gradient-based learning applied to document recognition. *Proc. IEEE* **1998**, *86*, 2278–2324. [CrossRef]

17. Zhang, Y.; Sung, W.J.; Mavris, D.N. Application of convolutional neural network to predict airfoil lift coefficient. In Proceedings of the 2018 AIAA/ASCE/AHS/ASC Structures, Structural Dynamics, and Materials Conference, Kissimmee, FL, USA, 8–12 January 2018; p. 1903. [CrossRef]

18. Fukami, K.; Nabae, Y.; Kawai, K.; Fukagata, K. Synthetic turbulent inflow generator using machine learning. *Phys. Rev. Fluids* **2019**, *4*, 064603. [CrossRef]

19. Jambunathan, K.; Hartle, S.; Ashforth-Frost, S.; Fontama, V. Evaluating convective heat transfer coefficients using neural networks. *Int. J. Heat Mass Transf.* **1996**, *39*, 2329–2332. [CrossRef]

20. Celik, N.; Kurtbas, I.; Yumusak, N.; Eren, H. Statistical regression and artificial neural network analyses of impinging jet experiments. *Heat Mass Transf.* **2009**, *45*, 599–611. [CrossRef]

21. Scalabrin, G.; Condosta, M.; Marchi, P. Modeling flow boiling heat transfer of pure fluids through artificial neural networks. *Int. J. Therm. Sci.* **2006**, *45*, 643–663. [CrossRef]

22. Ma, M.; Lu, J.; Tryggvason, G. Using statistical learning to close two-fluid multiphase flow equations for a simple bubbly system. *Phys. Fluids* **2015**, *27*, 092101. [CrossRef]

23. Hassanpour, M.; Vaferi, B.; Masoumi, M.E. Estimation of pool boiling heat transfer coefficient of alumina water-based nanofluids by various artificial intelligence (AI) approaches. *Appl. Therm. Eng.* **2018**, *128*, 1208–1222. [CrossRef]

24. Poletaev, I.; Pervunin, K.; Tokarev, M. Artificial neural network for bubbles pattern recognition on the images. *J. Phys. Conf. Ser.* **2016**, *754*, 072002. [CrossRef]

25. Bernardo, J.M.; Smith, A.F. *Bayesian Theory*; John Wiley & Sons: Hoboken, NJ, USA, 2009; Volume 405. [CrossRef]

26. Neal, R.M. *Bayesian Learning for Neural Networks*; Springer Science & Business Media: New York, NY, USA, 1996; Volume 118. [CrossRef]

27. Blundell, C.; Cornebise, J.; Kavukcuoglu, K.; Wierstra, D. Weight uncertainty in neural networks. *arXiv* **2015**, arXiv:1505.05424.

28. Graves, A. Practical variational inference for neural networks. In *Advances in Neural Information Processing Systems*; Curran Associates, Inc.: Red Hook, NY, USA, 2011; pp. 2348–2356.

29. MacKay, D.J. Bayesian Methods for Adaptive Models. Ph.D. Thesis, California Institute of Technology, Pasadena, CA, USA, 1992. [CrossRef]

30. Hernández-Lobato, J.M.; Adams, R. Probabilistic backpropagation for scalable learning of bayesian neural networks. In Proceedings of the International Conference on Machine Learning, Lille, France, 7–9 July 2015; pp. 1861–1869. [CrossRef]

31. Rasmussen, C.E.; Quinonero-Candela, J. Healing the relevance vector machine through augmentation. In Proceedings of the 22nd International Conference on Machine Learning, Bonn, Germany, 7–11 August 2005; pp. 689–696. [CrossRef]

32. Lakshminarayanan, B.; Pritzel, A.; Blundell, C. Simple and scalable predictive uncertainty estimation using deep ensembles. In *Advances in Neural Information Processing Systems*; Curran Associates Inc.: Long Beach, CA, USA, 2017; pp. 6402–6413. [CrossRef]

33. Breiman, L. Bagging predictors. *Mach. Learn.* **1996**, *24*, 123–140.:1018054314350. [CrossRef]

34. Hirt, C.W.; Nichols, B.D. Volume of fluid (VOF) method for the dynamics of free boundaries. *J. Comput. Phys.* **1981**, *39*, 201–225. [CrossRef]

35. Wu, J.; Dhir, V.K.; Qian, J. Numerical simulation of subcooled nucleate boiling by coupling level-set method with moving-mesh method. *Numer. Heat Transf. Part B Fundam.* **2007**, *51*, 535–563. [CrossRef]

36. Schiller, L.; Naumann, Z. VDI Zeitung 1935. *Drag Coeff. Correl.* **1935**, *77*, 318–320.

37. de Bertodano, M.A.L. Turbulent Bubbly Two-Phase Flow in a Triangular Duct. Ph.D. Thesis, Rensselaer Polytechnic Institute, Troy, NY, USA, 1992.

38. Lahey, R.T., Jr. The simulation of multidimensional multiphase flows. *Nucl. Eng. Des.* **2005**, *235*, 1043–1060. [CrossRef]

39. Del Valle, V.H.; Kenning, D. Subcooled flow boiling at high heat flux. *Int. J. Heat Mass Transf.* **1985**, *28*, 1907–1920. [CrossRef]

40. Benjamin, R.; Balakrishnan, A. Nucleation site density in pool boiling of saturated pure liquids: Effect of surface microroughness and surface and liquid physical properties. *Exp. Therm. Fluid Sci.* **1997**, *15*, 32–42. [CrossRef]

41. Ünal, H. Maximum bubble diameter, maximum bubble-growth time and bubble-growth rate during the subcooled nucleate flow boiling of water up to 17.7 MN/m$^2$. *Int. J. Heat Mass Transf.* **1976**, *19*, 643–649. [CrossRef]

42. Brooks, C.S.; Hibiki, T. Wall nucleation modeling in subcooled boiling flow. *Int. J. Heat Mass Transf.* **2015**, *86*, 183–196. [CrossRef]

43. Ranz, W.; Marshall, W.R. Evaporation from drops. *Chem. Eng. Prog.* **1952**, *48*, 141–146.

44. Al-Maeeni, L. *Sub-Cooled Nucleate Boiling Flow Cooling Experiment in a Small Rectangular Channel*; KTH, School of Engineering Sciences (SCI), Physics; KTH: Stockholm, Sweden, 2015.

45. Kromer, H.; Anglart, T.; Al-Maeeni, T.; Bel Fdhila, R. Experimental Investigation of Flow Nucleate Boiling Het Transfer in a Vertical Minichannel. In Proceedings of the First Pacific Rim Thermal Engineering Conference, Hawaii's Big Island, HI, USA, 13–17 March 2016; PRTEC-14973.

46. Tolubinsky, V.I.; Kostanchuk, D.M. Vapour bubbles growth rate and heat transfer intensity at subcooled water boiling. In *Proceedings of the International Heat Transfer Conference 4*; Begel House Inc.: Danbury, NY, USA, 1970; Volume 23.

47. Cole, R. A photographic study of pool boiling in the region of the critical heat flux. *AIChE J.* **1960**, *6*, 533–538. [CrossRef]

48. Kingma, D.P.; Ba, J. Adam: A method for stochastic optimization. *arXiv* **2014**, arXiv:1412.6980.

49. James, G.; Witten, D.; Hastie, T.; Tibshirani, R. *An Introduction to Statistical Learning*; Springer: New York, NY, USA, 2013; Volume 112. [CrossRef]

50. Baldi, P.; Sadowski, P.; Whiteson, D. Searching for exotic particles in high-energy physics with deep learning. *Nat. Commun.* **2014**, *5*, 1–9. [CrossRef] [PubMed]

51. Parish, E.J.; Duraisamy, K. A paradigm for data-driven predictive modeling using field inversion and machine learning. *J. Comput. Phys.* **2016**, *305*, 758–774. [CrossRef]

52. Angermueller, C.; Pärnamaa, T.; Parts, L.; Stegle, O. Deep learning for computational biology. *Mol. Syst. Biol.* **2016**, *12*, 878. [CrossRef]

53. Rasmussen, C.E. Gaussian processes in machine learning. In *Summer School on Machine Learning*; Springer: Berlin/Heidelberg, Germany, 2003; pp. 63–71._4. [CrossRef]

54. Srivastava, N.; Hinton, G.; Krizhevsky, A.; Sutskever, I.; Salakhutdinov, R. Dropout: A simple way to prevent neural networks from overfitting. *J. Mach. Learn. Res.* **2014**, *15*, 1929–1958. [CrossRef]

55. Gal, Y.; Ghahramani, Z. Dropout as a bayesian approximation: Representing model uncertainty in deep learning. In Proceedings of the International Conference on Machine Learning (ICML'16), New York, NY, USA, 20–22 June 2016; pp. 1050–1059. [CrossRef]
56. Quinonero-Candela, J.; Rasmussen, C.E.; Sinz, F.; Bousquet, O.; Schölkopf, B. Evaluating predictive uncertainty challenge. In *Machine Learning Challenges Workshop*; Springer: Berlin/Heidelberg, Germany, 2005; pp. 1–27. [CrossRef]

**Publisher's Note:** MDPI stays neutral with regard to jurisdictional claims in published maps and institutional affiliations.

 © 2020 by the authors. Licensee MDPI, Basel, Switzerland. This article is an open access article distributed under the terms and conditions of the Creative Commons Attribution (CC BY) license (http://creativecommons.org/licenses/by/4.0/).

Article

# Mathematical Modelling of Active Magnetic Regenerator Refrigeration System for Design Considerations

**Aref Effatpisheh [1], Amir Vadiee [2],* and Behzad A. Monfared [3]**

[1] Department of Mechanical and Aerospace Engineering, Shiraz University of Technology, Shiraz 71557, Iran; a.effatpishe@gmail.com

[2] School of Business Society and Engineering, Division of Civil Engineering and Energy Systems, Mälardalen University, 72123 Västerås, Sweden

[3] Department of Energy Technology, School of Industrial Engineering and Management, KTH Royal Institute of Technology, 11428 Stockholm, Sweden; behzadam@kth.se

* Correspondence: amir.vadiee@mdh.se

Received: 29 September 2020; Accepted: 26 November 2020; Published: 29 November 2020

**Abstract:** A magnetic refrigeration system has the potential to alternate the compression system with respect to environmental compatibility. Refrigeration systems currently operate on the basis of the expansion and compression processes, while active magnetic refrigeration systems operate based on the magnetocaloric effect. In this study, a single layer of Gd was used as the magnetocaloric material for six-packed-sphere regenerators. A one-dimensional numerical model was utilized to simulate the magnetic refrigeration system and determine the optimum parameters. The optimum mass flow rate and maximum cooling capacity at frequency of 4 Hz are 3 L·min$^{-1}$ and 580 W, respectively. The results show that the maximum pressure drop increased by 1400 W at a frequency of 4 Hz and mass flow rate of 5 L·min$^{-1}$. In this study, we consider the refrigeration system in terms of the design considerations, conduct a parametric study, and determine the effect of various parameters on the performance of the system.

**Keywords:** design considerations; magnetocaloric effect; coefficient of performance; refrigeration; capacity; mathematical modelling; energy systems

---

## 1. Introduction

Refrigeration systems are used in many areas, including domestic cooling systems, vehicles cooling systems, food storage cabinets, and hydrogen gas liquefaction. One of the most important issues that should be considered in the design of refrigeration systems is the ability to adapt to the environment. In recent years, extensive research has been performed on the use of various types of natural refrigerants such as ammonia (R717) and carbon dioxide (R744) in compressed air systems; however, these refrigerants have some drawbacks. Not only is there a shortage of components in small scale ammonia systems, but ammonia also has a pungent smell, and is both flammable and toxic. Carbon dioxide contributes significantly to global warming and also it is not compatible with all refrigeration system lubricants [1]. Despite the widespread use of these natural systems, issues such as the possibility of flammability and toxicity are serious barriers to the use of compressed air systems in different parts of industry.

Among the common refrigeration systems, compression refrigeration systems, which use chlorofluorocarbon refrigerators, are regarded as the most harmful to the environment; that is, they are responsible for global warming and ozone layer depletion.

In this regard, a magnetic refrigeration system can be developed as a replacement with similar efficiency, yet without any negative effects on the ozone layer, as compared with various compression refrigeration systems. Unlike liquid refrigerants, solid refrigerants do not enter the atmosphere and, therefore, have no direct environmental impact. The magnetic refrigeration system is based on the magnetocaloric effect. This property appears in the presence of a magnetic field. In magnetocaloric materials, the magnetic moments of the atoms are aligned with the external magnetic field. By applying the magnetic field, the magnetocaloric material (MCM) temperature increases and the entropy decreases as a result of the decreased disorder in the system. When the magnetic field is removed, the magnetic moment of the atoms returns to a random orientation, with the result that the entropy is increased and the MCM temperature is reduced.

The magnetocaloric effect is expressed in two forms: Either a temperature change if the magnetic field changes adiabatically or an entropy change if the magnetic field changes at a constant temperature. The maximum magnetocaloric effect occurs near the Curie point of the magnetocaloric material. The Curie point of MCM is the temperature at which the material changes from the ferromagnetic state to the paramagnetic state.

The efficiency of the magnetic refrigeration system is about 30–60% of the Carnot cycle, while the compression efficiency is between 5% and 10% of the Carnot cycle [2].

Numazawa et al. [3] compared two different regenerator geometries. The results showed the optimal geometries and dimensions of the regenerator at different frequencies. It was also shown that the flat parallel plates model can obtain a lower entropy and achieve a higher cooling capacity.

Aprea et al. [4] compared two different geometries of the regenerator—a porous medium and a flat plate. The results showed that for the flat plate regenerator, the coefficient of performance (COP) of the active magnetic regenerator (AMR) cycle is higher than that of the vapor compression plant only in the high-mass-flow-rate range.

Lozano et al. [5] published the experimental results of a laboratory-made magnetic refrigeration system. The material and prototype used were Gd and a rotary, respectively. The performance of the system was studied with fluid flow rate up to 600 L/h and the frequency up to 10 Hz.

In a study by Monfared et al. [6], the environmental effects of two magnetic cooling systems and a compression refrigeration system were compared. The effect of the electricity consumption for different operational alternatives, as an effective parameter in the life cycle assessment, was also investigated.

A study of regenerator geometries was performed by Lei et al. [7]. The results showed that the parallel plate and microchannel matrices had the highest theoretical efficiency, while the packed screen and packed sphere beds were possibly more appropriate from a practical point of view.

Another study examined the economic costs of the magnetic refrigeration system. In this study, the initial cost of the magnetocaloric and the magnetic materials and utility costs were taken into account in determining the final cost of the system. For a period of 15 years, the final cost of the device was estimated to be around $150 to $400, depending on the price of the magnetocaloric material and the magnetism. For a refrigeration system with a magnetic field of 1 T, frequency of around 4.5 Hz, and a COP of two, for generating a cooling power of 50 W, the minimum costs are around $100 and $40 for the magnet and magnetocaloric material, respectively [8].

In this study, the effects of various parameters on the active magnetic refrigeration system were investigated. The main objective of this study is to select various parameters and determine appropriate values in accordance with the design conditions. In this research, it is demonstrated how to design an appropriate refrigeration system according to the target and what criteria must be considered. In this study, an efficient numerical model is proposed that reduces computation time. In the following sections, the principles of active magnetic refrigeration are described along with the modeling.

## 2. Numeral Modeling

Different methods are used for the numerical modeling of the magnetic cooling system with the purpose of predicting the AMR performance in terms of cooling capacity, temperature span, and efficiency. Nielsen et al. [9] provided an overview of the numerical methods for the simulation of active magnetic refrigeration systems. Two approaches can be found in the modeling of AMR:

Steady-state and time-dependent models. On the one hand, steady-state AMR models are simple qualitative models for estimating AMR performance in terms of cooling capacity and COP versus temperature span and mass flow rate. The major disadvantage is that it does not provide information on the interactions among different parameters. To achieve a higher performance of the AMR cycle, a time-dependent approach is required. Time-dependent models describe the heat transfer between fluid and MCM which are coupled with the change in magnetic field and fluid flow distribution which is intrinsically time-dependent.

All the implemented numerical models were derived from a mathematical model describing the heat transfer between fluid and MCM. The magnetocaloric effect (MCE) occurs in the solid material due to the changing magnetic field.

Regarding the number of dimensions in the governing equations in AMR systems, one-dimensional, two-dimensional, and three-dimensional numerical models were selected depending on the purpose of the study. A notable difference between 1D and 2D or 3D models is the implementation of heat transfer between the solid and the fluid, which is a decisive process in the AMR cycle. The equations in 2D and 3D models for heat transfer between the fluid and solid material are coupled through an internal boundary condition, while in 1D models they are coupled via a heat transfer coefficient which is a crucial parameter of the AMR system.

The earlier studies have shown excellent qualitative and quantitative agreement between the 1D and 2D numerical simulation models. Because of this, 2D and 3D models were not used in this study. The additional computation time associated with the 2D model is not necessary and good results can be achieved with a 1D model [10,11]. This has an important effect on evaluating the performance of the AMR system using different parameters.

It is important to choose efficient numerical methods that reduce the computational time and minimize numerical errors. As mentioned earlier, the 1D numerical method appears to be suitable for this project. A 1D numerical method was employed to the model active magnetic refrigeration system, which has been widely used and validated by different researchers [10,12]. By applying the changes in of the properties of the developed magnetocaloric material to the numerical scheme [10], not only can higher solving speeds be achieved, but also the computational cost of AMR models can be reduced by employing an appropriate numerical method. The computation times of MCM properties in the new and pervious numerical models at different temperatures are shown in the Table 1. The magnetic field was assumed to be 1.2 (T).

**Table 1.** The computation time of magnetocaloric material (MCM) properties in the numerical model.

| Temperature | 270 (K) | 280 (K) | 290 (K) | 300 (K) | 310 (K) |
| --- | --- | --- | --- | --- | --- |
| New method | 0.09 (s) | 0.06 (s) | 0.07 (s) | 0.06 (s) | 0.08 (s) |
| [10] | 11.44 (s) | 11.29 (s) | 11.32 (s) | 11.55 (s) | 11.38 (s) |

The computation time of the new and previous numerical models during a cycle at different mass flows are shown in the Table 2. The model assumptions are as follows: The magnetic field was assumed to be 1.2 (T) and temperature span 1 (K).

**Table 2.** The computation time of numerical models during a cycle.

| Mass Flow Rate | 0.2 L·min$^{-1}$ | 0.4 L·min$^{-1}$ | 0.6 L·min$^{-1}$ | 0.8 L·min$^{-1}$ | 1 L·min$^{-1}$ |
| --- | --- | --- | --- | --- | --- |
| New method | 770 (s) | 287 (s) | 155 (s) | 80 (s) | 37 (s) |
| [10] | 1997 (s) | 787 (s) | 395 (s) | 223 (s) | 120 (s) |

In the following sections, the basic components of an AMR model will be described. These include the theory of AMR systems, governing equations and how the modeling assumptions are implemented.

### 2.1. The Theory and Basis of the Active Magnetic Regenerator Refrigeration System Performance with the Active Regenerator

The magnetic refrigeration cycle consists of four stages—similar to the thermodynamic cycles of the compression refrigeration system. In Figure 1, an outline of the active magnetic cooling system cycle is shown with a cold heat exchanger (CHX) and hot heat exchanger (HHX).

A. The magnetocaloric material with a constant initial temperature is exposed adiabatically to the external magnetic field, and its temperature increases according to the magnetocaloric effect $(T + \Delta T_{ad})$.

B. The heat transfer fluid enters from the cold side and absorbs the heat of the solid refrigerant, thereby reducing the temperature of the solid refrigerant $(T)$. The temperature of the fluid is increased due to the heat absorbed from solid refrigerant, so that the outlet temperature of the fluid in the hot end of the regenerator is increased and the heat is transferred to a hot heat exchanger or the environment.

C. The bed of the regenerator is demagnetized adiabatically, resulting in a decrease in the MCM temperature $(T - \Delta T_{ad})$.

D. The working fluid exits from the cold heat exchanger and flows in the opposite direction in the regenerator bed (from the hot end to the cold end). The temperature of the fluid leaving the cold end is less than the refrigeration load temperature, which can absorb the cooling load from the cold reservoir $(T)$. The temperature of the regenerator backing into it is the original temperature (the zero magnetic field), thereby the cycle is completed and commence from stage A.

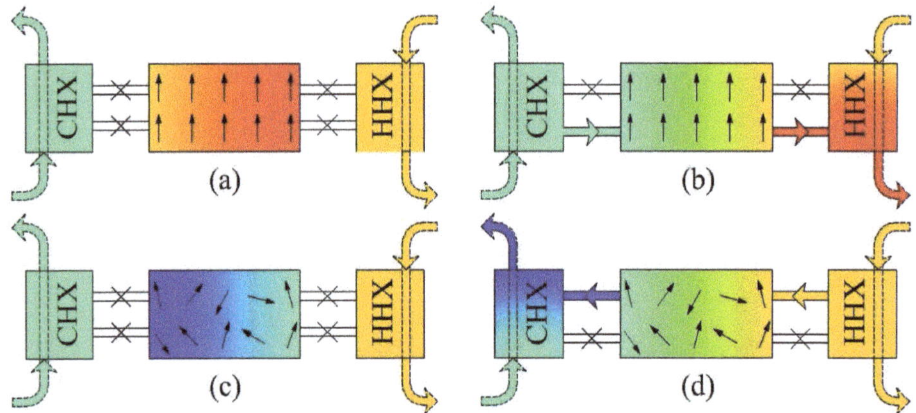


**Figure 1.** Schematic description of a magnetic refrigeration cycle: (**a**) magnetization, (**b**) hot heat transfer, (**c**) demagnetization, (**d**) cold heat transfer with a cold heat exchanger (CHX) and hot heat exchanger (HHX).

The temperature-entropy diagram is shown in Figure 2. The processes of applying the on magnetic field and removing the magnetic field are assumed to be isentropic.

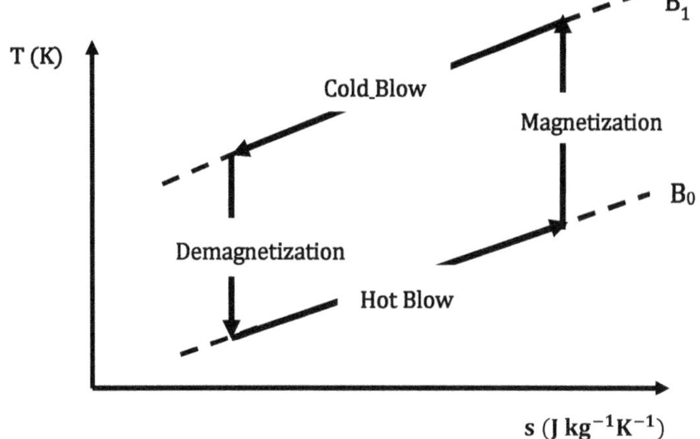


**Figure 2.** Schematic demonstrating temperature relative to entropy for the magnetic refrigeration cycle with the active regenerator.

### 2.2. Governing Equations

By applying the first law of thermodynamics for the one-segment solid refrigerants (MCM) and heat transfer fluid (energy conservation equation), the energy equations of the regenerator and heat transfer fluid are derived as follows:

$$A_c \frac{\partial}{\partial x}\left(K_{feff}\frac{\partial T_f}{\partial x}\right) - \dot{m}_f C_f \frac{\partial T_f}{\partial x} - h_{fs}A_c a_s\left(T_f - T_s\right) + \left|\frac{\partial p}{\partial x}\frac{\dot{m}_f}{\rho_f}\right| = \rho_f \varepsilon A_c C_f \frac{\partial T_f}{\partial t} \tag{1}$$

$$h_{fs}A_c a_s\left(T_f - T_s\right) + A_c \frac{\partial}{\partial x}\left(K_{seff}\frac{\partial T_s}{\partial x}\right) + (1-\varepsilon)A_c T_s \frac{\partial s}{\partial B}\frac{\partial B}{\partial t} = (1-\varepsilon)\rho_s A_c C_s \frac{\partial T_s}{\partial t} \tag{2}$$

where $k$, $T$, $\rho$, $c$, and $s$ are the thermal conductivity, temperature, density, specific heat, and specific entropy, respectively, and $A_c$, $x$, $t$, $\dot{m}_f$, $B$, $h_{fs}$, and $\varepsilon$ are the cross-sectional area, axial position, time, mass flow rate, magnetic field, heat transfer coefficient, and porosity, respectively. The subscripts $f$ and $s$ represent the fluid and solid refrigerant, respectively. The term on the right-hand side of Equation (1) represents the energy stored in the fluid. On the left-hand side of Equation (1), the first term describes the axial conduction, the second is the advection term, the third accounts for the convective heat transfer between the fluid and the solid, and the fourth accounts for the heat generated due to viscous dissipation. Similarly, in Equation (2), the right-hand-side term describes the energy storage in the regenerator bed. On the left-hand side, the first term represents the convective heat transfer between the fluid and the solid, the second term accounts for the axial conduction, and the third accounts for the MCE.

### 2.3. Modeling Conditions and Assumptions

In general, for analyzing of AMR mathematical models following assumptions had been considered:

#### 2.3.1. Properties of the Magnetocaloric Materials

The properties of the magnetocaloric materials are of great importance in a magnetic refrigeration system. An inappropriate choice of the magnetocaloric material increases the cost and reduces the system efficiency. The magnetocaloric material also has a crucial impact on the MCE. Therefore, it is important to use the MCM material that is suitable for the particular application. Some of the most important materials used in near-room-temperature applications in the magnetic cooling system cycle have been found through research [13]. A comprehensive review on a wide variety of magnetocaloric materials was reported in [14]. In this study, it was shown that for low-temperature applications, Laves phases are employed for hydrogen liquefaction, and for room-temperature applications, Gd (and its compounds), $La(FeSi)_{13}$ type phases and MnFePSi type phases are typically employed. It is notable that some materials (Heusler alloys) demonstrate giant MCE, but they have drawbacks such as significant magnetic hysteresis and slow transformation kinetics, which have rendered them less attractive for use in magnetic refrigeration devices. These problems could be eliminated, or their impact reduced by materials engineering processes. Good examples of these methods are the following: Multiphase materials and composites, fabrication techniques, powder metallurgy, nanostructures, and special treatment conditions (hydrogenation, annealing) [14].

Magnetocaloric materials with high MCE are still in the development stage, requiring more investigation and the design of solid refrigerants suitable over a wide working temperature. Rare earth (RE)-based intermetallic compounds are promising materials which manifest a significant MCE. The properties and MCE of RE-based intermetallic compounds depend on the crystal structure, magnetic properties, and magnetic phase transition. The RE-based intermetallic compounds can be stratified into three categories: Binary, ternary, and quaternary compounds. It was indicated that by modification of the concentrations and physical properties of these compounds, a reasonable MCE with a wide range of working temperatures is achievable [15].

RE-based intermetallic compounds ($Er_{40}Zn_{60}$) are practical at low temperatures and can, therefore, be used in cryogenic applications. A number of magnetic refrigerants with excellent cryogenic magnetocaloric properties were summarized in [16], which also briefly reviewed RE-based intermetallic compound materials with a substantial MCE in low magnetic fields. In addition, it was shown that magnetic properties and magnetic entropy changes could be modified by applying hydrostatic pressure.

In this study, a single layer of Gd was used as a refrigerant. Gadolinium is one of the MCM materials which has widely been used in magnetic refrigeration for near-room temperature applications. The MCE of pure gadolinium near room temperature is around 293 K. In addition, the properties of gadolinium are fairly acceptable ($\Delta T_{ad} = 3.3$ K, $C_H = 300$ J·kg$^{-1}$·K$^{-1}$, $\Delta s_M = 3.1$ J·kg$^{-1}$·K$^{-1}$) at a magnetic field change of 1 T [17]. Its properties have made it an ideal reference candidate to compare to other different MCM materials in magnetic refrigeration applications. However, two of the main drawbacks when choosing Gd are its purity and price, which could limit its applications. In spite of these challenges, its MCE properties and lack of magnetic hysteresis make it the first choice for room-temperature magnetic refrigeration. The properties of MCM were obtained with respect to the mean field theory (MFT); a more complete description of the mean field theory was presented in a study by Petersen [18]. According to the mean field theory the magnitude of the entropy changes with magnetization is predictable. The thermodynamic properties of the MCM can be obtained by the MFT for a wide range of temperatures and magnetic fields.

Figure 3a shows the isothermal entropy change in the MCM during the magnetization of pure gadolinium when it is magnetized from 0 to 1.5 Tesla. Entropy changes are shown in two manners: Experimentally [10] and the MFT. As can be seen, there are differences between the MFT and experimental data. The Curie point was predicted by the MFT to be sharp point and well-defined, while in the experimental curve, it is smooth. The entropy changes at temperatures below the Curie temperature in the MFT are more than the experimental values, while above the Curie point a good agreement is found between the experimental results and the outcome of the MFT.

These discrepancies can be explained by the following reasons. Firstly, due to the impurities in gadolinium which are observed in experimental tests, its entropy changes are less than those predicted by the mean field theory. Secondly, the MFT over-estimates the values and is too idealized. Thirdly, it was assumed that the MCM material was uniformly distributed throughout the material, while in the real model this was not possible. Although there is a difference the between numerical and experimental results, the MFT method is an efficient for modeling the properties of a magnetocaloric material. Not only does it predict the properties of magnetocaloric materials in the steady state at wide temperature ranges and magnetic fields, but it also removes some of the issues associated with experiments. Figure 3b shows the entropy changes at different temperatures and magnetic fields from zero to 1 T and compares these to the results reported in [18]. As can be seen, there is a good agreement between the two works. The properties of MCM were obtained in [18] by MFT.

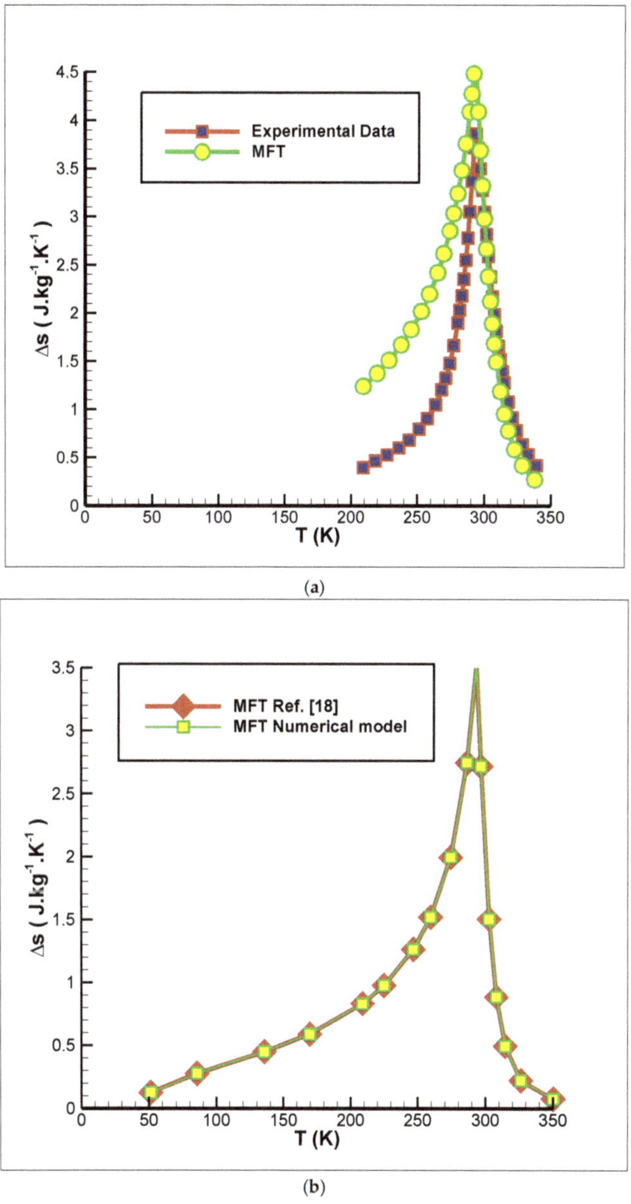

(a)

(b)

**Figure 3.** (**a**) the comparison of the isothermal entropy changes of Gd during the magnetization (0–1.5 T) and (**b**) the comparison of the isothermal entropy changes of Gd (0–1 T) in [18] which were obtained by mean field theory (MFT).

### 2.3.2. The Type of Magnetic Refrigeration System

The first magnetic prototype was built in 1976 by Brown [17], and a wide range of different prototypes have been designed and built over the 40 years since. There is a one feature that plays an important role for all these prototypes. This feature classifies devices into two groups which are a reciprocating (linear motion) manner and a rotary manner. The reciprocating motion means that the AMR system or the magnet moves in a reciprocal direction back and forth, and in the rotary system, the AMR or magnet is rotating. Each of these methods has its advantages and disadvantages, and the appropriate choice depends on the purpose of the prototype. One of the considerable drawbacks of reciprocating systems is that the operating frequencies are limited to ≤1 Hz, which could influence the refrigeration capacity, there is no frequency limitation in the rotary system. However, reciprocating devices are still practical as experimental testing devices compared to rotary system.

On the other hand, there are two main problems associated with the rotary system. The first issue is in the assembly of AMR and magnet systems, and the second, in the auxiliary components, such as the heat exchanger and heat-transfer fluid system. However, most rotary devices, as can be expected, work more efficiently as they operate continuously compared to a linear device, which does not.

In a study performed by Yu et al. [19], different types of magnetic refrigerators and heat pumps with different geometries were investigated. In the present study, the rotary system that was published by Zimm et al. [20] was chosen.

### 2.3.3. The Other Simplifying Assumptions

- It was assumed that there is no phase change occurs in the heat transfer fluid, to prevent the freezing of water, a mixture of water and 10% ethylene glycol was used as a heat transfer fluid.
- The fluid heat transfer was assumed to be incompressible (constant density). Among the properties of the heat transfer fluid used in the modeling of the magnetic cooling system were the viscosity, thermal conductivity coefficient, and heat capacity. Fluid properties were considered to be a function of temperature; therefore, the properties of the fluid were considered as a polynomial function of the temperature.
- There is no flow leakage during the cycle which is an ideal assumption. It may be difficult to control the flow leakage during experiments.
- The radiation heat transfer is negligible compared to the convective and conductive heat transfer which is a good approximation in room temperature applications.
- It was assumed that the solid material is distributed uniformly in the regenerator. This assumption is not applicable in the experimental models.
- One possible way to predict the initial distribution temperature of the fluid and the regenerator is to extrapolate them from the linear profile.
- Regardless of magnetic hysteresis, the on and off magnetic fields are assumed to be adiabatic and reversible processes. When the magnetic field is removed completely or reaches the zero value, the temperature of the magnetocaloric material returns to its original (initial) state.

### 2.4. Selecting Regenerator Geometry

Selection of the regenerator geometry, namely, the main parameter that determines the performance of the magnetic refrigeration system, was considered. In a study on various geometries of the magnetic refrigeration system [7], parallel flat plates and microchannel matrices were found, theoretically, to have the highest efficiency; however, it was found that it is possible to use a packed sphere bed in applications. The spherical particle bed characteristic is the spherical particle diameter, in which the same or different particle diameters can be considered. In this study, the sphere particle diameters were considered to be the same. A schematic of the various regenerator geometries is shown in Figure 4. In Figure 4, the dark-colored regions represent solid matter and white areas indicate the heat transfer fluid that is normal for the case of a sheet of paper.

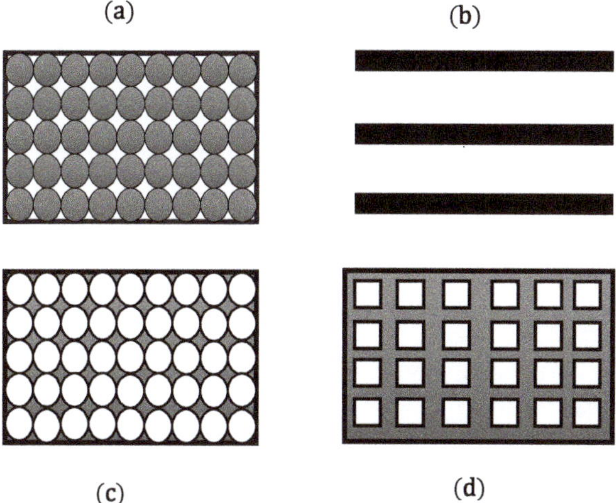

**Figure 4.** Different regenerators geometries: (**a**) spherical particle bed, (**b**) parallel flat plates, (**c**) spherical microchannel matrices, and (**d**) square and rectangular microchannel matrices.

The correction of the pressure drop for spherical particles is as follows [21]:

$$\frac{dp}{dx} = \frac{180(1-\varepsilon)^2 \mu}{d_p^2 \varepsilon^3} V + \frac{1.8\rho(1-\varepsilon)}{d_p \varepsilon^3} V^2. \tag{3}$$

In the pressure drop equation for spherical particles, the velocity of the fluid is obtained by dividing the volumetric flow rate (Q) of the fluid over the cross-sectional area of the regenerator. In addition, $d_p$ is the particle diameter, and the porosity ($\varepsilon$) is considered to be 0.362. The specific cross-sectional area is given in Equation (4):

$$a_s = 6\frac{(1-\varepsilon)}{d_p}. \tag{4}$$

The Nusselt number for packed sphere nodes was presented by Wakao and Kaguei [22]:

$$h = \frac{Nu_{fs}k_f}{d_p} = \frac{k_f}{d_p}\left(2 + 1.1Re^{0.6}Pr^{\frac{1}{3}}\right). \tag{5}$$

To calculate the thermal resistance of spherical particles, Equation (5) was modified to give Equation (6) [23,24]:

$$\frac{1}{h^*} = \frac{1}{h} + \frac{d_p}{\beta k_s} \tag{6}$$

where $k_{feff}$ and $k_{seff}$ are the effective thermal conductivity for the fluid and magnetocaloric material, which are presented by Equations (7) and (8) [25]:

$$k_{feff} = k_f(\varepsilon + 0.5RePr) \tag{7}$$

$$k_{seff} = (1-\varepsilon)k_s. \tag{8}$$

The temperature range $\Delta T = T_H - T_C$ was calculated according to the difference between the hot- and cold-source temperatures. $T_C$ is the temperature of the cold source and $T_H$ is the hot source temperature. The refrigeration capacity and heating load were calculated according to Equations (9) and (10), and the COP was calculated according to Equation (11):

$$\dot{Q}_H = ( \int_0^\tau |\dot{m}_f| \left( e_{f,x=L} - e_{f,T_H} \right) dt ) / \tau \tag{9}$$

$$\dot{Q}_C = ( \int_0^\tau |\dot{m}_f| \left( e_{f,T_C} - e_{f,x=0} \right) dt ) / \tau \tag{10}$$

$$COP = \frac{\dot{Q}_C}{\dot{Q}_H - \dot{Q}_C} \tag{11}$$

where $\tau$, $t$, and $\dot{m}_f$ are the cycle period, time, and mass fluid flow rate, respectively.

$e_{f,x=L}$ and $e_{f,x=0}$ are the enthalpy of the output fluid from the cold and hot ends of the regenerator at any time step. The temperature of the fluid at the time it entered the cold and hot ends of the regenerator was assumed to be equivalent to the cold- and hot-source temperatures, respectively. The heat conduction in the ends was neglected. The positive mass flow rate is defined when the fluid flows from the cold end to the hot end of the regenerator and the negative mass fluid flow rate is defined as the reverse. The boundary conditions of the regenerator are shown in Equations (12) and (13) in accordance with the fluid flow direction:

$$if \; \dot{m}(t) > 0 \; then \; T_f(x = 0, t) = T_C \tag{12}$$

$$if \; \dot{m}(t) < 0 \; then \; T_f(x = L, t) = T_H \tag{13}$$

The governing Equations (1) and (2) are solved numerically using the finite differences method. The fluid and regenerator energy balance from Equations (1) and (2) are discretized and provided for each control volume. It is assumed that the fluid and MCM properties in each time step are constant. In other words, there are no significant changes in the fluid and MCM properties in each time step. The fluid and MCM temperatures at each point in the time step will be obtained by solving the discretized equations for the previous time step. The discretized forms of the fluid and regenerator energy equations are solved by assuming an initial temperature, and the fluid and regenerator temperatures are obtained at the next time step. Discretized equations were iteratively solved, and the iteration process continued until the fluid and regenerator temperatures between repetitions remained unchanged. Since in numerical solutions, the contiguous space becomes discrete, one of the important advantages of this method is that it is independent from computational networks. It is essential to choose a grid which is not dependent on the number grids to accurately predict the performance of the AMR. To determine the sensitivity of the numerical model from the number of time steps and axial nodes, grid sensitivity analysis was performed. The independence of the results from the number of computational nodes and the disjoint of temporal steps are indicated in Figures 5 and 6, respectively. These figures represent refrigeration capacity (Qc) based on the number of axial nodes and time steps. In the numerical model, the number of time steps and axial points in the longitudinal direction (the spatial node) were considered to be 8000 and 60, respectively. The parameters used in the grid study were a frequency of 4 Hz, mass flow rate of 3 L·min$^{-1}$, porosity of 0.362, and sphere diameter of 0.5 mm. The hot and cold reservoirs temperatures were 286 and 283 (K), respectively.

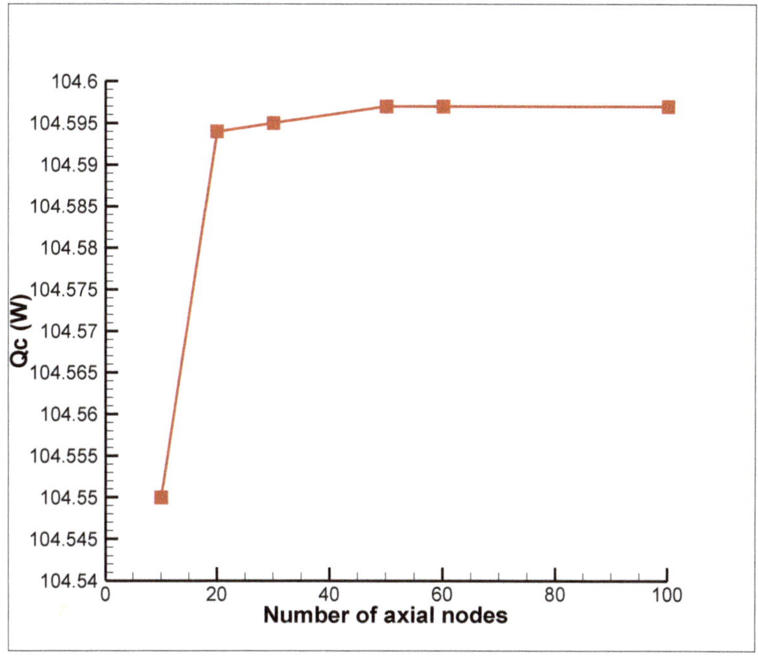

**Figure 5.** Independence of the results from the number of computational nodes.

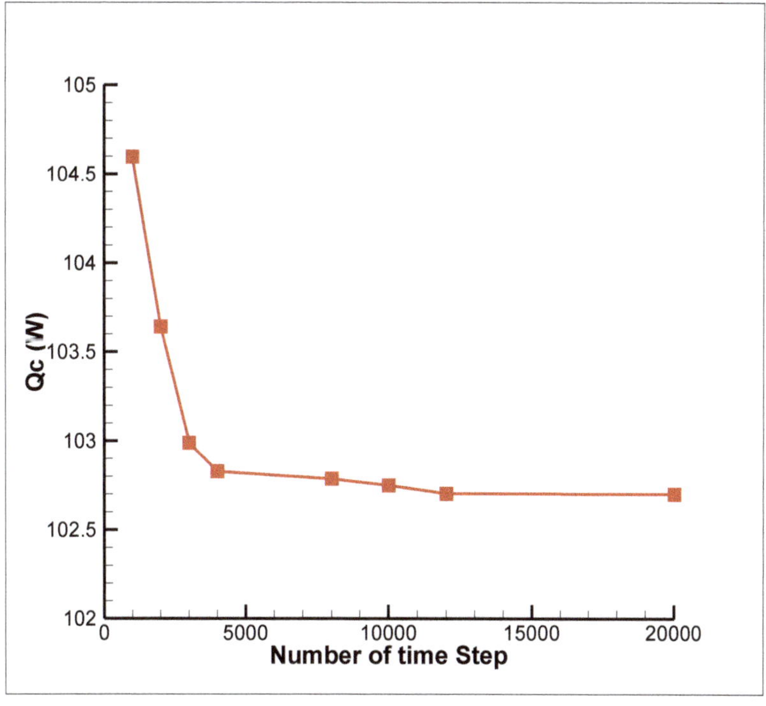

**Figure 6.** Independence of the results from the temporal step disjoining.

### 2.5. The Heat Transfer Fluid

The heat transfer fluid has an important role in the performance of the AMR. It should have acceptable thermodynamic properties, i.e., high thermal conductivity and low viscosity. These characteristics are significant at high operation frequencies. The majority of AMR systems use water as a heat-transfer fluid. The advantage of water is that it has good properties, is non-toxic and highly available. The disadvantage of using water is that it can cause corrosion of the magnetocaloric materials and also turn into the solid phase below 0 °C. A solution to this is the use of a mixture of anti-corrosives and anti-freeze with water (e.g., at a ratio of 20:80), this can not only prevent corrosion, but also decrease the freezing point of the mixture to below 0 °C. The impact of the heat-transfer fluid on the AMR performance is undeniable. Numerical, experimental, and theoretical studies have been performed on AMR performance [17]. In [17], they compared four types of heat-transfer fluid: Water, liquid metals, different alcohols, and different mixtures of water and ethanol. The results showed that liquid metals had the best cooling properties, while pure water yielded the best AMR performance. In spite of the fact liquid metals (e.g., gallium) have a better performance compared to the other fluids, they may be highly toxic and, therefore, cannot be used as the working fluid in a magnetic refrigeration system. They also often have high densities or viscosities, which results in viscous dissipation.

Generally, the distribution of fluid flow in all numerical models is assumed to be a periodic fluid flow as a function of time. There are two methods to assume the mass flow rate profile, a discrete mass flow rate profile through a ramping method (an appropriate function) and a continuous flow curve (sinusoidal or hyperbolic tangent) [9]. Discrete mass flow rate profiles were determined to be the best choice because they eliminate some of the errors which originate from specific experimental devices [9]. The function of the mass fluid flow is shown in Figure 7a.

### 2.6. The Magnetic Field

With respect to the magnetic field, the same method that was used for the mass flow rate profile can be considered for the magnetic field distribution. For the numerical model to resemble the experimental model, the function of the magnetic field could be performed through a number of time steps. Figure 7b shows the function of the magnetic field during a cycle.

The parameters considered in the numerical model are presented in Table 3. In this model, the intensity of the magnetic field and the dimensions of the regenerator geometry were considered to be constant. The other parameters were changed to evaluate the effect on the performance of the magnetic refrigeration system. The refrigeration capacity, heating load, and COP were defined as positive.

**Table 3.** Modeling parameters.

| Parameter | Value |
| --- | --- |
| Hot reservoir temperature $T_H$ | 290–305 (K) |
| Cold reservoir temperature $T_C$ | −273–300 (K) |
| Frequency $f$ | 0.67-1-2-4 (Hz) |
| Porosity $\varepsilon$ | 0.15-0.25-0.362-0.45-0.55 |
| Sphere diameter $d_p$ | 0.1-0.2-0.3-... (mm) |
| Volumetric flow rate Q | 0.2-0.3-0.4-... (L·min$^{-1}$) |
| Regenerator volume $V_r$ | 33 cm$^3$ |
| Number of beds | 6 |
| Intensity of magnetic field B | 1.5 T |
| MCM density $\rho_s$ | 7900 kg·m$^{-3}$ |
| MCM thermal conductivity $k_s$ | 11 W·m$^{-1}$·K$^{-1}$ |

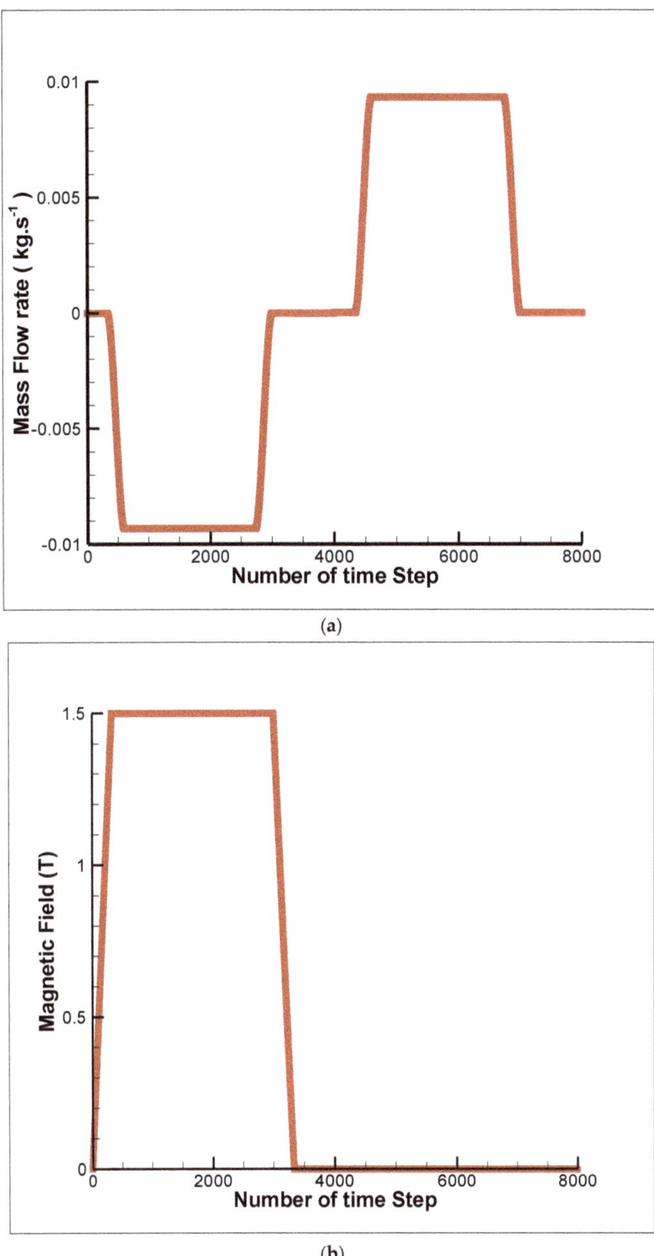

**Figure 7.** (**a**) mass flow rate and (**b**) magnetic field variations for each bed during a cycle in a rotary device.

## 2.7. Validation

After the grid study for the numerical model described in the previous section was performed, the numerical model was validated by the experimental data. The results of this study were compared with those of a laboratory study [10]. The outlet temperature of fluid from the cold end of the regenerator ($T_{outlet}$) was obtained at the demagnetization step at a frequency of 1 Hz. The inlet temperature was equal to the temperature of the cold reservoir. The diameter of the spherical particles was 0.5 mm, and the porosity was 0.362 for different mass flow rates (Figure 8).

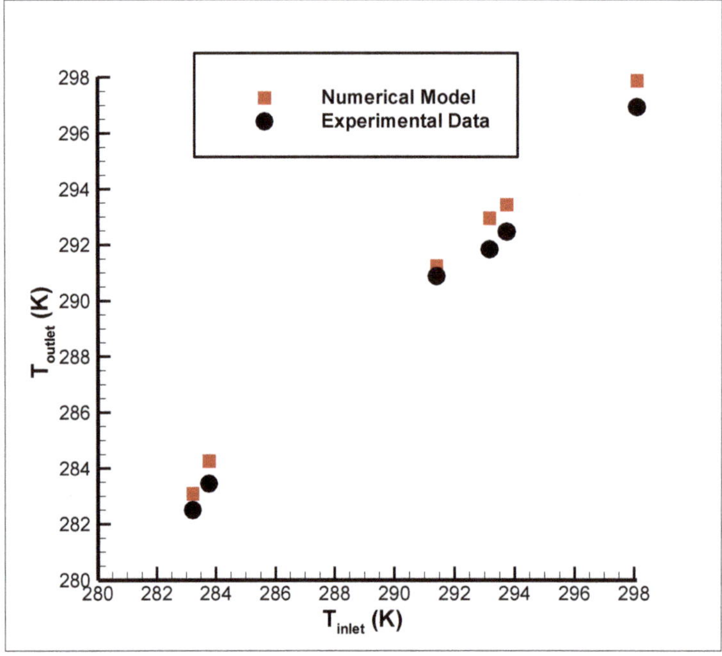


**Figure 8.** The comparison of experimental and numerical results.

## 3. A Study of the Parameters That Effect System Performance

Initially, the effect of the mass flow rate on the output of the model, including the refrigeration capacity and coefficient of performance, was investigated in different working conditions. In addition, the effect of different parameters on the performance of the magnetic refrigeration system was shown. Design charts were based on the coefficient of the performance and refrigeration capacity of the magnetic refrigeration system.

### 3.1. Fluid Flow Rate

The effect of the mass fluid flow rate parameter was investigated, and the other parameters were assumed to be constant. By increasing the mass flow rate of the fluid, the heat transfer between the fluid and the magnetocaloric material increased because more heat is absorbed by the fluid. Therefore, the coefficient of performance and refrigeration capacity increased. Furthermore, by increasing the mass flow rate of the fluid, the viscous dissipation increased and produced heat in the fluid. Increasing the heat transfer rate between the fluid and the magnetocaloric material compensated for this so that the mass flow rate of the fluid could achieve an optimum value. From this point on, by increasing the mass flow rate of the fluid, the viscous dissipation had a greater effect than the increase in the heat transfer rate between the fluid and the solid refrigerant, causing the coefficient of performance and refrigeration capacity to decrease overall (Figure 9). By plotting a horizontal line parallel to the horizontal axis, the graph was disconnected at two different points. This meant that the refrigeration capacity and coefficient of performance were proportional to the two different mass fluid flow rates.

It is important to note that the viscous dissipation of the fluid for the lower mass fluid flow rate was lower than that in the higher mass flow rate. Thus, it is important to ensure that the fluid flow rate is either too low or too high.

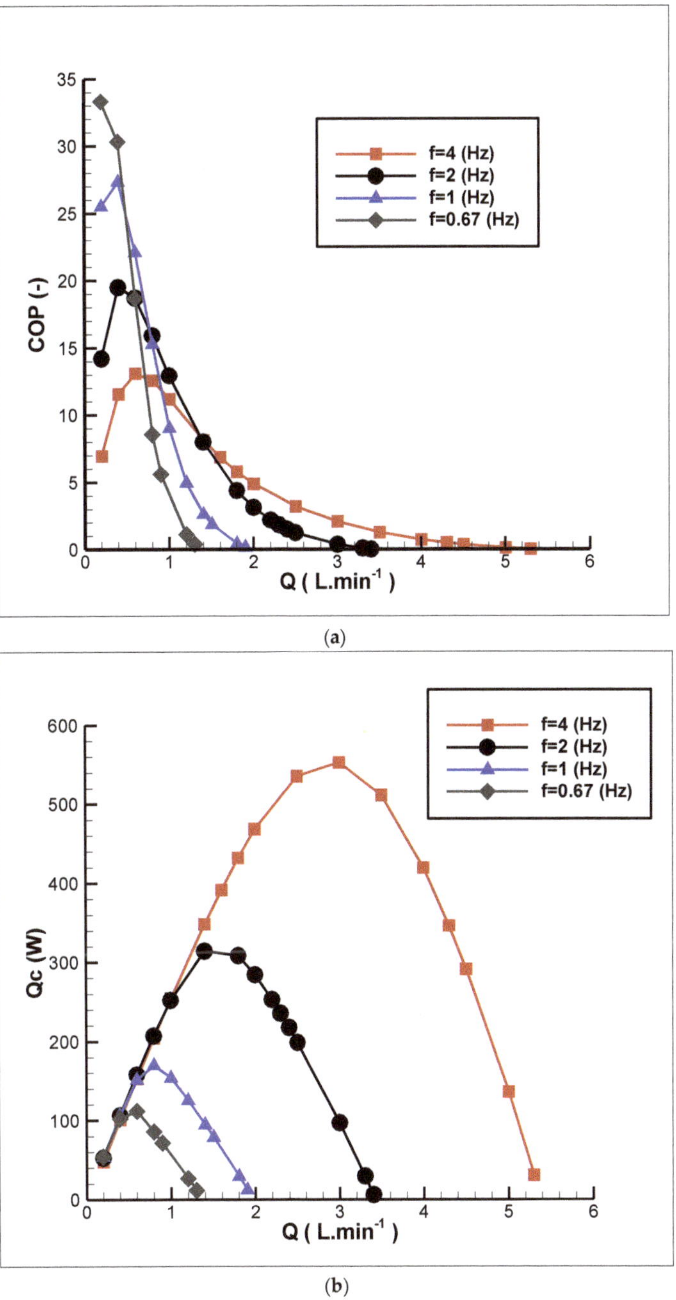

(a)

(b)

**Figure 9.** (**a**) the coefficient of performance and (**b**) refrigeration capacity based on the volumetric flow rate for a temperature span of 1 K at different frequencies.

Table 4 shows the parameters that were considered to evaluate the effect of the volumetric flow rate.

**Table 4.** Applied parameters to evaluate the effect of the volumetric flow rate.

| Temperature Span (K) | Porosity (-) | Frequency (Hz) | Sphere Diameter (mm) $T_H = 291, T_C = 290$ |
|---|---|---|---|
| 1 | 0.362 | 0.67-1-2-4 | 0.5 |

### 3.2. Frequency

As shown in Figure 9, the refrigeration capacity and coefficient of performance increased with decreasing frequency at a low flow rate because the heat transfer time between the fluid and the magnetocaloric material increased. At high volumetric flow rates, by increasing the frequency, the number of cycles completed at the time step increased, which increased the refrigeration capacity and the coefficient of performance. As it observed, higher frequencies are achievable in higher mass flow rates and, therefore, it would result in higher viscous dissipation and consequently, higher input pump work was required that reduced the overall performance.

### 3.3. Temperature Span (Hot- and Cold-Source Temperature Difference)

The temperature span is another important design parameter. In this study, the effects of the temperature span variation on the refrigeration capacity and coefficient of performance were evaluated and the results are shown in Figure 10. According to Figure 10, it can be concluded that the refrigeration capacity and coefficient of performance were inversely proportional to the temperature range. In the low-temperature span, due to the fact that less power and energy are needed to transfer heat from a cold source to a hot supplier, it was expected that the refrigeration capacity and coefficient of performance would be higher than in the higher temperature span. Hence, the refrigeration capacity and the coefficient of performance decreased by increasing the temperature span. As the temperature span increased due to the axial heat conductivity from the hot end to the cold end of the regenerator, heat loss is increased, and it would result in reducing the overall efficiency.

### 3.4. Spherical Particle Diameter

One of the key parameters of the packed sphere bed regenerator is the size of the spherical particle diameter, as shown in Figure 11. By increasing the diameter of the spherical particles, the refrigeration capacity and the coefficient of performance first increase and then decreased meaning that there will be an optimal diameter for the spherical particles. Below the optimal diameter of spherical particles, increasing it reduces the viscous dissipation, thereby increasing the refrigeration capacity. On the other hand, beyond the optimal diameter of spherical particles, increasing it results in a decrease in the coefficient of the heat transfer and, therefore, the coefficient of performance and the refrigeration capacity decrease.

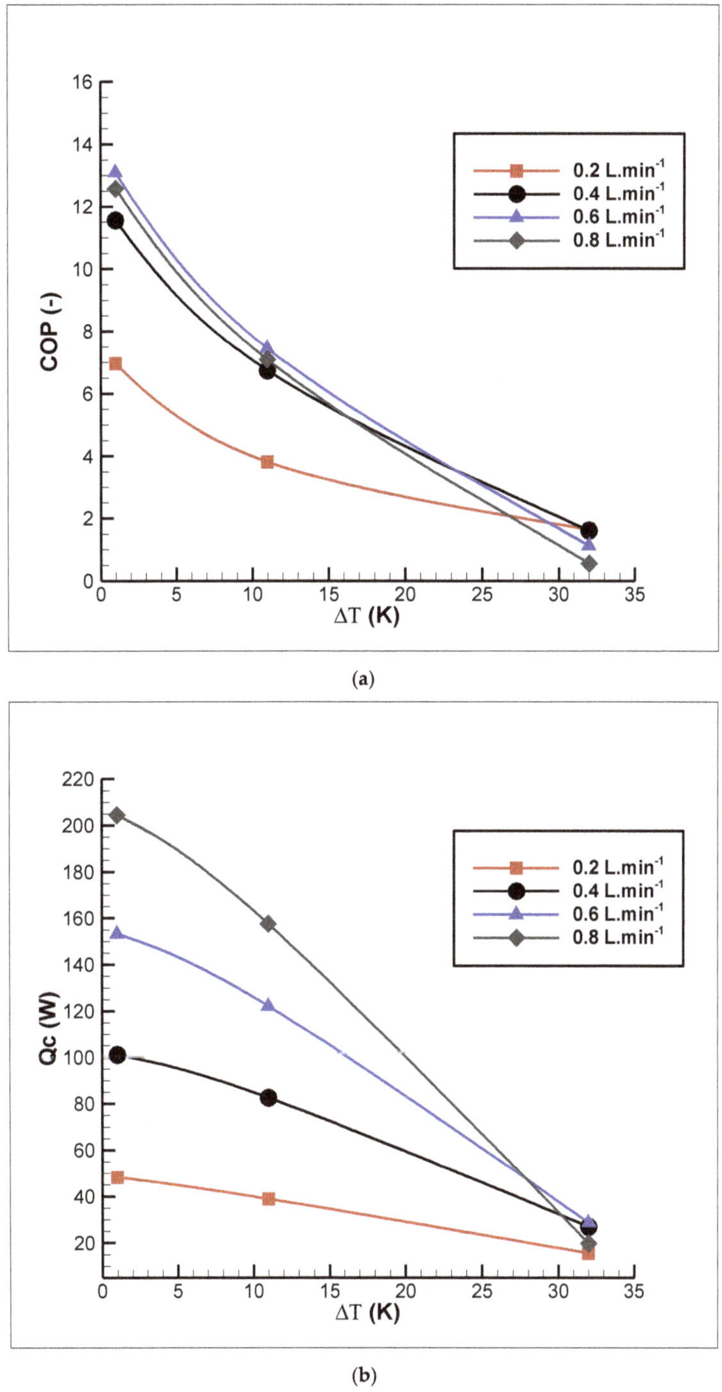



**Figure 10.** The effects of the span temperature variation on the coefficient of performance (**a**) and refrigeration capacity (**b**).

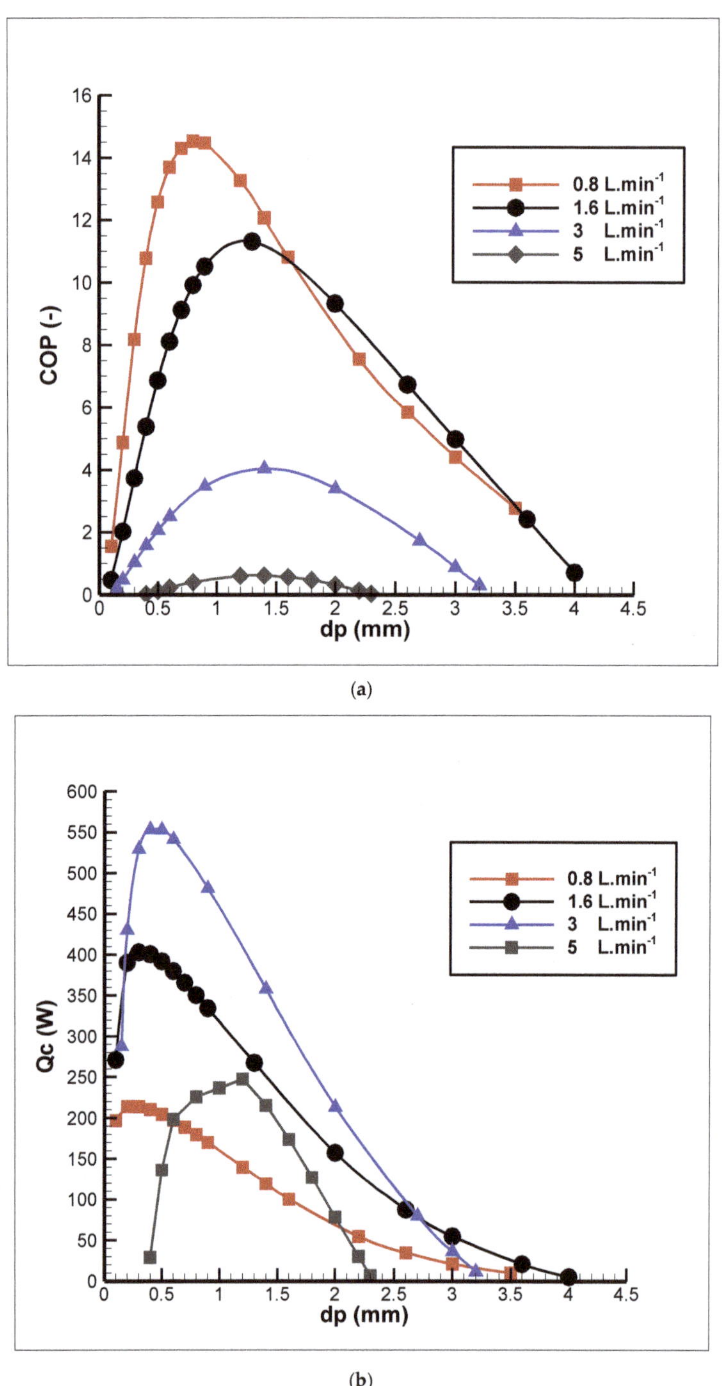

**Figure 11.** (**a**) the coefficient of performance and (**b**) refrigeration capacity based on the sphere particle diameter for a temperature span of 1 K and a frequency of 4 Hz at different volumetric flow rates.

### 3.5. Porosity

Porosity is defined as the ratio of the volume of empty space to the total regenerator volume. By increasing the porosity, the free space of regenerator increases, therefore more fluid can pass through the regenerator bed and more heat is absorbed from the solid refrigerant, leading to an increase in the refrigeration capacity and the coefficient of performance. On the other hand, an increase in porosity reduces the amount of magnetocaloric material and the magnetocaloric effect, thereby reducing the refrigeration capacity and coefficient of performance. In this case, a more intense magnetic field can be used to increase the refrigeration capacity and coefficient of performance. Furthermore, it should be noted that in a very-low-porosity condition, the viscous dissipation is increased, so that a lower refrigeration capacity and coefficient of performance can be expected (Figure 12).

### 3.6. Pump Power

One of the most important parameters influencing the performance of the magnetic refrigeration system is the pump power, that is, the viscous dissipation. Viscous dissipation in the fluid is the irreversible process which causes mechanical energy to transform into heat and may increase the heat losses. The impact of the viscous dissipation is included in the AMR model via a friction factor, as shown in Equation (3). The effect of viscosity loss at high frequencies will increase and, in some cases, become significant in the models of compact AMRs, because the small geometries require higher fluid flow to maintain the same cooling capacity at a large scale. Excessive pressure drops (viscous dissipation) increase the work required to pump the fluid through the AMR.

In this study, it was assumed that there is no leakage in the system, and the mechanical parts of the system such as the piping system were not considered in the numerical model. The impact of these parameters on the performance of the AMR could be considered as a correction factor to the pump power.

An important parameter in the viscous dissipation is the spherical particle diameter. As shown in Figure 13, the work of the pump was increased due to the increased pressure drop by reducing the diameter of the spherical particles and increasing the mass flow rate of the fluid.

### 3.7. Design Analysis

Figures 9–13 show the effect of different parameters on the performance of the magnetic refrigeration system. The performance of the AMR considerably depends on the operational parameters. Figure 9 shows that low or high mass flow rate is not desirable for the AMR and by increasing the frequency cooling power will increase. The results which are presented in Figure 10 show that there is a linear dependency of COP and refrigeration capacity on the temperature span. Figure 11 shows the refrigeration capacity and COP as a function of sphere diameters. It is evident that there is an optimum point for sphere diameter for each mass flow rates. The effect of diameter of spherical particles and mass flow rate simultaneously on AMR performance in are presented in Figure 13.

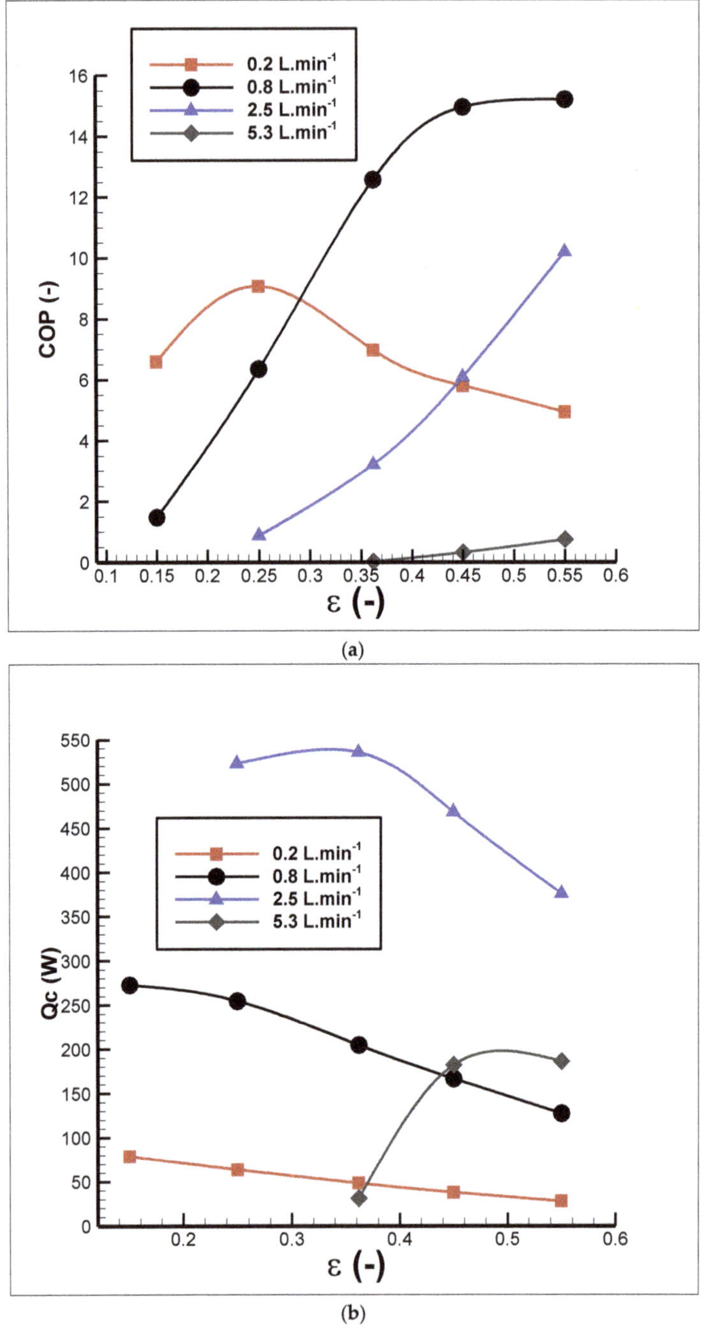

**Figure 12.** Chart of (**a**) the coefficient of performance and (**b**) refrigeration capacity based on porosity for a temperature range of 1 K, frequency of 4 Hz, and a sphere diameter of 0.5 mm at different volumetric flow rates.

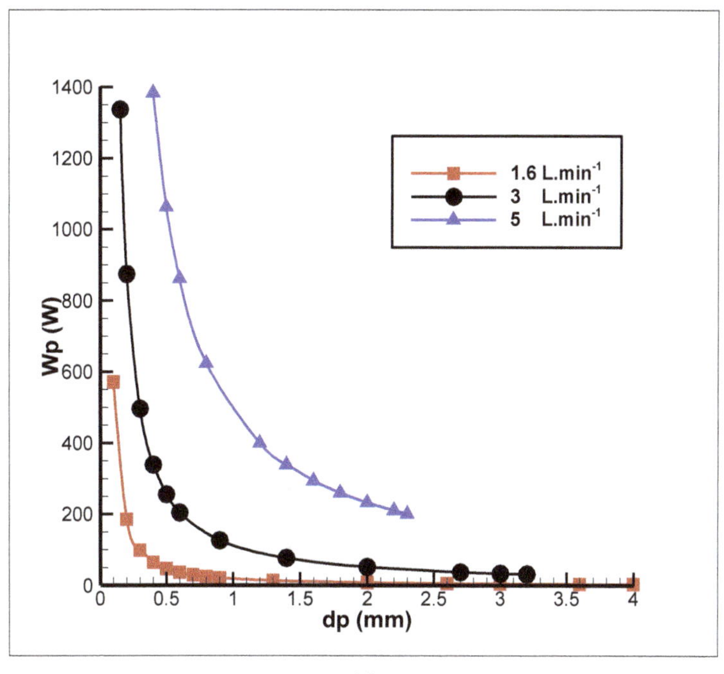

(**a**)

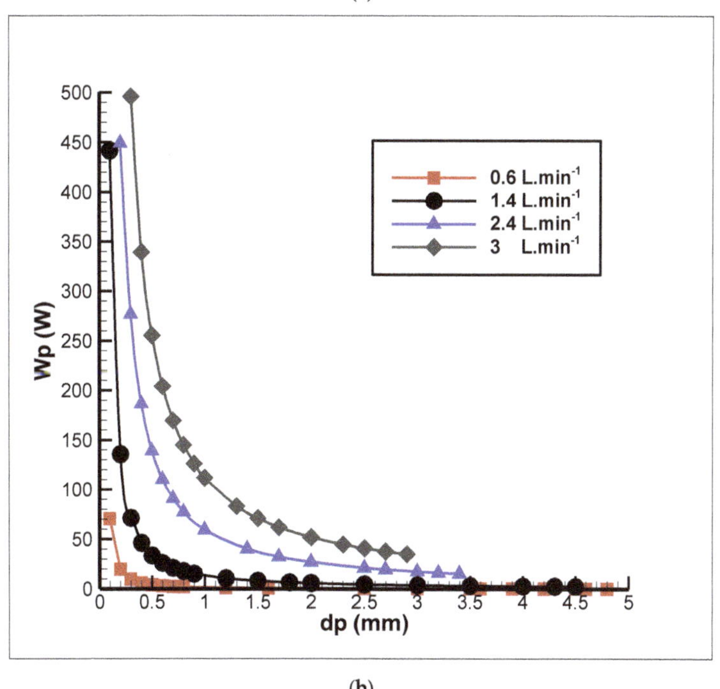

(**b**)

**Figure 13.** *Cont.*

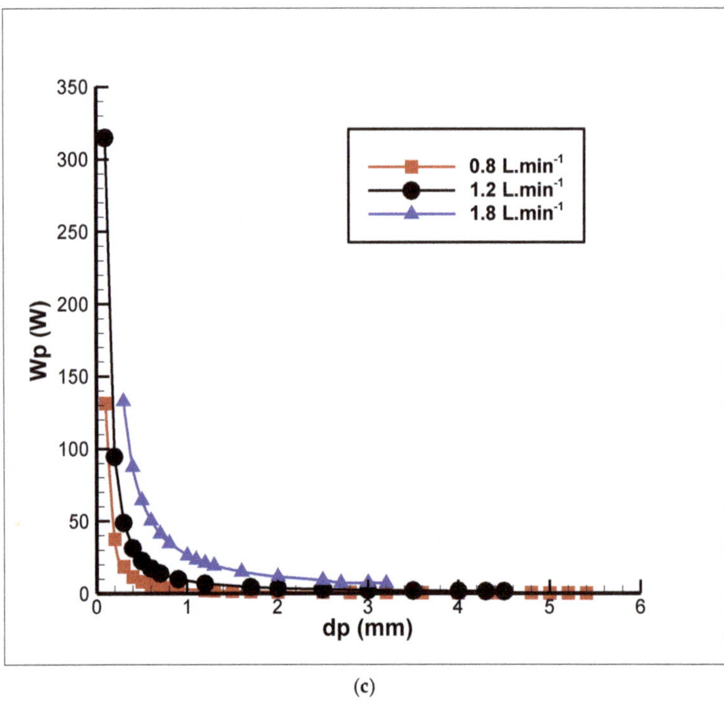



(c)

**Figure 13.** The power pump based on the of the sphere particle diameter at a different volumetric flow rate in the temperature span of 1 K: (**a**) frequency of 4 Hz, (**b**) frequency of 2 Hz, and (**c**) frequency of 1 Hz.

The designer of a new magnetic refrigeration system can select the parameters that are appropriate for the working conditions by using the diagrams presented in this study. According to the parameters reported in Table 3, it is possible to predict the third parameter by having two parameters. For example, as shown in Figure 9, the refrigeration capacity and the coefficient of performance can be calculated by using the flow rate and operating frequency. Furthermore, using the frequency and the refrigeration capacity, it can be predicted how much refrigeration capacity is being met at a specific rate of mass fluid flow. In the same way, like the other parameters, the porosity, temperature range, and diameter of the spherical particles can be calculated. Design charts can be categorized into different groups: The design diagram based on the operational parameters, the price of the magnetocaloric material, the dimensions, and geometry of the regenerator. Design charts can be considered as a tool for developing a magnetic refrigeration system without performing mathematical calculations that leading to time-saving. The designer of the cooling system must be aware of the application of the system and understand what the purpose and operating conditions related to the system are and consider all the aspects of their designs, including the limitations, designers must also offer all possible options for the client's system requirements. Some of the losses in the magnetic refrigeration system that affect the AMR performance are the insufficient heat transfer between the heat transfer fluid and the magnetocaloric material, the magnetic hysteresis, insufficient heat transfer in heat exchangers, and the pressure drops in the piping and heat exchanger. Another important point that should be considered in the system design is the desired economics of the refrigeration system, that is, having the lowest cost and the highest efficiency. The basic information generally required in order to design an active-reactive magnetic refrigeration system is shown in Table 5.

**Table 5.** Parameters required for the design of a magnetic refrigeration system.

---

- Magnetic refrigeration cycle processes
- Type of magnetic refrigeration
- Magnetocaloric material
- Properties of magnetocaloric material
- The lifetime of the system
- The price of magnetocaloric material
- The heat transfer fluid
- Magnetic refrigeration system operating frequency and the type of system: Reciprocating or rotary
- The regenerator geometry
- The maximum temperature and pressure operating system
- Refrigeration capacity and the temperature of cold source
- Preventing system losses (energy recovery)
- Equipment details and the type of mechanical system
- The standard equipment used in the system
- Equipment insulation
- Safety of equipment and potential hazards in the system
- Restrictions on equipment maintenance
- Auxiliary equipment (if needed)
- The final cost of the device
- Preventing the corrosion of equipment and oxidation of materials
- Awareness of environmental conditions (e.g., room temperature)
- Application of magnetic refrigeration system (domestic use, food shops, food storage in cold stores, liquefaction of gases, and vehicles)

---

## 4. Conclusions

In this study, the effect of different parameters, such as the fluid flow rate, porosity, spherical particle diameter, temperature span, and frequency, on the active magnetic refrigeration system was evaluated. As a result, the optimal parameter was obtained in each working condition. In this research, it was shown that the temperature span is inversely related to the refrigeration capacity and coefficient of performance. At high-fluid-flow rates, increasing the frequency makes it possible to increase the coefficient of performance and refrigeration capacity. The spherical particle diameter is one of the parameters that influence the performance of the magnetic refrigeration system, which is inversely related to the pressure drop. Therefore, by conducting a parametric study on the mass flow rate of the fluid and the diameter of spherical particles in each working condition, the pressure drop can be controlled and evaluated. In this study, an efficient numerical method is proposed that reduce the computational time and minimize numerical errors.

The study showed that the magnetic refrigeration system efficiency is highly dependent on the selected parameters. According to the refrigeration capacity and coefficient of performance, the designer of a magnetic refrigeration system can extract the required parameters from the design charts. Design charts and tables are of particular importance for the design of a magnetic refrigeration system because of their time-saving capacity. Furthermore, without having to make complex calculations and creating additional costs, the desired parameter can be selected from the tables and design charts. Some of the items to be monitored that should be considered in designing a magnetic refrigeration system are presented in Table 5.

The limitations of this study include the lack of laboratory equipment to accurately measure the properties of the magnetocaloric material and build a prototype of AMR model. Some errors in system modeling are due to the assumptions of a uniform distribution of fluid flow in all regenerators, that there is no leakage in the system and there is no detectable magnetic hysteresis. These items are difficult to implement in the experimental model, which results in a discrepancy between the numerical model and experimental data, and leads to an overestimation of the outputs of the numerical model. AMR modeling is an immature field and requires further detailed research. Using new methods to calculate the actual magnetic field would result in customer demands being met with higher accuracy.

**Author Contributions:** Methodology, software, validation, formal analysis, writing—original draft preparation, resources A.E.; Conceptualization, review, editing and project administration A.V.; supervision and technical support, B.A.M. All authors have read and agreed to the published version of the manuscript.

**Funding:** This research received no external funding.

**Conflicts of Interest:** The authors declare no conflict of interest.

## References

1. Makumbi, T. Investigating the Application of Environmentally Friendly Solutions in Refrigeration Applications of Uganda. Master's Thesis, University of Gävle, Gävle, Sweden, 2013.
2. Yu, B.; Gao, Q.; Zhang, B.; Meng, X.; Chen, Z. Review on research of room temperature magnetic refrigeration. *Int. J. Refrig.* **2003**, *26*, 622–636. [CrossRef]
3. Numazawa, T.; Mastumoto, K.; Yanagisawa, Y.; Nakagome, H. A modeling study on the geometry of active magnetic regenerator. *AIP Conf. Proc.* **2012**, *1434*, 327–334. [CrossRef]
4. Aprea, C.; Greco, A.; Maiorino, A. Magnetic refrigeration: A promising new technology for energy saving. *Int. J. Ambient. Energy* **2014**, *37*, 294–313. [CrossRef]
5. Lozano, J.; Engelbrecht, K.; Bahl, C.R.; Nielsen, K.K.; Barbosa, J.J.; Prata, A.; Pryds, N. Experimental and numerical results of a high frequency rotating active magnetic refrigerator. *Int. J. Refrig.* **2014**, *37*, 92–98. [CrossRef]
6. Monfared, B.; Furberg, R.; Palm, B. Magnetic vs. vapor-compression household refrigerators: A preliminary comparative life cycle assessment. *Int. J. Refrig.* **2014**, *42*, 69–76. [CrossRef]
7. Lei, T.; Engelbrecht, K.; Nielsen, K.R.; Veje, C.T. Study of geometries of active magnetic regenerators for room temperature magnetocaloric refrigeration. *Appl. Therm. Eng.* **2017**, *111*, 1232–1243. [CrossRef]
8. Bjørk, R.; Bahl, C.R.; Nielsen, K.K. The lifetime cost of a magnetic refrigerator. *Int. J. Refrig.* **2016**, *63*, 48–62. [CrossRef]
9. Nielsen, K.K.; Tusek, J.; Engelbrecht, K.; Schopfer, S.; Kitanovski, A.; Bahl, C.R.H.; Smith, A.; Pryds, N.; Poredos, A. Review on numerical modeling of active magnetic regenerators for room temperature applications. *Int. J. Refrig.* **2011**, *34*, 603–616. [CrossRef]
10. Engelbrecht, K. A Numerical Model of an Active Magnetic Regenerator Refrigerator with Experimental Validation. Ph.D. Thesis, University of Wisconsin-Madison, Madison, WI, USA, 2008.
11. Jaime, B.T. Modelling and Analysis of an Air-Conditioning System for Vehicles Based on Magnetocaloric Refrigeration. Ph.D. Thesis, Universitat Politècnica de València, Valencia, Spain, 2016.
12. Aprea, C.; Greco, A.; Maiorino, A. A dimensionless numerical analysis for the optimization of an active magnetic regenerative refrigerant cycle. *Int. J. Energy Res.* **2013**, *37*, 1475–1487. [CrossRef]
13. Engelbrecht, K.L.; Nellis, G.F.; Klein, S.A.; Zimm, C.B. Recent Developments in Room Temperature Active Magnetic Regenerative Refrigeration. *HVAC R Res.* **2007**, *13*, 525–542. [CrossRef]
14. Franco, V.; Blázquez, J.S.; Ipus, J.J.; Law, J.Y.; Moreno-Ramírez, L.M.; Conde, A. Magnetocaloric effect: From materials research to refrigeration devices. *Prog. Mater. Sci.* **2018**, *93*, 112–232. [CrossRef]
15. Zhang, Y. Review of the structural, magnetic and magnetocaloric properties in ternary rare earth RE2T2X type intermetallic compounds. *J. Alloy. Compd.* **2019**, *787*, 1173–1186. [CrossRef]
16. Li, L.; Yan, M. Recent progresses in exploring the rare earth based intermetallic compounds for cryogenic magnetic refrigeration. *J. Alloy. Compd.* **2020**, *823*, 153810. [CrossRef]
17. Kitanovski, A.; Tušek, J.; Tomc, U.; Plaznik, U.; Ozbolt, M.; Poredoš, A. *Magnetocaloric Energy Conversion*; Springer: Cham, Switzerland, 2016.
18. Petersen, T.F. Numerical Modelling and Analysis of a Room Temperature Magnetic Refrigeration System. Ph.D. Thesis, Technical University of Denmark, Roskilde, Denmark, 2007.
19. Yu, B.; Liu, M.; Egolf, P.W.; Kitanovski, A. A review of magnetic refrigerator and heat pump prototypes built before the year 2010. *Int. J. Refrig.* **2010**, *33*, 1029–1060. [CrossRef]
20. Zimm, C.; Boeder, A.; Chell, J.; Sternberg, A.; Fujita, A.; Fujieda, S.; Fukamichi, K. Design and performance of a permanent-magnet rotary refrigerator. *Int. J. Refrig.* **2006**, *29*, 1302–1306. [CrossRef]
21. Macdonald, I.F.; El-Sayed, M.S.; Mow, K.; Dullien, F.A.L. Flow through Porous Media-the Ergun Equation Revisited. *Ind. Eng. Chem. Fundam.* **1979**, *18*, 199–208. [CrossRef]
22. Wakao, N.; Kagei, S. *Heat and Mass Transfer in Packed Beds*; Taylor & Francis: Abingdon, UK, 1982.

23. Nield, D.A.; Bejan, A. *Convection in Porous Media*; Springer: New York, NY, USA, 2006.
24. Dixon, A.G.; Cresswell, D.L. Theoretical prediction of effective heat transfer parameters in packed beds. *AIChE J.* **1979**, *25*, 663–676. [CrossRef]
25. Amiri, A.; Vafai, K. Transient analysis of incompressible flow through a packed bed. *Int. J. Heat Mass Transf.* **1998**, *41*, 4259–4279. [CrossRef]

**Publisher's Note:** MDPI stays neutral with regard to jurisdictional claims in published maps and institutional affiliations.

 © 2020 by the authors. Licensee MDPI, Basel, Switzerland. This article is an open access article distributed under the terms and conditions of the Creative Commons Attribution (CC BY) license (http://creativecommons.org/licenses/by/4.0/).

*Article*

# Research on the Prediction Method of Centrifugal Pump Performance Based on a Double Hidden Layer BP Neural Network

**Wei Han [1,2,*], Lingbo Nan [1,*], Min Su [3], Yu Chen [1], Rennian Li [1,2] and Xuejing Zhang [4]**

[1]  Department of Energy and Power Engineering, Lanzhou University of Technology, Lanzhou 730050, China
[2]  Key Laboratory of Fluid Machinery and Systems, Lanzhou University of Technology, Lanzhou 730050, China
[3]  Department of Electrical and Information Engineering, Lanzhou University of Technology, Lanzhou 730050, China
[4]  Department of Architecture and Transportation Engineering, Guilin University of Electronic Technology, Guilin 541004, China
*  Correspondence: hanwei@lut.cn (W.H.); LingboNan19951118@126.com (L.N.)

Received: 31 May 2019; Accepted: 12 July 2019; Published: 15 July 2019

**Abstract:** With the aim of improving the shortcomings of the traditional single hidden layer back propagation (BP) neural network structure and learning algorithm, this paper proposes a centrifugal pump performance prediction method based on the combination of the Levenberg–Marquardt (LM) training algorithm and double hidden layer BP neural network. MATLAB was used to establish a double hidden layer BP neural network prediction model to predict the head and efficiency of a centrifugal pump. The average relative error of the head between the experimental and prediction obtained by the double hidden layer BP neural network model was 4.35%, the average relative error of the model prediction efficiency and the experimental efficiency was 2.94%, and the convergence time was 1/42 of that of the single hidden layer. The double hidden layer BP neural network model effectively solves the problems of low learning efficiency and easy convergence into local minima—issues that were common in the traditional single hidden layer BP neural network training. Furthermore, the proposed model realizes hydraulic performance prediction during the design process of a centrifugal pump.

**Keywords:** centrifugal pump; double hidden layer; Levenberg–Marquardt algorithm; performance prediction

---

## 1. Introduction

The performance prediction of centrifugal pumps has become an indispensable part of the optimization design of centrifugal pumps. At present, the traditional prediction methods mainly include computational fluid dynamics (CFDs) numerical simulation methods, and empirical formulas [1]. When the CFDs method is used to predict the performance of a centrifugal pump, the performance prediction error of centrifugal pumps is usually more than 5%, due to the fact that the mechanical loss and partial volume loss of the centrifugal pump are ignored. Meanwhile, empirical formulas can be used to predict the efficiency of a centrifugal pump under the design parameters, but they cannot effectively predict the actual efficiency of an impeller with the same design parameters and different structural parameters (blade number, blade angle, etc.). Table 1 shows the empirical relationship of the disc loss with different specific speeds of a centrifugal pump. The disc loss is nearly inversely proportional to the specific speed [2]. Otherwise, the disc loss is difficult to test and apply in the pump system.

**Table 1.** Disc loss with different specific speeds of a centrifugal pump.

| Specific Speed | 30 | 40 | 50 | 60 | 70 | 80 | 90 | 100 | 150 | 200 |
|---|---|---|---|---|---|---|---|---|---|---|
| Disc Loss/% | 28.5 | 20.4 | 15.7 | 12.7 | 10.6 | 9.1 | 7.9 | 7.0 | 4.4 | 3.1 |

Testing the performance of a centrifugal pump by the experimental method includes three steps: The design, manufacturing, and performance test of the prototype pump. However, this method has some shortcomings, such as tedious measurement process, high cost, and a long research and development cycle. Therefore, it is generally not used as the preferred method to predict the performance during the design process of pumps nowadays.

In recent years, the artificial neural network (ANN) method has often been used to solve uncertain or nonlinear control problems owing to its advantages of strong adaptability, high accuracy, precise function approximation ability, and nonlinear mapping ability. In ANN, the nonlinear mapping relationship is built between the input and output of sample data through the self-learning ability of each neuron. It has the characteristics of large-scale parallel processing, fault tolerance, self-organization and self-adaptation ability, and strong associative function. Therefore, it has been applied in many fields for performance prediction, target recognition, intrusion detection, on-line supervisory control, fault diagnosis, etc. Li et al. [3] proposed a method of combining a chaos algorithm with a genetic algorithm to overcome the shortcomings of the neural network for gesture recognition, and used the optimal result as the initial weight and threshold of the BP neural network to identify gestures. Simulation and experimental results showed that this method has better real-time accuracy for gesture recognition. Peng et al. [4] proposed a vehicle and personnel identification method based on an improved neural network—that is, the seismic signals of moving objects are processed and analyzed to obtain eigenvectors—so as to realize good classification. Experimental results showed that the method has better recognition accuracy and effectiveness.

The ANN method has a shorter prediction cycle and lower cost compared with the experimental method. Moreover, it can predict the hydraulic performance during the design process of a centrifugal pump, as it is forward-looking and time-sensitive, and provides a new idea for the efficient performance prediction of centrifugal pumps. Ne et al. [5] first applied the BP algorithm in ANN for the performance prediction of a centrifugal pump. In the training process, the method constantly modified the weight value using the gradient fastest descent method, and the weight value changed along the negative gradient direction of the error function. The maximum predicted the deviation of the head and efficiency to be 7% and 8%, respectively, which shows the feasibility of pump performance prediction based on the ANN method. Yao et al. [6] used the BP neural network to predict the centrifugal pump head, and the training function was learned by the gradient descent method. The highly nonlinear mapping between the geometric parameters' input of the transition parts and the performance output of the centrifugal pump was realized through the self-application and learning function of the neural network. Cong et al. [7] used the Bayesian regularization algorithm to predict the performance of a single stage centrifugal pump hydraulic model, in which the regularization method improved the generalization ability by modifying the training performance function of the neural network. They found that the error of performance prediction with the improved neural network was lower than 6%. Jiang et al. [8] considered that the CFD numerical simulation method directly used to predict the performance of a centrifugal pump will greatly increase the calculation cost, and the design cycle will also become longer. Therefore, they used the combination of CFDs simulation and the BP neural network to predict the efficiency of the centrifugal pump impeller. BP neural network technology was used to establish an approximate proxy model between the influencing factors and the response value, and the optimized hydraulic model of the centrifugal pump was given. Otherwise, the method was limited to the optimization design of the centrifugal pump impeller with a lower specific speed.

The above studies on the performance prediction of centrifugal pumps are all based on the single hidden layer BP neural network. Due to the simple structure of the single hidden layer BP neural network, it is impossible to extract more characteristic information of input parameters, and the prediction of centrifugal pump performance faces problems, such as small coverage and low robustness. Considering that a deeper network depth will result in network redundancy and the convergence time of the network will become longer, this will lead to an inaccurate performance prediction of the centrifugal pump. Therefore, it is necessary to establish an appropriate method to increase the network depth and accelerate the convergence of the deep BP neural network to improve the prediction accuracy of centrifugal pump performance.

In recent years, the LM algorithm has been favored by domestic and overseas scholars [9–12]. It is a classical nonlinear numerical optimization algorithm that combines the advantages of the gradient descent method and the Gauss-Newton method. Moreover, it has the local convergence of the Gauss-Newton method and the global characteristics of the gradient descent method. The fast convergence of the Gauss-Newton method is adopted when the solution is near to the optimal solution, and the convergence characteristic is achieved by adaptively adjusting the damping factor. The approximate second-order derivative information is used to achieve the ability of superlinear convergence near the optimal solution, which has a higher iterative convergence speed and accelerates the convergence of the neural network. The robustness of the gradient descent method is adopted when the solution is far from the optimal solution. That is, when the input information or the neural network has finite perturbation, the neural network can still maintain the stability characteristics from the input to the output relationship to obtain reliable solutions, thus solving problems, such as the poor numerical stability of the neural network. This approach has thus been widely applied to various research fields.

Wang [13] designed a new intrusion detection model based on the LM-BP neural network to address the issues of the traditional BP neural network, such as slow convergence speed, the tendency to fall into local minima, and high cost of computation. Through further research of the neural network and intrusion detection systems, the comparative experimental results showed that the new model combines the advantages of anomaly and misuse detection. It can quickly detect new intrusions as well as reduce the false alarm rate and missed alarm rate. Zhao et al. [14] applied the LM algorithm neural network with the learning rate for on-line supervisory control. Compared with the traditional forward neural network BP, the new control strategy can improve the operation speed and the local minima. It can improve the tracking performance of the servo system with an unknown load disturbance. The search direction can be optimized by the LM-BP neural network and has been applied to transformer fault diagnosis [15]. Through test and analysis, not only was the convergence speed accelerated, but the accuracy was also greatly improved. The study results showed that the positive rate of the fuzzy fault diagnosis reached 92.5%; the verified method could improve the transformer fault diagnosis performance.

To sum up, the nonlinear relationship between the hydraulic performance and the structural parameters of the impeller is so complex that the actual efficiency of the impeller is difficult to predict with different structural parameters, whether by CFDs or empirical methods. On the other hand, the double hidden layer BP neural network can enhance the mapping ability of the complex relationship between the input and output of the system, and has stronger approximation and fault tolerance abilities than the single hidden layer BP neural network [16]. So, the combination of the LM algorithm and double hidden layer BP neural network can be used to predict the performance of centrifugal pumps rapidly and precisely for use in strong nonlinear engineering prediction fields, such as intrusion detection, on-line supervisory control, and transformer fault diagnosis.

## 2. Structural Design of the Centrifugal Pump Performance Prediction Model

### 2.1. Double Hidden Layer BP Neural Network Structure

Multi-layer BP neural networks can be used to predict different engineering problems for the forward propagation of parameters and the back propagation of errors. In theory, it has the ability to approximate any nonlinear continuous map, so it is very suitable for the modeling and performance prediction of nonlinear systems [17]. In this paper, a double hidden layer BP neural network with multiple inputs and two outputs is constructed. It is assumed that the input layer is composed of m neurons, which is used to complete the input of m predicted parameters of the system. The first hidden layer is composed of k neuron nodes, which are used to complete the spatial weighted aggregation of input signals and the excitation output. The second hidden layer is composed of p neuron nodes, which are used to improve the nonlinear mapping capability of the network for a complex relationship between the input and output of the system. The output layer consists of two process neuron nodes that are used to complete the system output, as shown in Figure 1.

The transfer function of each layer in the BP neural network has a significant impact on the performance of the model. These transfer functions are usually determined experimentally. In the BP neural network, the commonly used transfer functions are linear transfer function ("purelin"), tangent transfer function ("tansig"), logarithmic sigmoid transfer function ("logsig"), etc. When the transfer function of the input layer and the first hidden layer is "tansig", the transfer function of the first hidden layer and the second hidden layer is "tansig", and the transfer function of the second hidden layer and the output layer is "purelin", which are shown in the MATLAB neural network toolbox [18–20].

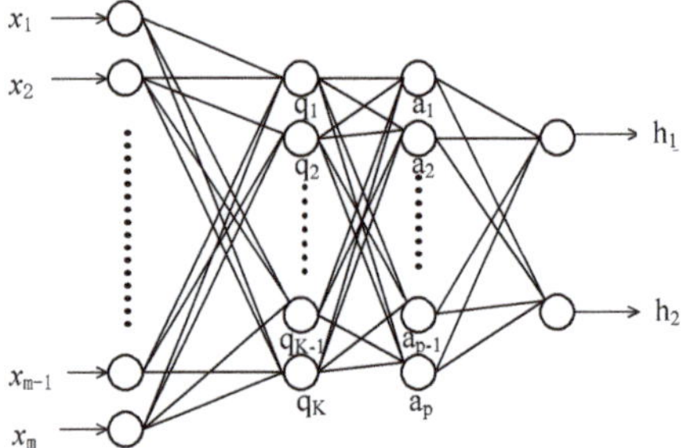


**Figure 1.** Double hidden layer back propagation (BP) neural network structure.

Assuming that the signal that passes from the input layer to the first hidden layer is $m_j$, this can be expressed as in Equation (1):

$$m_j = \sum_{k=1}^{m} x_k w_{kj} + b_j, \tag{1}$$

where $x_k$ is the input neuron and $x_k$ represents each design parameter of the centrifugal pump in this paper. $w_{kj}$ represents the input layer to the first hidden layer weight and $b_j$ represents the first hidden layer bias.

The first hidden layer output signal is denoted as $y_j$, shown as in Equation (2):

$$y_j = \mathrm{tansig}(m_j). \tag{2}$$

The signal transmitted from the first hidden layer to the second hidden layer is denoted as $n_i$, and the calculation of $n_i$ is shown in Equation (3):

$$n_i = \sum_{j=1}^{k} y_j w'_{ji} + b'_i, \tag{3}$$

where $w'_{ji}$ represents the weight of the first hidden layer of the second hidden layer and $b'_i$ is the bias of the second hidden layer. The output signal of the second hidden layer is denoted as $z_i$, which is expressed as in Equation (4):

$$z_i = \mathrm{tansig}(n_i). \tag{4}$$

The signals transmitted from the second hidden layer to the output layer are, respectively, recorded as $s_1$ and $s_2$, shown as in Equations (5) and (6):

$$s_1 = \sum_{i=1}^{p} z_i w_{i1}'' + b_1'', \tag{5}$$

$$s_2 = \sum_{i=1}^{p} z_i w_{i2}'' + b_2'', \tag{6}$$

where $w_{i1}''$ and $w_{i2}''$ represent the weights from the second hidden layer to the output layer, and $b_1''$ and $b_2''$ represent the biases of the output layer.

During the process of centrifugal pump performance prediction, the output signals of the head (H) and the efficiency ($\eta$) of the output layer are denoted as $h_1$ and $h_2$, respectively, which can be expressed as in Equations (7) and (8), wherein $a_1$, $a_2$, $b_1$, and $b_2$ represent the coefficients of the linear transfer function "purelin":

$$h_1 = a_1 \left( \sum_{i=1}^{p} \text{tansig} \left( \sum_{j=1}^{k} \text{tansig} \left( \sum_{k=1}^{m} x_k w_{kj} + b_j \right) w_{ji}' + b_1'' \right) w_{i1}'' + b_1'' \right) + b_1, \tag{7}$$

$$h_2 = a_2 \left( \sum_{i=1}^{p} \text{tansig} \left( \sum_{j=1}^{k} \text{tansig} \left( \sum_{k=1}^{m} x_k w_{kj} + b_j \right) w_{ji}' + b_i'' \right) w_{i2}'' + b_2'' \right) + b_2. \tag{8}$$

### 2.2. LM Algorithm

As the traditional gradient descent method is used to train the deep network, the convergence speed is greatly reduced because of the greater network depth. In order to solve this problem, the improved form of the Gauss-Newton method combined with the LM training algorithm is used to train the network in this paper. The proposed method can accelerate the network training and convergence speed effectively [21,22]. The LM algorithm is a second-order algorithm. So, when the gradient of the error surface is small, the LM algorithm is similar to the gradient descent method. When the gradient of the error surface is large, the LM algorithm is similar to the Gauss-Newton method [23]. In addition, the LM algorithm can estimate the learning rate (LR) in each gradient direction of the error surface according to the Hessian matrix, so compared with a first-order algorithm, the LM algorithm is more effective than others for training neural networks at present [24]. The following is a brief description of the LM algorithm: Firstly, the error index function of the neural network is set as E(x), as shown in Equation (9). Here, N is the number of samples, $Y_i$ represents the expected network output vector, $Y_i'$ denotes the actual network output vector, and $r_i(x)$ represents the current error:

$$E(x) = \frac{1}{2} \sum_{i=1}^{N} \left\| Y_i - Y_i' \right\|^2 = \frac{1}{2} \sum_{i=1}^{N} r_i^2(x). \tag{9}$$

If the weight vector of the t-th iteration of the neural network is $x_t$, the new weight vector, $x_{t+1}$, of the Newton algorithm can be obtained by Equation (10). Here, $\Delta x_t$, $H_e$, and $g_t$ represent the updated value of the weight, Hessian matrix, and current gradient, respectively. $H_e$ and $g_t$ can be obtained from Equations (11) and (12), where J and r(x) represent the Jacobian matrix and the error, respectively:

$$x_{t+1} = x_t + \Delta x_t = x_t = H_{et}^{-1} g_t, \tag{10}$$

$$H_e = J^T J, \tag{11}$$

$$g_t = \nabla E(x)\big|_{x=x(t)} = J^T(x)r(x). \tag{12}$$

However, the matrix, $H_e$, is not always reversible. In order to solve this problem, considering the introduction of a coefficient, $\lambda_t$, the LM algorithm after updating the weight is as shown in Equation (13), where I is an identity matrix:

$$x_{t+1} = x_t - \left[J^T(x_t)J(x_t) + \lambda_t I\right]^{-1} J^T(x_t)r(x_t). \tag{13}$$

Obviously, when $\lambda_t$ is very large, the LM algorithm approximates the gradient descent method; when $\lambda_t$ is close to zero, it is equivalent to the Gauss-Newton algorithm. Since the approximate second-order derivative information is used in the calculation, the algorithm is much faster than the gradient descent method. In other words, the LM algorithm is superior to the Gauss-Newton method. Therefore, it is feasible to train the double hidden layer BP neural network based on the LM algorithm to improve the convergence speed of deep network training.

## 3. Establishment of Sample Data Sets

In the design of a centrifugal pump, the design parameters of the impeller have an important effect on the efficiency, cavitation performance, and characteristic curve of the centrifugal pump [25]. Therefore, it is necessary to investigate the influence of the design parameters of the impeller, such as flow rate ($Q$), rotating speed ($n$), impeller inlet diameter ($D_j$), hub diameter ($d_h$), impeller outlet diameter ($D_2$), specific speed ($n_s$), blade outlet width ($b_2$), and blade numbers ($Z$), on the centrifugal pump design and performance prediction. In the prediction of centrifugal pump performance, the sample data used in this paper are representative data obtained from [26], which construct 44 groups of a training sample set. These sample data are from Dalian Hongze Pump Industry, Zhejiang Institute of Mechanical, and Electrical Design, Jiangsu University Fluid Machinery Engineering Technology Research Center, China. In order to ensure the universality of the model, low, medium, and high values of $n_s$ as well as large and small values of $Q$ are considered in the sample data. Some of these shown in Table 2.

**Table 2.** Centrifugal pump performance prediction model sample data.

| Serial Number | $n_s$ | $Q$ (m³/h) | $N$ (r/min) | $D_j$ (mm) | $d_h$ (mm) | $D_2$ (mm) | $b_2$ (mm) | $Z$ | $H$ (m) | $\eta$ (%) |
|---|---|---|---|---|---|---|---|---|---|---|
| 1 | 23.1 | 12.5 | 2900 | 52 | 0 | 242 | 4 | 4 | 80.78 | 42.21 |
| 2 | 30 | 21 | 2900 | 60 | 0 | 245 | 10 | 10 | 80 | 41 |
| 3 | 33 | 12.5 | 2900 | 48 | 0 | 200 | 6 | 5 | 50.34 | 51.09 |
| 4 | 47.2 | 12.5 | 2900 | 44 | 0 | 160 | 5.6 | 5 | 31.25 | 56.32 |
| 5 | 48 | 300 | 1450 | 175 | 45 | 547 | 17 | 7 | 100 | 75 |
| 6 | 58 | 75 | 2950 | 100 | 50 | 290 | 11 | 6 | 80 | 70 |
| 7 | 73 | 148 | 2900 | 110 | 25 | 278 | 15 | 6 | 90 | 80.5 |
| 8 | 81 | 140 | 1450 | 138 | 32 | 317 | 19 | 6 | 30 | 79.7 |
| 9 | 90 | 200 | 2900 | 110 | 25 | 255 | 18 | 7 | 84 | 80 |
| 10 | 103 | 130 | 1450 | 140 | 38 | 262 | 23 | 6 | 21 | 82 |
| 11 | 131 | 400 | 1450 | 190 | 0 | 345 | 35 | 6 | 32 | 83 |
| 12 | 151 | 243 | 1450 | 162 | 35 | 262 | 34 | 6 | 19 | 86.8 |
| 13 | 205 | 2600 | 740 | 450 | 0 | 640 | 120 | 5 | 25 | 88.9 |
| 14 | 225 | 6300 | 490 | 700 | 0 | 965 | 186 | 5 | 23 | 88 |
| 15 | 302 | 6650 | 660 | 625 | 0 | 775.5 | 187 | 4 | 24 | 85.8 |

## 4. Prediction Results and Analysis

The flow diagram of predicting the performance of the centrifugal pump in this paper is shown in Figure 2. It can be seen from the figure that the whole process of centrifugal pump performance prediction is divided into three parts. The first step is the construction of a double hidden layer BP neural network. In order to ensure the correctness of the training direction in the training process and accelerate the speed of the network convergence, all the data are initialized before the training starts, known as normalization processing. The normalization process here is achieved by the "prestd" function in the MATLAB toolbox. Then, the number of neurons in each hidden layer of the double hidden layer BP neural network is determined through training. The second step is the training process of the double hidden layer BP neural network. In this case, the training parameters were first set and the double-hidden BP neural network was built by MATLAB software. In the process of training, the LM algorithm was used to calculate the error value of each iteration and compare it with the set mean squared error (MSE) value. If the iterative error was less than the set value, the training would be completed; otherwise, the next iteration would continue until network convergence was achieved. The third step is the evaluation of the double hidden layer BP neural network model. From the second step, the centrifugal pump back propagation (CPBP) performance prediction model will be established, then the test data will be input into this model for the prediction of the head and efficiency.

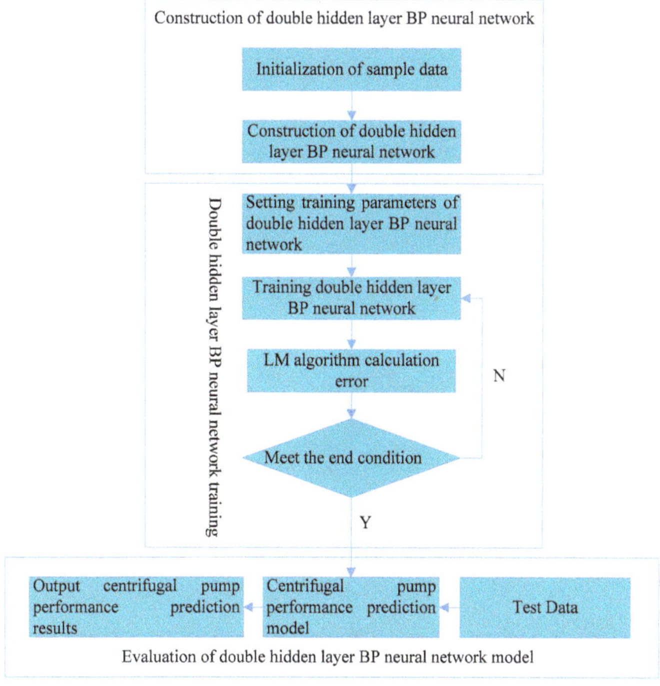


**Figure 2.** Centrifugal pump performance prediction flow diagram with the centrifugal pump back propagation (CPBP) model.

### 4.1. Parameter Selection of the CPBP Model

With the aim of overcoming the shortcomings of predicting the performance of a centrifugal pump with a single hidden layer BP neural network, a double hidden layer BP neural network was constructed in this paper, including one input layer, two hidden layers, and one output layer. In order to obtain a better nonlinear mapping relationship of the related parameters through network learning, the S-type tangent function and linear function were used for the transfer functions of the two hidden layers and the output layer, respectively. Similar to the traditional BP neural network, the performance of the double hidden layer BP neural network is also affected by the number of neurons in the input layer and hidden layer. In general, the number of input layer neurons is determined by the number of input sample variables. So, the number of neurons in the input layer was set as eight in this paper. Similarly, the number of neurons in the output layer needs to be equal to the number of the predicted parameters. The number of neurons in the output layer of the network was set as two, including the head and efficiency of the centrifugal pump. With regard to the determination of the number of neurons in the hidden layer, this paper adopted the design method of the number of neurons in the hidden layer [27,28], as shown in Equation (14):

$$c = \sqrt{d + u} + f, \tag{14}$$

where c, d, and u represent the number of neurons in the hidden layer, input layer, and output layer, respectively. The values of the constant "f" ranged from 1 to 10, so the range of neuron numbers in the hidden layer was 4 to 13.

The number of hidden layer neurons is input into the corresponding model and then the network is trained once again. Figure 3 shows the MSE values corresponding to different networks. The horizontal axis is the ratio ($n_2/n_1$) of the number of neurons ($n_2$) in the first hidden layer to the number of neurons ($n_1$) in the input layer; the longitudinal axis is the ratio ($n_4/n_3$) of the number of neurons ($n_4$) in the output layer to the number of neurons ($n_3$) in the second hidden layer; and the vertical axis is the training MSE corresponding to each network. As the numbers of neurons in the first hidden layer and the second hidden layer were both six, the corresponding MSE was the minimum. Therefore, the final structure of the double hidden layer BP neural network suitable for the prediction of centrifugal pump performance was 8-6-6-2. Figure 4 is the structure diagram of the CPBP network.

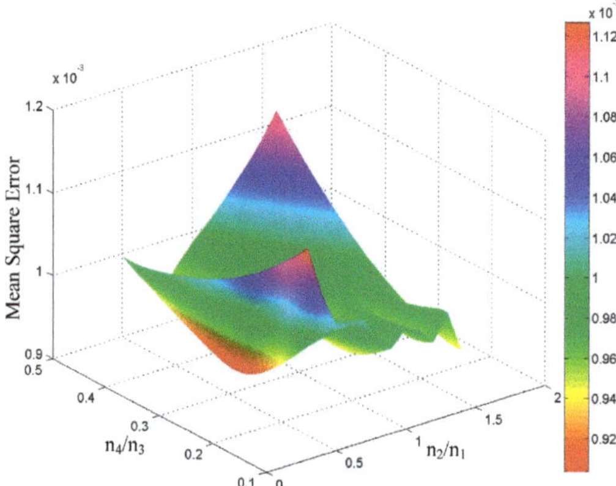

**Figure 3.** Comparison of mean squared errors of back propagation (BP) networks with different double hidden layers.

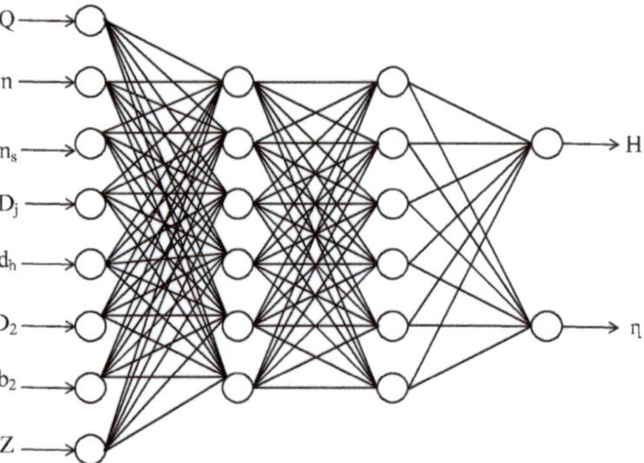

**Figure 4.** Centrifugal pump back propagation (CPBP) neural network structure diagram.

### 4.2. Training Process of CPBP Model

The training parameters of neural networks have an important influence on the performance prediction of centrifugal pumps, so it is necessary to choose the appropriate training parameters. The parameters used in the CPBP network training include the maximum number of iterations: Epochs, LR, momentum factor (MF), and MSE. These training parameters have a direct impact on the prediction accuracy of the final centrifugal pump performance. The CPBP network model can be established with the help of the neural network toolbox in MATLAB to determine the design values of the network training parameters, as shown in Table 3.

The training network is started after the CPBP network is built. During the training process, the back propagation of error is used to iterate repeatedly until the convergence of the CPBP network is achieved. Finally, the prediction model of the centrifugal pump can be established by the CPBP network.

**Table 3.** Training parameter settings.

| Parameter | Epochs | LR | MF | MSE |
|---|---|---|---|---|
| Setting | 550 | 0.04 | 0.95 | 0.001 |

### 4.3. Prediction Results of CPBP Model

In order to verify the accuracy of the prediction method, five sets of test data provided by Shanghai Kaiquan Pump Co., Ltd. China, were input into the trained model to obtain the prediction values of the head and efficiency. The corresponding relative error can be calculated between the experimental and the predictive values. Simultaneously, the prediction accuracy of the CPBP network model can be obtained. The comparison between the experimental and the predicted results is shown in Table 4. Among them, $H^{**}$ and $\eta^{**}$ are the head and efficiency predicted by the CPBP model, respectively.

The maximum relative error of the head predicted by the CPBP model compared with the experimental results was 11.87%, the minimum was 0.13%, and the average relative error was 4.35%. From these error values, it can be found that the error distribution was relatively average. The maximum relative error between the model prediction and the experimental efficiency was 4.99%, the minimum was 0.47%, and the average relative error was 2.94%, so the error distribution was also uniform. The average relative error of the prediction values on the head and efficiency indicate the reliability of the CPBP model.

**Table 4.** Comparison of prediction results and experimental results of five test sets.

| Serial Number | Input Value | | | | | | | | Experimental Value | | Predictive Value | |
|---|---|---|---|---|---|---|---|---|---|---|---|---|
| | $n_s$ | $Q$ (m³/h) | $n$ (r/min) | $D_j$ (mm) | $d_h$ (mm) | $D_2$ (mm) | $b_2$ (mm) | $Z$ | $H$ (m) | $H$ (%) | $H^{**}$ (m) | $\eta^{**}$ (%) |
| 1 | 180 | 620 | 1450 | 225 | 50 | 340 | 54 | 5 | 28 | 85 | 24.6756 | 84.6602 |
| 2 | 114 | 775 | 1450 | 240 | 48 | 440 | 42 | 6 | 60 | 88.2 | 58.2233 | 81.3399 |
| 3 | 246 | 3500 | 740 | 500 | 0 | 650 | 137 | 5 | 24 | 88.9 | 23.9672 | 89.1542 |
| 4 | 85.6 | 100 | 2900 | 90 | 0 | 210 | 16 | 6 | 56.5 | 81.25 | 54.1095 | 80.2150 |
| 5 | 128.1 | 100 | 2900 | 100 | 0 | 178 | 17 | 6 | 33 | 74.2 | 32.1484 | 77.9264 |

### 4.4. Comparison of Results

In order to compare the prediction accuracy of the CPBP network with the traditional single hidden layer BP neural network, the traditional single hidden layer BP neural network with the best prediction effect was constructed to predict the performance of the centrifugal pump. Firstly, Equation (14) was used to determine the value range of the number of hidden layer neurons when the number of input neurons was eight and the number of output neurons was two. After calculation, the value range of hidden layer neurons was also 4 to 13. Then, the single hidden layer BP networks corresponding to 4 to 13 hidden layer neurons were set up in MATLAB and trained. The MSE values corresponding to different networks are shown in Figure 5, where the abscissa is the ratio ($n_2/n_1$) of the number of neurons ($n_2$) in the hidden layer to the number of neurons ($n_1$) in the input layer and the ordinate is the training MSE corresponding to each network.

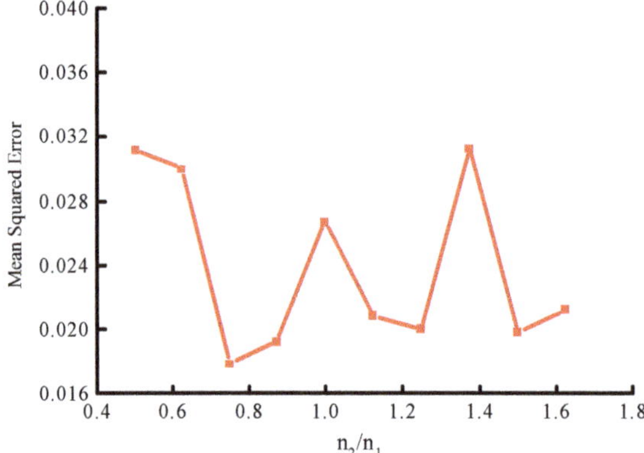


**Figure 5.** Comparison of the mean squared errors of different single hidden layer back propagation (BP) networks.

As can be seen from Figure 5, when the number of neurons in the hidden layer was six, the corresponding MSE was the smallest. That is, the single hidden layer BP neural network composed of six neurons in the hidden layer had the best prediction effect compared with other single hidden layer BP neural networks. Its structure is shown in Figure 6.

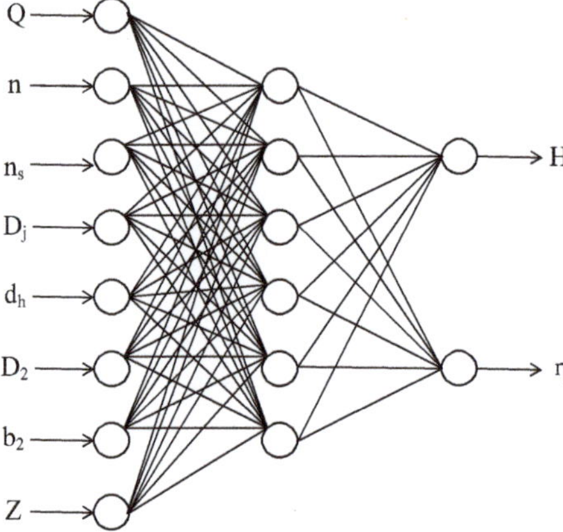


**Figure 6.** Single hidden layer back propagation (BP) network structure diagram.

Five groups of identical test data with the CPBP were input into the centrifugal pump prediction model obtained by the single hidden layer BP neural network training to obtain the predicted results. The results of the prediction of centrifugal pump performance between the single hidden layer BP network and the double hidden layer BP network were compared, as is shown in Table 5 and Figure 7. Among them, H* and $\eta$* are the head and efficiency of the single hidden layer BP neural network prediction, respectively. $\Delta\eta$* and $\Delta$H* are the relative errors of the efficiency and head of the single hidden layer BP neural network, respectively. $\Delta\eta$** and $\Delta$H** are the relative errors of the efficiency and head of the CPBP model, respectively. Compared with the CPBP network, the maximum, the minimum, and the average relative error of the head value of the single hidden layer BP network increased by

11.41%, 0.23%, and 3.40%, respectively. Meanwhile, the maximum, the minimum, and the average relative error of the efficiency prediction value increased by 4.99%, 0.47%, and 2.37%, respectively. It can be concluded that the CPBP network has a higher prediction accuracy than the single hidden layer BP network for the performance prediction of centrifugal pumps.

**Table 5.** Comparison of the results of centrifugal pump performance prediction between the single hidden layer back propagation (BP) network and the double hidden layer back propagation (BP) network.

| Serial Number | Experimental Value | | Predicted Value and Relative Error of BP Network with Single Hidden Layer | | | | Predicted Value and Relative Error of Double Hidden Layer (CPBP Model) | | | |
|---|---|---|---|---|---|---|---|---|---|---|
| | $H$/m | $\eta$/% | $H^{*}$/m | $\eta^{*}$/% | $\Delta\eta^{*}$/% | $\Delta H^{*}$/% | $H^{**}$/m | $\eta^{**}$/% | $\Delta\eta^{**}$% | $\Delta H^{**}$/% |
| 1 | 28 | 85 | 23.5700 | 83.5183 | 1.7400 | 15.8200 | 24.6756 | 84.6602 | 0.3990 | 11.8700 |
| 2 | 60 | 88.2 | 58.0873 | 80.9351 | 8.2400 | 3.1880 | 58.2233 | 81.3399 | 7.7700 | 2.9600 |
| 3 | 24 | 88.9 | 21.2275 | 87.6447 | 1.4000 | 11.5500 | 23.9672 | 89.1542 | 0.2860 | 0.1360 |
| 4 | 56.5 | 81.25 | 52.0179 | 76.9955 | 5.2400 | 7.9000 | 54.1095 | 80.2150 | 1.2700 | 4.2300 |
| 5 | 33 | 74.2 | 33.1127 | 81.6154 | 9.9900 | 0.3400 | 32.1484 | 77.9264 | 5.0000 | 2.5800 |

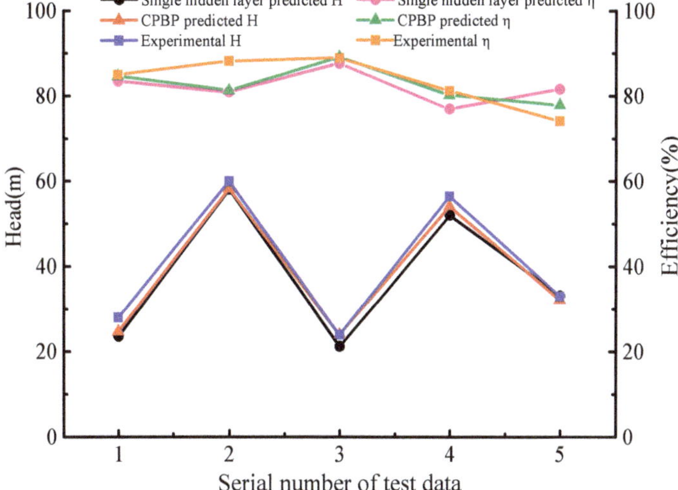


**Figure 7.** Comparison between the prediction results and experimental results of five test sets.

As can be seen from Figure 7, the predicted and the experimental values exhibited the same change trend; the small deviation may be caused by the accuracy of the experimental data and the deficiencies of the training samples.

Figure 8 shows the variation of the MSE during the training of the BP network with a single hidden layer and the CPBP network. It can be seen that the traditional single hidden layer BP network had difficulty in achieving a high convergence accuracy quickly compared with the CPBP network. Therefore, the CPBP network is more effective in predicting the performance of centrifugal pumps compared to the traditional single hidden layer BP neural network.

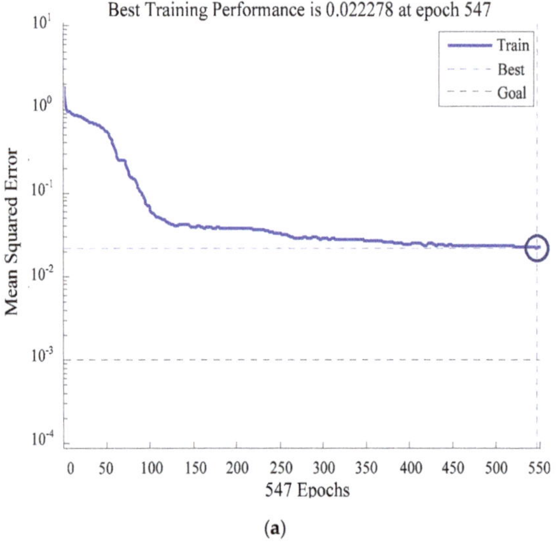

(**a**)

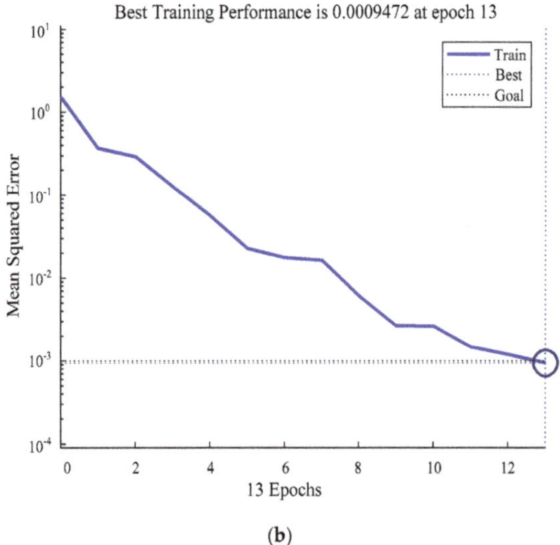

(**b**)

**Figure 8.** Change of the mean squared errors on the back propagation (BP) neural network with different hidden layers. (**a**) Back propagation (BP) network with a single hidden layer; (**b**) centrifugal pump back propagation (CPBP) network.

## 5. Conclusions

(i) In this paper, a new CPBP model for the performance prediction of centrifugal pumps based on design and structure parameters was established. The LM algorithm was used to accelerate the convergence rate of the double hidden layer BP network. The proposed model improves the mapping ability of the complex relationship between the input and output of the nonlinear system, known as the centrifugal pump performance, with its structure and operation parameters.

(ii) The CPBP network could easily achieve a high convergence accuracy quickly compared with the traditional single hidden layer BP network.

(iii) Compared with the traditional single hidden layer BP network, the average relative error of the head and efficiency prediction values of the CPBP network decreased by 3.40% and 2.37%, respectively, and the training time of CPBP model was only 1/42 of that of the traditional BP network. This indicates that the CPBP network model is more suitable for the rapid and efficient determination of the optimum design of a centrifugal pump.

**Author Contributions:** W.H. conceived and designed the research, L.N. drafted the article, M.S. completed the neural network architecture and verification, Y.C. collected and collated the test sample data, R.L. is the leader of the research team. X.Z. completed the English check.

**Funding:** This study was supported by the National Key R&D Program Projects of China (Grant no. 2018YFB0606100) and the National Natural Science Foundation of China (Grant no. 51669012, 51579125).

**Acknowledgments:** The authors would like to acknowledge the support in the form of experimental data provided by Shanghai Kaiquan Pump Co., Ltd. China.

**Conflicts of Interest:** There are no known conflicts of interest.

## References


1. Kaewnai, S.; Chamaoot, M.; Wongwises, S. Predicting performance of radial flow type impeller of centrifugal pump using CFD. *J. Mech. Sci. Technol.* **2009**, *23*, 1620–1627. [CrossRef]
2. Ran, J. *Research and Implementation of Performance Prediction of Centrifugal Pump Based on Neural Network*; University of Electronic Science and Technology of China: Chengdu, China, 2014.
3. Li, D.J.; Li, Y.Y.; Li, J.X. Gesture Recognition Based on BP Neural Network Improved by Chaotic Genetic Algorithm. *Int. J. Autom. Comput.* **2018**, *15*, 21–30. [CrossRef]
4. Peng, Z.Q.; Cao, C.; Huang, J.Y. Seismic signal recognition using improved BP neural network and combined feature extraction method. *J. Cent. South Univ.* **2014**, *21*, 1898–1906. [CrossRef]
5. Nie, S.B.; Guan, X.F.; Liu, H.L. Exploration of Predicting the Performance of Centrifugal Pumps by Artificial Neural Network. *Pump Technol.* **2002**, *5*, 16–18.
6. Yao, Y.F.; Jiang, Q.Y.; He, R.G. Performance Prediction of Centrifugal Pump Based on BP Neural Network. *Mech. Des. Manuf.* **2008**, *10*, 73–74.
7. Cong, X.Q.; Yuan, S.Q.; Yuan, D.Q. Performance Prediction of Centrifugal Pump Based on Improved BP Neural Network. *J. Agric. Mach.* **2006**, *37*, 56–59.
8. Jiang, W.Z.; Duan, S.Q.; Yang, C.M. Optimization design of centrifugal pump impeller based on CFD simulation and BP neural network. *Mach. Tool Hydraul.* **2016**, *44*, 67–70.
9. Zayani, R.; Bouallegue, R.; Roviras, D. Adaptive Predistortions Based on Neural Networks Associated with Levenberg-Marquardt Algorithm for Satellite Down Links. *EURASIP J. Wirel. Commun. Netw.* **2008**, *2008*, 132729. [CrossRef]
10. Camargo, A.; He, Q.; Palaniappan, K. Performance evaluation of optimization methods for super-resolution mosaicking on UAS surveillance videos. In Proceedings of the SPIE Defense, Security, and Sensing 2012, Baltimore, MD, USA, 24–26 April 2012; Volume 8355, p. 30.
11. Jebur, A.A.; Atherton, W.; AlKhaddar, R.M. Settlement Prediction of Model Piles Embedded in Sandy Soil Using the Levenberg–Marquardt (LM) Training Algorithm. *Geotech. Geol. Eng.* **2018**, *36*, 2893–2906. [CrossRef]
12. Comon, P.; Luciani, X.; De Almeida, A.L.F. Tensor decompositions, alternating least squares and other tales. *J. Chemom. Soc.* **2010**, *23*, 393–405. [CrossRef]
13. Wang, T.Y. *Research and Implementation of LM-BP Neural Network Intrusion Detection System*; Hunan University: Changsha, China, 2007.
14. Zhao, H.; Zhou, R.X. Neural Network Supervisory Control Based on Levenberg-Marquardt Algorithm. *J. Xi'an Jiaotong Univ.* **2002**, *36*, 523–527.
15. Song, Z.J.; Wang, J. Fuzzy clustering and LM algorithm to improve BP neural network for transformer fault diagnosis. *High. Volt. Electr. Appl.* **2013**, *5*, 54–59.
16. Cui, D.W. Comprehensive evaluation of water resources vulnerability in Wenshan Prefecture of Yunnan Province based on improved BP neural network model. *J. Yangtze River Sci. Res. Inst.* **2013**, *30*, 1–7.

17. Ding, H.; Dong, W.Y.; Wu, D.M. Water Level Prediction of Double Hidden Layer BP Neural Network Based on LM Algorithm. *Stat. Decis.* **2014**, *5*, 16–19.

18. Khayet, M.; Cojocaru, C. Artificial neural network model for desalination by sweeping gas membrane distillation. *Desalination* **2013**, *308*, 102–110. [CrossRef]

19. Choi, H.K.Y.; Lin, W.; Loon, S.C. Facial Scanning with a Digital Camera: A Novel Way of Screening for Primary Angle Closure. *J. Glaucoma* **2015**, *24*, 522–526. [CrossRef] [PubMed]

20. Cai, Z.L.; Chen, X.L.; Shi, W.R. Improvement of Comprehensive Evaluation Method of Learning Effect Based on BP Neural Network. *J. Chongqing Univ.* **2007**, *30*, 96–99.

21. Gao, J.; Zhang, Y.; Du, Y. Optimization of the tire ice traction using combined Levenberg–Marquardt (LM) algorithm and neural network. *J. Braz. Soc. Mech. Sci. Eng.* **2019**, *41*, 40. [CrossRef]

22. An, R.; Li, W.J.; Han, H.G. An improved Levenberg-Marquardt algorithm with adaptive learning rate for RBF neural network. In Proceedings of the 2016 35th Chinese Control Conference (CCC), Chengdu, China, 27–29 July 2016.

23. Zhang, H.H.; Tao, Y.R.; Hu, J. Prediction model of photosynthetic rate of cucumber seedlings fused with chlorophyll content. *J. Agric. Mach.* **2015**, *46*, 259–263.

24. Wilamowski, B.M.; Yu, H. Improved Computation for Levenberg–Marquardt Training. *IEEE Trans. Neural Netw.* **2010**, *21*, 930–937. [CrossRef] [PubMed]

25. Wang, Y.Y. *Effect of Blade Wrap Angle and Exit Angle on Impeller Performance*; Lanzhou University of Technology: Lanzhou, China, 2017.

26. Guan, X.F. *Modern Pump Theory and Design*; China Aerospace Publishing House: Beijing, China, 2010.

27. Jiao, B.; Ye, M.X. Method for determining the number of hidden layer units in BP neural network. *J. Shanghai Dianji Univ.* **2013**, *16*, 113–116.

28. Wang, R.B.; Xu, H.Y.; Li, B. Research on the Method of Determining the Number of Nodes in BP Neural Network. *Comput. Technol. Dev.* **2018**, *28*, 31–35.

© 2019 by the authors. Licensee MDPI, Basel, Switzerland. This article is an open access article distributed under the terms and conditions of the Creative Commons Attribution (CC BY) license (http://creativecommons.org/licenses/by/4.0/).

*Article*

# Influence of Critical Wall Roughness on the Performance of Double-Channel Sewage Pump

Xiaoke He [1], Yingchong Zhang [2], Chuan Wang [3,*], Congcong Zhang [3], Li Cheng [3,*], Kun Chen [4] and Bo Hu [5]

[1] School of Electric Power, North China University of Water Resources and Electric Power, Zhengzhou 450045, China; hexiaoke@ncwu.edu.cn
[2] National Research Center of Pumps, Jiangsu University, Zhenjiang 212013, China; zhangyingchong1994@163.com
[3] College of Hydraulic Science and Engineering, Yangzhou University, Yangzhou 225009, China; Conniezcc@163.com
[4] Ningbo Jushen Pumps Industry Co., Ltd., Ningbo 315100, China; skyckun2009@163.com
[5] Deputy Energy & Power Engineering, Tsinghua University, Beijing 100084, China; hubo@mail.tsinghua.edu.cn
* Correspondence: wangchuan@ujs.edu.cn (C.W.); chengli@yzu.edu.cn (L.C.)

Received: 22 November 2019; Accepted: 13 January 2020; Published: 17 January 2020

**Abstract:** The numerical method on a double-channel sewage pump was studied, while the corresponding experimental result was also provided. On this basis, the influence of wall roughness on the pump performance was deeply studied. The results showed that there was a critical value of wall roughness. When the wall roughness was less than the critical value, it had a great influence on the pump performance, including the head, efficiency, and shaft power. As the wall roughness increased, the head and efficiency were continuously reduced, while the shaft power was continuously increased. Otherwise, the opposite was true. The effect of wall roughness on the head and hydraulic loss power was much smaller than that on the efficiency and disk friction loss power, respectively. With the increase of wall roughness, mechanical efficiency and hydraulic efficiency reduced constantly, leading to the decrement of the total efficiency. With the increase of flow rate, the effect of wall roughness on the head and efficiency gradually increased, while the influence on the leakage continuously reduced. The influence of the flow-through component roughness on the pump performance was interactive.

**Keywords:** double-channel sewage pump; critical wall roughness; numerical calculation; external characteristics

---

## 1. Introduction

Pumps are classified as general machinery with varied applications [1–5]. Sewage pump, as important equipment in the sewage treatment project, is widely used in chemical, municipal, and other industries. It is mainly used to transport production and domestic sewage containing a large amount of solid particles or fibrous solid substances. The double-channel sewage pump has the characteristics of compact structure, high efficiency, and good anti-winding and anti-clogging performance, while it has a wide application range, which has been studied by a large number of scholars [6–10].

It is well known that pumps have huge energy consumption, and wall roughness has an important effect on pump efficiency [11–16]. In the process of using the double-channel sewage pump, the abrasion of the wall surface caused by the impurities in the transport medium and the damage of the blade surface caused by erosion will cause the change of the wall roughness, and roughness $Ra$ is one of the important factors affecting the performance of the pump.

## 2. Literature Overview

In the past years, many scholars studied the effect of wall roughness on the flow in pipes, fans, compressors, microchannels. In order to study the effect of wall roughness in turbulent pipe flow, Hemeida [17] developed an equation for estimating the thickness of the laminar sublayer in turbulent pipe flow of pseudoplastic fluids and found that the turbulent pipe flow could be divided into two regions: smooth wall and rough wall turbulence. The roughness Reynolds number was used to determine the smooth wall turbulence and rough wall turbulence regions. Kandlikar [18] studied the roughness effects at microscale—reassessing Nikuradse's experiments on liquid flow in rough tubes, and found that Nikuradse's work was revisited in light of the recent experimental work on roughness effects in microscale flow geometries. Li et al. [19] studied the influence of the internal surface roughness of the nozzle on cavitation erosion characteristics of submerged cavitation jets from the aspects of erosion intensity and erosion efficiency; it could be concluded that excessive smooth surface was not conducive to the formation of cavitation bubbles, leading to an attenuated intensity of cavitation erosion, while excessive rough surface caused much energy dissipation and led to divergent jets, resulting in a significant reduction of erosion intensity. According to the experimental results, there existed an optimum inner surface roughness value to achieve the strongest aggressive cavitation erosion capability for submerged cavitating jets. Tang et al. [20] analyzed the existing experimental data in the literature on the friction factor in microchannels. The friction factors in stainless steel tubes were much higher than the theoretical predictions for tubes of conventional size. This discrepancy resulted from the large relative surface roughness in the stainless steel tubes. From the literature review and the present test data, it is suggested that for gaseous flow in microchannels, with relative surface roughness less than 1%, the conventional laminar prediction should still be applied. Gamrat et al. [21] used three different approaches in the present study to predict the influence of roughness on laminar flow in microchannels. The numerical simulations, the rough layer model, and the experiments agreed to show that the Poiseuille number $Po$ increased with the relative roughness and was independent of Re in the laminar regime ($Re < 2000$). The increase in $Po$ observed during the experiments was predicted well both by the three-dimensional simulations and the rough layer model. Li et al. [22] studied the effect of the roughness of the compressor blade on the performance based on the equivalent Reynolds number correction principle and found that when the surface roughness increased, the main characteristic parameters of the compressor were reduced to varying degrees, making the compressor overall performance degraded. Li et al. [23] discussed the formula for calculating friction loss of fluid flow in similar fan ducts. The calculation formulas between the model and the physical flow efficiency under different relative surface roughness of the flow channel were introduced. Li et al. [24] studied the effect of surface roughness on the micro-gap leakage of oil-free lubrication scroll compressors and found that under the condition of the same micro-gap size and inlet pressure, a larger rough element height and distribution density could effectively increase the leakage. The flow resistance of the gas reduced the leakage flow rate, thereby reducing the amount of leakage. Gao [25] used low-speed compressor plane cascade experiments to study the effect of different roughness positions on cascade performance. It was found that cascade performance was more sensitive to leading-edge roughness and suction front roughness. In terms of roughness, the total pressure loss value was reduced by 23.1% compared to the smooth blade. Han et al. [26] used numerical methods to study the influence of blade surface roughness on the internal flow field and characteristic parameters of the compressor under the compressor-level environment and found that the blade roughness had a significant effect on the main performance parameters of the compressor; with the blade roughness with the increase of the compressor, the performance degradation of the compressor stage was intensified, and the energy loss of the internal airflow was increased, especially the energy loss of the airflow near the middle and upper part of the leading edge of the impeller was severe, and the temperature of the blade end wall was increased.

With the rapid development of computer technology, numerical simulation is increasingly widely used in the fluid flow [27–33]. Guelich et al. [34] studied the effect of wall roughness on the efficiency of centrifugal pumps. It was found that the effect of wall roughness in the volute was stronger than the roughness inside the impeller. The magnitude of hydraulic loss depended on the wall roughness, turbulent flow near the wall, and actual velocity distribution in the flow channel. Zhu et al. [35] simulated the influence of wall roughness of flow parts, including impeller, diffuser, pump cavity, and hub, on the performance of the axial flow pump and found that reducing the roughness of the flow surface could effectively improve the head and efficiency. The efficiency of the axial flow pump was more sensitive to changes in the roughness of the blade surface. Yamazaki et al. [36] studied the effect of wall roughness on the performance of jet pumps. It was found that the wall roughness near the throat inlet had the greatest influence on the efficiency of the jet pump. The optimum efficiency decreased linearly with increasing wall roughness. Aldas et al. [37] used CFD numerical calculation to determine the effect of pump absolute roughness and relative roughness on pump efficiency. It was found that under certain absolute roughness, the efficiency was significantly improved until a certain degree. Feng et al. [38] found that under the optimal condition, compared with no roughness, when the roughness was 3.2 μm, 6.3 μm, and 12.5 μm, the head decreased by 0.3 m, 0.5 m, and 0.7 m, while the efficiency decreased by 4.7%, 5.7%, and 6.8%. Pan et al. [39] obtained the performance difference of the axial flow pump under the influence of different rough wall through simulation calculation. It was found that under the influence of wall roughness, the pump device showed a decrease in the head and the efficiency, and the efficiency decreased most. Deshmukh et al. [40] analyzed the turbulent flow energy and eddy viscosity characteristics of electric submersible pumps with different roughness, and the influence of wall disturbance on the pressure distribution and velocity field. It was found that the roughness effect of the high viscosity oil was the most significant relative to water. Under non-design conditions, the Reynolds number affects the overall roughness effect. Gu et al. [41] used the experimental design method, based on the numerical simulation technology to test the design and simulation of the head, shaft power, and efficiency of the axial flow pump. It was found that the impeller wall roughness had the greatest influence on the hydraulic performance, the influence coefficient on the head was –0.265, while efficiency was −0.283, and the shaft power was 0.099. Lim [42] studied the effect of wall roughness of the double-suction centrifugal pump components on performance. It was found that the impeller roughness had the greatest influence on the performance of the double suction centrifugal pump, while the suction chamber had the least impact. Pump performance had a strong correlation with the wall roughness of the impeller cover. To sum up, many scholars only study the effect of wall roughness on pump efficiency, but there are few manuscripts on critical wall roughness.

In this paper, numerical and experimental studies were made in a double-channel sewage pump. The influence of critical wall roughness on the performance of the pump was analyzed, which had practical guiding significance for the production, maintenance, and daily management of the double-channel sewage pump.

## 3. Numerical Calculation

### 3.1. Calculation Model

The design parameters of the double-channel sewage pump were as follows: rated flow $Q_d = 12$ m$^3$/h, rated head $H = 13$ m, impeller speed $n = 2800$ r/min, impeller blade number $Z = 2$, specific speed $n_s = 3.65nQ^{0.5}/H^{0.75} = 86$. Impeller and volute were the core components of the pump. Due to the requirement of non-blocking performance, the impeller of the sewage pump adopted a curved pipe shape, and the volute section adopted a rectangular section, as shown in Figure 1. The geometric parameters of the impeller and volute were calculated by the velocity coefficient method, as shown in Table 1. The calculation model was the basis of numerical calculation, and the integrity of the model had a significant impact on the accuracy of the calculation results. If only the impeller and volute were considered, the disk friction loss and volume leakage loss were neglected. Therefore, to study the influence of wall roughness on the performance of the pump more accurately, the calculation model of the pump included the ring clearance, the impeller, the pump cavity, the volute, the inlet, and outlet sections whose length were 5 times and 10 times of impeller outlet diameter, as shown in Figure 2.

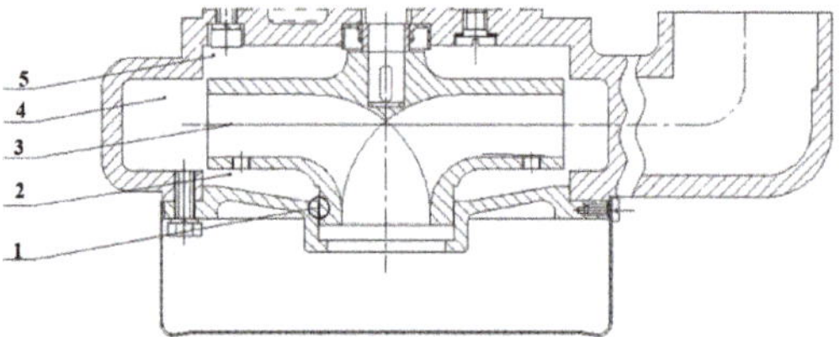

**Figure 1.** Schematic diagram of the double-channel sewage pump. 1. Ring clearance; 2. Front pump cavity; 3. Impeller; 4. Volute; 5. Rear pump cavity.

**Table 1.** Basic geometric parameters of the pump.

| Geometric Parameters | Numerical Value | Geometric Parameters | Numerical Value |
|---|---|---|---|
| Impeller inlet diameter $D_j$ (mm) | 30 | Blade outlet angle $\beta_2$ (°) | 22.5 |
| Impeller outlet diameter $D_2$ (mm) | 107 | Volute base diameter $D_3$ (mm) | 124 |
| Impeller outlet width $b_2$ (mm) | 20 | Volute inlet width $b_3$ (mm) | 30 |
| Blade inlet angle $\beta_1$ (°) | 68 | Volute outlet diameter $D_d$ (mm) | 44 |

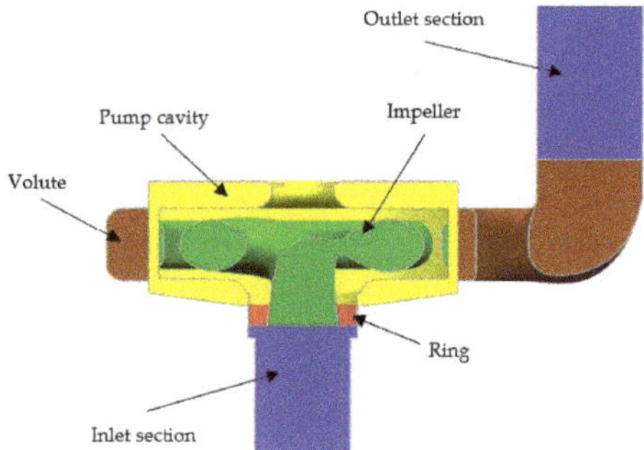


**Figure 2.** Calculation model of the double-channel sewage pump (most of the inlet and outlet sections are hidden).

### 3.2. Grid Independence

Compared with the common centrifugal pump impeller, the impeller of the double-channel sewage pump had the characteristics of severe distortion, large wrap angle, and special profile. Therefore, in the ICEM (Integrated Computer Engineering and Manufacturing), the hexahedral structured grid was generated for the total calculation domain of the double-channel sewage pump, as shown in Figure 3. In order to determine the appropriate number of grids, five grid sizes ($G$ values) were selected for the grid-independent analysis under the same settings, as shown in Figure 4. It could be seen that the influence of the grid size $G$ on the pump performance was within 2%. When the grid size was large, that is, the grid number was small, the efficiency $\eta$ and head $H$ were relatively high. When $G \leq 1.5$ mm, the efficiency and head were basically stable. Considering the coordination of calculation accuracy and time, $G = 1.5$ mm was selected for meshing, as shown in Figure 3. Moreover, the grid quality of the calculation model can be seen in Table 2. The total grid quality was more than 0.37, which could meet the requirement of the numerical calculation.

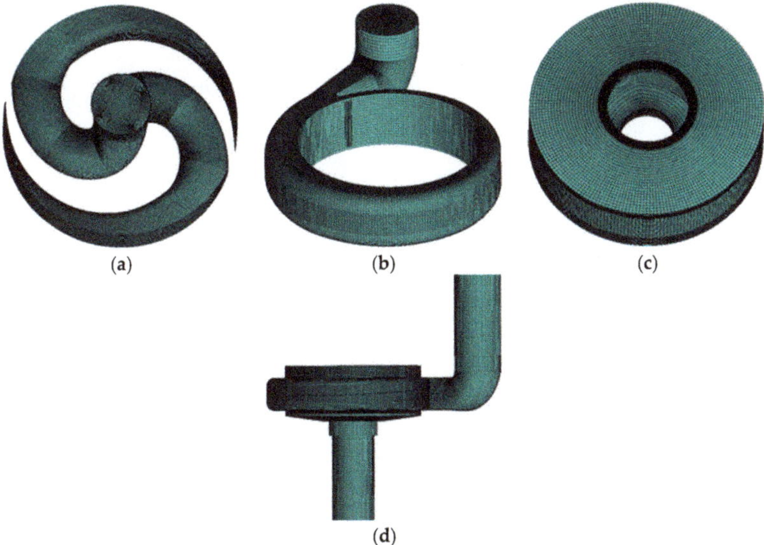

**Figure 3.** Structure meshing. (**a**) Impeller, (**b**) Volute, (**c**) Pump cavity, (**d**) Overall domain.

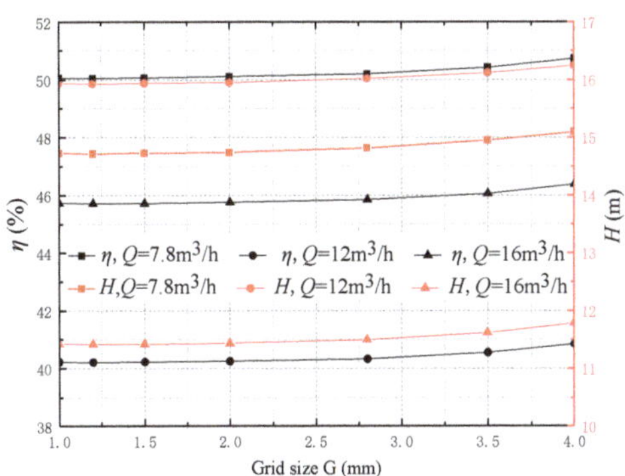

**Figure 4.** Grid independence analysis.

**Table 2.** Grid quality.

| Parts | Grid Number | Node Number | Impeller | Volute | Pump Cavity | Inlet Section | Outlet Section |
|---|---|---|---|---|---|---|---|
| Grid quality | 4,839,030 | 4,289,372 | 0.37 | 0.43 | 0.54 | 0.65 | 0.63 |

### 3.3. Turbulent Model

Turbulence models for pumps in ANSYS CFX (Computational Fluid X) include standard $k$-$\varepsilon$ ($k$ and $\varepsilon$ represent the pulsating kinetic energy of the turbulent flow and its dissipation rate respectively), RNG (Renormalization-group) $k$-$\varepsilon$, SST (shear stress transport), $k$-$\omega$ ($\omega$ represents another kind of dissipation rate) and others. The calculation results of different turbulence models were quite different and needed to be selected according to the actual situation. Therefore, the numerical calculations with different turbulence modes under the rated flow condition were carried out and compared with the experimental results, as shown in Table 3. It was found that the numerical results of the standard $k$-$\varepsilon$ turbulence model agreed well with the experimental results. Therefore, the standard $k$-$\varepsilon$ turbulence model was selected.

**Table 3.** Experimental and numerical results with different turbulent models under the rated flow condition.

| Turbulent Models | Efficiency $\eta$ (%) | Head $H$ (m) |
| --- | --- | --- |
| $k$-$\varepsilon$ | 50.06 | 14.73 |
| RNG $k$-$\varepsilon$ | 51.12 | 14.77 |
| SST | 62.89 | 14.92 |
| $k$-$\omega$ | 62.97 | 14.93 |
| Experimental value | 48.09 | 14.34 |

### 3.4. Boundary Setting

The impeller and the shroud in the pump cavity were based on the rotating reference frame, whereas the other sub-domains were based on the stationary reference frame throughout the entire calculation domains. The interfaces between the impeller and its adjacent sub-domains were set to "frozen rotor" mode, and the other interfaces were set to "general connection" mode. Moreover, the non-slip walls were selected as the wall boundaries. The open inlet and mass outflow were selected as the inlet and outlet boundaries.

### 3.5. Equivalent Sand Model

In actual production, the wall roughness has peaks and valleys, and its shape and size are different. The arithmetic average deviation of the profile ($R_a$) and unevenness ten-point height ($R_z$) are the two typical parameters that can illustrate the value of the surface roughness. For simplicity, $R_a$, the arithmetic average deviation of the profile, was selected in the manuscript.

The wall function method used by CFX is an extension of the method proposed by Launder and Spalding. In the logarithmic regular region, the near-wall tangential velocity of the fluid was logarithmically related to the wall shear stress, and the empirical formula was used to connect the near-wall boundary conditions of the average flow and the turbulent transport equation. The logarithmic relationship of the near-wall velocity was as follows [43]:

$$u^+ = \frac{U_t}{u_\tau} = \frac{1}{\kappa}\ln(y^+) + C \tag{1}$$

$$y^+ = \frac{\rho \Delta y u_\tau}{\mu} \tag{2}$$

$$u_\tau = \sqrt{\frac{\tau_w}{\rho}} \tag{3}$$

where $u^+$ is the near-wall velocity (in m/s), $u_\tau$ is the friction velocity (in m/s), $U_t$ is the tangential velocity at a distance from the wall surface $\Delta y$ (in m/s), $y^+$ is the dimensionless distance from the wall, $\tau_w$ is the wall shear stress (in N), $k$ is the Von Karman constant, $C$ is a constant associated with wall roughness.

Surface roughness increased the generation of turbulence near the wall, which, in turn, led to a significant increase in wall shear stress, destroying the viscous sub-layer in the turbulent flow. The logarithmic velocity profile near the wall moved down:

$$u^+ = \frac{1}{\kappa}\ln(y^+) + B - \Delta B \tag{4}$$

where $B$ take 5.2, offset $\Delta B$ is a function of dimensionless roughness $h^+$ ($h^+ = hu_\tau/v$); $v$ is kinematic viscosity (in m$^2$/s).

For grit roughness, the offset $\Delta B$ could be expressed in the following form using the dimensionless sand roughness $h_s{}^+$:

$$\Delta B = \frac{1}{\kappa}\ln(1 + 0.3h_s^+) \tag{5}$$

There are usually two methods for measuring roughness in production and life. One is to measure the roughness of the surface of an object by means of a roughness measuring instrument [44,45]. The second is to compare the roughness samples [46], compare the surface roughness of the object with the standard surface roughness samples, and estimate the size of the object surface roughness. The contour arithmetic mean deviation $R_a$ was used as a measurement parameter of roughness, which was defined as follows:

$$R_a = \frac{1}{N}\sum_{i=1}^{n}|z_i - z|$$ (6)

In actual production, the roughness has peaks and valleys, the shape and size are different. In CFX, the equivalent sand roughness $k_s$ was described [47,48], that is, a tightly arranged ball of equal diameter $h_s$ is placed on the smooth plane. Simulating the undulating wall surface, the equivalent sand grain roughness only affected the fluid in the upper half of the ball, as shown in Figure 5. Therefore, the effect of the same surface roughness and the equivalent sand roughness on the fluid was quite different.

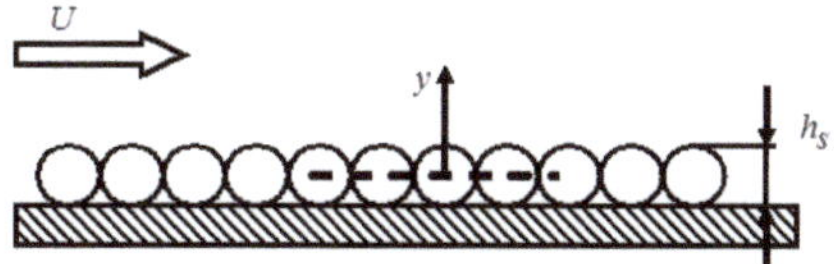


**Figure 5.** Schematic diagram of CFX equivalent sand model.

Colebrook et al. [49] used the equivalent sand model to experimentally study the gas in rough pipes and found the relationship between "smooth" law and "rough" law. The sand-grain roughness values required for use with the moody chart were not derived from any direct measure of roughness using modern surface characterization equipment, such as an optical profilometer. Using direct measurements of surface roughness in fluid flow calculations might, therefore, result in significant error.

For the equivalent sand model, it could be obtained from formula (6):

$$R_a = \frac{1}{h_s}\sum_{i=1}^{h_s}|y_i - y|$$ (7)

When the number of points tended to infinity, formula (7) became the integral formula:

$$R_a = \frac{1}{h_s}\int_{x=0}^{h_s}|y - \overline{y}|dx$$ (8)

and

$$\overline{y} = \frac{\pi h_s}{8}$$ (9)

Taking Equations (8) and (9) into Equation (7) gave:

$$h_s = 11.03 R_a$$ (10)

Equation (8) showed that if a roughness meter was used to measure the roughness $R_a$ of a surface composed of a sphere layer with a diameter of $h_s$, the final value of the measured roughness $R_a$ was an order of magnitude smaller than a suitable sand grain roughness.

Adams et al. [50] proposed an algorithm to convert the measured surface roughness parameters into equivalent sand grain roughness and used this algorithm to convert the surface roughness into equivalent sand grain roughness so that the experimental results and numerical calculation of fluid flow had better consistency. Adams et al. obtained through deduction and experiment:

$$k_s = [(1.2 \pm 0.1)/0.204]R_a = (5.392 \sim 6.372)R_a$$ (11)

Combining the research of Deshmukh [51] and so on, the conversion coefficient of the equivalent roughness $k_s$ and the average deviation Ra of the contour arithmetic was selected as:

$$k_s = 5.863 R_a \tag{12}$$

The conversion between the wall roughness $R_a$ and the equivalent sand roughness $k_s$ was utilized in CFX, and as shown in Table 4, the numerical calculation regarding the roughness of the double-channel sewage pump was performed.

**Table 4.** Conversion between the wall roughness Ra and the equivalent sand roughness $k_s$.

| Wall Roughness $R_a$ (µm) | Equivalent Sand Roughness $k_s$ (µm) |
|:---:|:---:|
| 0 | 0 |
| 20 | 117.26 |
| 40 | 234.52 |
| 50 | 293.15 |
| 60 | 351.78 |
| 80 | 469.04 |
| 100 | 586.30 |

## 4. Influence of Wall Roughness on the Pump Performance

### 4.1. Influence of Wall Roughness on the Pump Performance

Six wall roughness of 0 µm, 20 µm, 40 µm, 50 µm, 60 µm, 80 µm, and 100 µm were selected for the numerical calculation of double-channel sewage pump. From Figure 6 and Table 5, it could be seen that there was a critical wall roughness ($R_a$ = 50 µm) for the influence of roughness on the performance of the pump. When 0 µm $\leq R_a \leq$ 50 µm, with the increase of $R_a$, the head $H$ and efficiency $\eta$ gradually decreased, and the decreasing rate gradually reduced, while the shaft power $P$ gradually increased, and the increasing rate gradually decreased. When 50 µm $\leq R_a \leq$ 100 µm, with the increase of $R_a$, $H$ and $\eta$ gradually increased, and the increasing rate gradually decreased, while $P$ decreased gradually, and the decreasing rate gradually reduced. When $R_a$ increased from 0 µm to 50 µm, the value of $P$ increased by 43.39%, and the reductions of $H$ and $\eta$d were 33.18% and 4.12%, respectively. It showed that when the wall roughness was less than the critical wall roughness, the wall roughness had a great influence on the performance of the pump, and the influence on the shaft power and efficiency was much greater than that on the head. When the wall roughness was larger than the critical wall roughness, the wall roughness had little effect on the performance of the pump.

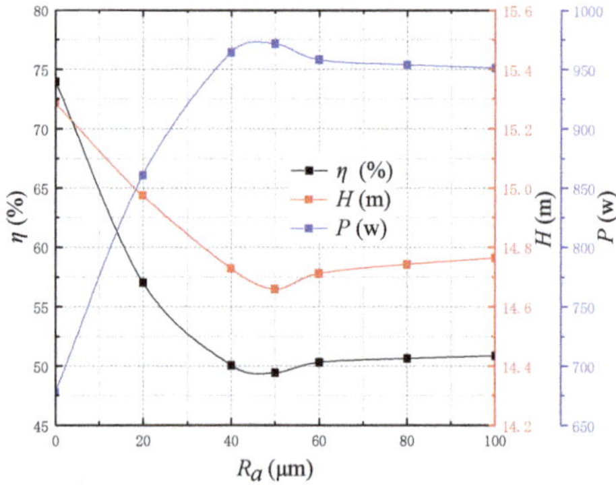


**Figure 6.** Pump performance of the pump with different wall roughness under the rated flow condition.

**Table 5.** Pump performance with different roughness under the rated flow condition.

| Roughness $R_a$ (µm) | 0 | 20 | 40 | 50 | 60 | 80 | 100 |
|---|---|---|---|---|---|---|---|
| Efficiency $\eta$ (%) | 73.95 | 57.02 | 50.06 | 49.45 | 50.33 | 50.65 | 50.87 |
| Head $H$ (m) | 15.29 | 14.98 | 14.73 | 14.66 | 14.71 | 14.74 | 14.76 |
| Shaft power $P$ (w) | 677.54 | 860.83 | 964.28 | 971.55 | 958.14 | 953.97 | 951.17 |

*4.2. Influence of Wall Roughness on the Internal Flow of the Pump*

The influence of wall roughness on the internal flow of the pump could be expressed by the turbulent energy. Estimating the turbulent kinetic energy using turbulence intensity, the formula was as follows:

$$k = \frac{3}{2}(ul)^2 \tag{13}$$

$$l = 0.16 \times R_e^{(-1/8)} \tag{14}$$

where $u$ is the average velocity (in m/s), $l$ is the turbulence intensity, and $R_e$ is the Reynolds number. Figure 7 shows the turbulent energy distribution in the middle section of the pump with different wall roughness under the rated flow condition ($Q = 12$ m$^3$/h). It could be seen that due to the influence of the impeller's rotation, the turbulent energy was gradually reduced from the shroud edge to the shroud center. There was the largest turbulent energy at the clearance ring, so the domain of the clearance ring could not be ignored in the numerical calculation. As the wall roughness increased, the turbulent flow energy $k$ gradually increased, and the increasing rate gradually reduced.

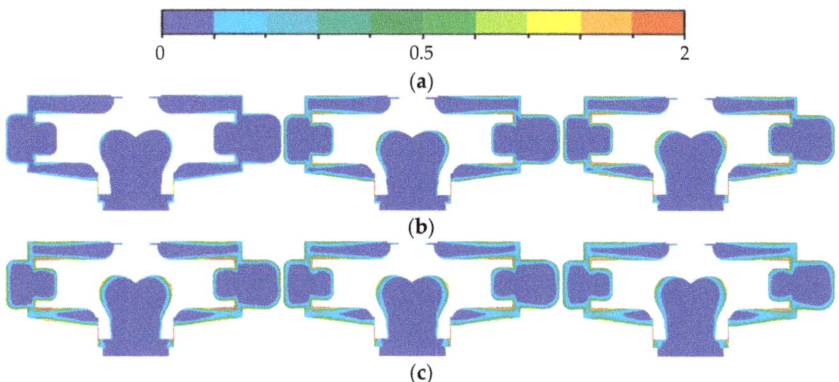

**Figure 7.** Turbulence energy distribution in the middle section of the pump with different wall roughness under the rated flow condition. (**a**) Turbulence kinetic energy (m$^2$/s$^2$), (**b**) $R_a = 0$ µm, $k = 0.096$ m$^2$/s$^2$; $R_a = 20$ µm, $k = 0.217$ m$^2$/s$^2$; $R_a = 40$ µm, $k = 0.278$ m$^2$/s$^2$, (**c**) $R_a = 60$ µm, $k = 0.307$ m$^2$/s$^2$; $R_a = 80$ µm, $k = 0.327$ m$^2$/s$^2$; $R_a = 100$ µm, $k = 0.346$ m$^2$/s$^2$.

The variation of the hydraulic loss distribution inside the impeller was related to the pressure distribution. Figure 8 shows the pressure distribution in the middle section of the pump with different wall roughness under the rated flow condition. It could be seen that the pressure inside the impeller was center-symmetric. As the wall roughness increased, the pressure inside the impeller gradually decreased, mainly at the edge of the impeller, and the decreasing rate gradually reduced.

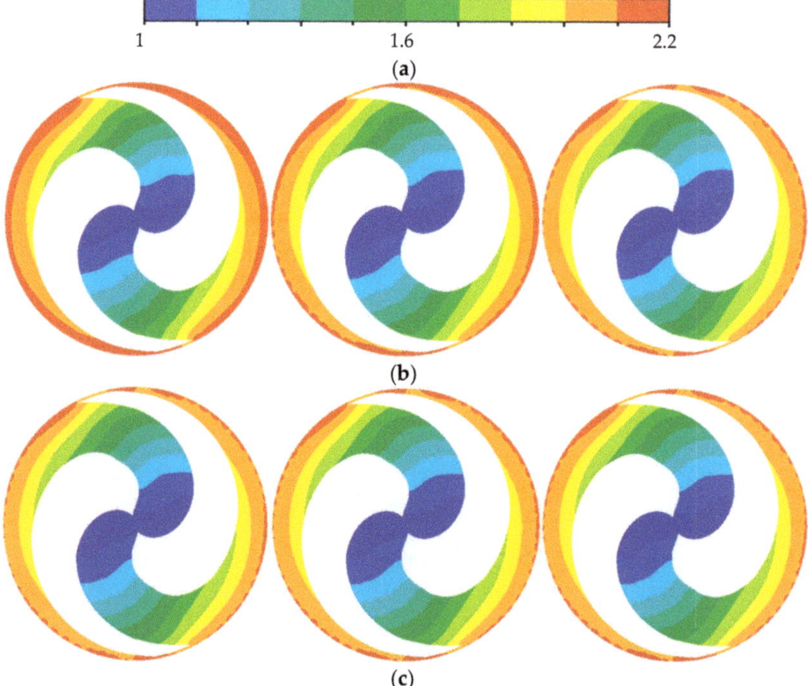

**Figure 8.** Pressure distribution in the middle section of the pump with different wall roughness under the rated flow condition. (**a**) Static pressure ($10^5$ pa), (**b**) $R_a = 0$ μm, Pressure = 113,484 Pa; $R_a = 20$ μm, Pressure = 96,037 Pa; $R_a = 40$ μm, Pressure = 95,091 Pa, (**c**) $R_a = 60$ μm, Pressure = 94,068 Pa; $R_a = 80$ μm, Pressure = 93,088 Pa; $R_a = 100$ μm, Pressure = 94,248 Pa.

*4.3. Influence of Wall Roughness on the Components of the Efficiency and Shaft Power*

In order to study why the influence of wall roughness on the performance of the pump is so obvious, this paper further subdivided the external characteristic of the pump. The components of its shaft power and efficiency are shown in Table 6. Without considering the mechanical loss at the bearing shaft seal, the formulae were as follows:

$$\eta = \frac{\rho g Q H}{P} \tag{15}$$

$$P = P_m + P_h \tag{16}$$

$$P_m = M\omega \tag{17}$$

The three sub-efficiencies of the pump were obtained from the following formulae:

$$\eta_m = 1 - \frac{P_m}{P} \tag{18}$$

$$\eta_v = \frac{Q}{Q + q} \tag{19}$$

$$\eta_h = \frac{\eta}{\eta_m \eta_v} \tag{20}$$

where $P$ is the shaft power (in W), $P_m$ is the disk friction loss power (in W), $P_h$ is the hydraulic power (in W), $q$ is the average ring leakage (in m$^3$/h), $\eta_m$ is the mechanical efficiency, $\eta_v$ is the volumetric efficiency, and $\eta_h$ is the hydraulic efficiency.

**Table 6.** Pump's performance with different wall roughness under the rated flow condition.

| $R_a$ | $P_m$ | $P_h$ | $P$ | $Q$ | $q$ | $\eta_m$ | $\eta_v$ | $\eta_h$ | $\eta$ |
|---|---|---|---|---|---|---|---|---|---|
| (μm) | (W) | (W) | (W) | (m$^3$/h) | (m$^3$/h) | (%) | (%) | (%) | (%) |
| 0 | 46.53 | 631.01 | 677.54 | 12 | 0.82 | 93.13 | 93.59 | 84.84 | 73.95 |
| 20 | 155.40 | 705.43 | 860.83 | 12 | 0.99 | 81.95 | 92.37 | 75.33 | 57.02 |
| 40 | 216.01 | 748.27 | 964.28 | 12 | 1.03 | 77.60 | 92.10 | 70.05 | 50.06 |
| 50 | 224.65 | 746.90 | 971.55 | 12 | 1.06 | 76.88 | 91.90 | 70.00 | 49.45 |
| 60 | 217.53 | 740.60 | 958.14 | 12 | 1.07 | 77.30 | 91.87 | 70.87 | 50.33 |
| 80 | 212.25 | 741.73 | 953.97 | 12 | 1.07 | 77.75 | 91.83 | 70.94 | 50.65 |
| 100 | 208.80 | 742.37 | 951.17 | 12 | 1.08 | 78.05 | 91.77 | 71.03 | 50.87 |

Figure 9 illustrates the three kinds of power $P_m$, $P_h$, and $P$ with different wall roughness $R_a$ under the rated flow conditions ($Q = 12$ m$^3$/h). When 0 μm $\leq R_a \leq$ 50 μm, as $R_a$ increased, $P_m$, $P_h$, and $P$ increased together, but the increasing rate gradually reduced; when 50 μm $\leq R_a$, as $R_a$ increased, $P_m$, $P_h$ and $P$ decreased slightly. When $R_a$ changed from 0 μm to 50 μm, $P_m$ and $P_h$ increased to 382.81% and 18.37%, respectively. When $R_a$ changed from 50 μm to 100 μm, $P_m$ and $P_h$ increased to 7.06% and 0.61%, respectively, indicating that when the wall roughness did not reach the critical wall roughness, the influence of the wall roughness was relatively large. The effect of the wall roughness on $P_h$ was much smaller than that on $P_m$.

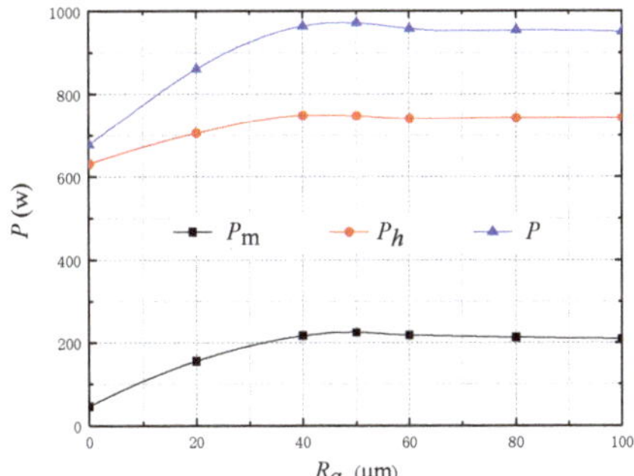


**Figure 9.** Three kinds of power of the pump with different wall roughness under the rated flow condition.

Figure 10 shows the pump mechanical efficiency $\eta_m$, hydraulic efficiency $\eta_h$, and volumetric efficiency $\eta_v$ with different roughness $R_a$ under the rated flow condition. It could be seen from Table 5 that as $R_a$ increased, $\eta_v$ gradually decreased, but the decreasing rate gradually reduced. When $R_a$ did not reach the critical wall roughness ($R_a = 50$ μm), with the increase of $R_a$, $\eta_m$ and $\eta_h$ gradually decreased, and the decreasing rate gradually reduced. When $R_a$ exceeded the critical wall roughness, as $R_a$ increased, $\eta_m$ and $\eta_h$ gradually increased, but the increasing rate gradually reduced. When $R_a$ changed from 0 μm to 50 μm, $\eta_m$, $\eta_h$, and $\eta_v$ reduced by 16.25%, 14.84%, and 1.69%, respectively. When $R_a$ increased from 50 μm to 100 μm, $\eta_m$ and $\eta_h$ increased by 1.17% and 1.03%, and $\eta_v$ reduced by 0.13%. It showed that when the wall roughness did not reach the critical wall roughness, the effect of wall roughness on the efficiency was large; when the roughness exceeded the critical roughness, the effect of wall roughness on the efficiency was small. With the increase of wall roughness, mechanical efficiency and hydraulic efficiency reduced constantly, leading to the decrement of the total efficiency.

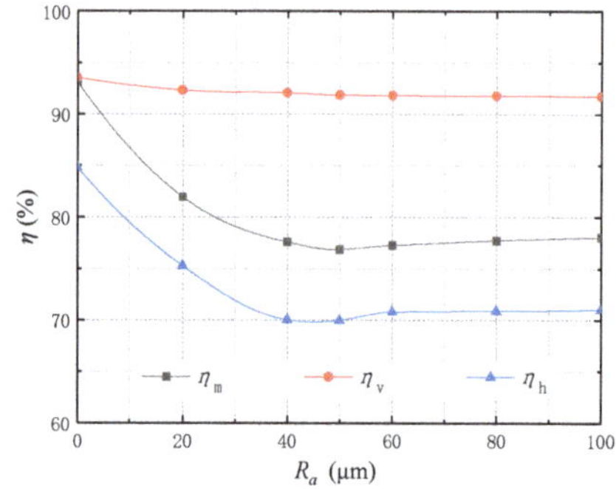

**Figure 10.** Three kinds of power of the pump with different wall roughness under the rated flow condition.

### 4.4. Influence of Wall Roughness on the Pump Performance Under Five Flow Conditions

Because of complicated use situations, the pump's operating conditions of the pump were constantly changing. The above research was mainly to study the influence of wall roughness of the pump under the rated flow condition. Therefore, the influence of different wall roughness under five flow conditions ($Q$ = 7.8, 10, 12, 14, and 16 m$^3$/h) was numerically calculated, as shown in Table 7.

**Table 7.** Pump's performance with five flow conditions and wall roughness.

| $Q$ | $R_a$ | $H$ | $P_m$ | $P_h$ | $q$ | $\eta_m$ | $\eta_v$ | $\eta_h$ | $\eta$ |
|---|---|---|---|---|---|---|---|---|---|
| (m$^3$/h) | (μm) | (m) | (W) | (W) | (m$^3$/h) | (%) | (%) | (%) | (%) |
| 7.8 | 0 | 16.13 | 43.50 | 505.41 | 0.84 | 92.08 | 90.33 | 75.04 | 62.41 |
| 10 | 0 | 15.90 | 43.95 | 568.54 | 0.83 | 92.82 | 92.32 | 82.44 | 70.65 |
| 12 | 0 | 15.29 | 46.53 | 631.01 | 0.82 | 93.13 | 93.59 | 84.84 | 73.95 |
| 14 | 0 | 14.22 | 47.77 | 694.73 | 0.81 | 93.57 | 94.52 | 82.50 | 72.96 |
| 16 | 0 | 12.93 | 49.15 | 756.15 | 0.79 | 93.90 | 95.30 | 78.13 | 69.91 |
| 7.8 | 50 | 15.89 | 221.50 | 626.10 | 1.11 | 73.87 | 87.50 | 61.59 | 39.81 |
| 10 | 50 | 15.61 | 222.74 | 686.09 | 1.10 | 75.49 | 90.13 | 68.74 | 46.77 |
| 12 | 50 | 14.66 | 224.65 | 746.90 | 1.06 | 76.88 | 91.90 | 70.00 | 49.45 |
| 14 | 50 | 13.13 | 226.20 | 808.30 | 1.00 | 78.13 | 93.32 | 66.32 | 48.36 |
| 16 | 50 | 11.31 | 228.90 | 865.62 | 0.93 | 79.09 | 94.52 | 60.19 | 45.00 |
| 7.8 | 100 | 15.92 | 205.32 | 619.97 | 1.14 | 75.12 | 87.30 | 62.46 | 40.96 |
| 10 | 100 | 15.67 | 206.90 | 680.28 | 1.11 | 76.68 | 89.99 | 69.67 | 48.07 |
| 12 | 100 | 14.76 | 208.80 | 742.37 | 1.08 | 78.05 | 91.77 | 71.03 | 50.87 |
| 14 | 100 | 13.43 | 210.56 | 804.93 | 1.02 | 79.27 | 93.20 | 68.21 | 50.39 |
| 16 | 100 | 11.80 | 212.18 | 864.33 | 0.94 | 80.29 | 94.45 | 62.96 | 47.75 |

Figure 11 shows the pump head $H$ and efficiency $\eta$ under five flow conditions. When $R_a$ changed from 0 μm to 50 μm, $H$ and $\eta$ were reduced, while the leakage amount $q$ increased. When $Q$ = 7.8m$^3$/h, the increase of $R_a$ led to a decrease of $H$ and $\eta$ of 0.24 m and 22.6%, respectively, and $q$ increased by 0.27 m$^3$/h. When $Q$ = 16 m$^3$/h, the increase of $R_a$ led to a decrease of $H$ and $\eta$ of 1.62 m and 24.91%, respectively, and $q$ increased by 0.14 m$^3$/h; When $R_a$ increased from 50 μm to 100 μm, $H$, $\eta$, and $q$ increased together. When $Q$ = 7.8 m$^3$/h, with the increase of $R_a$, $H$, $\eta$, and $q$ increased by 0.03 m, 1.15%, and 0.03 m$^3$/h, while they increased by 0.49 m, 2.75%, and 0.01 m$^3$/h at $Q$ = 16 m$^3$/h. It showed that with the increase of flow rate, the influence of wall roughness on the head and efficiency increased gradually, but the influence on the leakage amount decreased gradually.

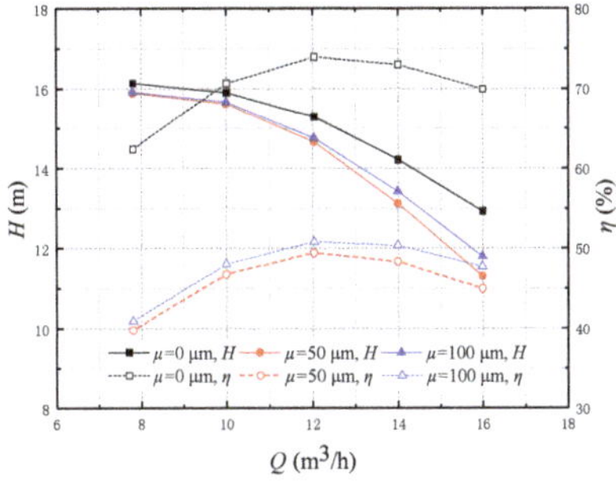

**Figure 11.** Pump's performance with three kinds of wall roughness under different flow conditions.

Figure 12 shows the three kinds of power $P_m$, $P_h$, and $P$, with three kinds of wall roughness under five flow conditions. As $Q$ increased, $P_h$ and $P$ showed an increasing trend, while $P_m$ kept basically unchanged. With the increase of $R_a$, $P_m$, $P_h$, and $P$ increased jointly, but the increasing rate kept the same with $Q$, indicating that the effect of wall roughness on the power was independent of Reynolds number.

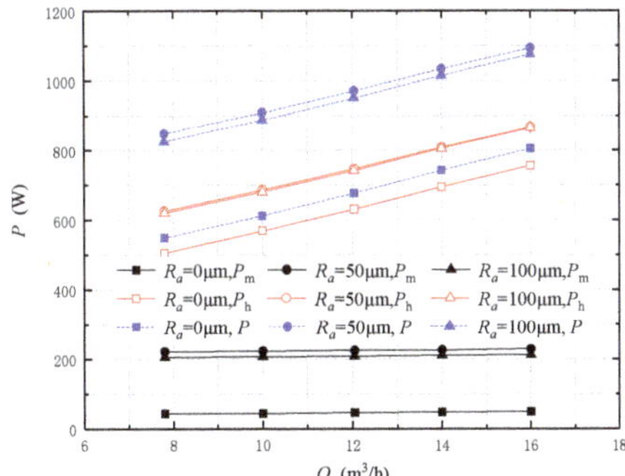

**Figure 12.** Three kinds of power of the pump under five flow conditions.

Figure 13 indicates the three kinds of efficiency $\eta_m$, $\eta_h$, and $\eta_v$ under five flow conditions. With the increase of $Q$, $\eta_m$ and $\eta_v$ increased, while $\eta_h$ first increased and then decreased. The reason why $\eta_m$ increased was that $P_m$ kept basically unchanged with the increase of $Q$, while $P_h$ and $P$ increased together, so the proportion of $P_m$ in $P$ gradually decreased. The reason why $\eta_v$ increased was that the pump's leakage amount decreased as with $Q$. The reason why $\eta_h$ first increased and then decreased was that 12 m³/h was the design condition (optimal flow condition) of the pump. Under different roughness conditions, with the increase of flow rate, the variation amplitude of mechanical efficiency and volumetric efficiency decreased gradually, while the variation amplitude of hydraulic efficiency increased gradually. The reason why the variation amplitude of mechanical efficiency decreased gradually was that with the increase of flow rate, the variation amplitude of disk friction loss power and the roughness caused by disk friction loss power and hydraulic power basically remain unchanged, while the hydraulic power increased. The reason why the variation amplitude of volumetric efficiency decreased was that with increasing the flow rate, the influence of wall roughness on leakage also decreased gradually. When the roughness was less than the critical wall roughness, the reason for the increase of hydraulic efficiency was that with the increase of flow rate, the resistance loss along the pump path caused by roughness increased gradually. When the roughness was greater than the critical wall roughness, the reason for the increase of hydraulic efficiency was that with the increase of roughness, the energy loss in the pump was close to a constant value [28], and the proportion decreased gradually.

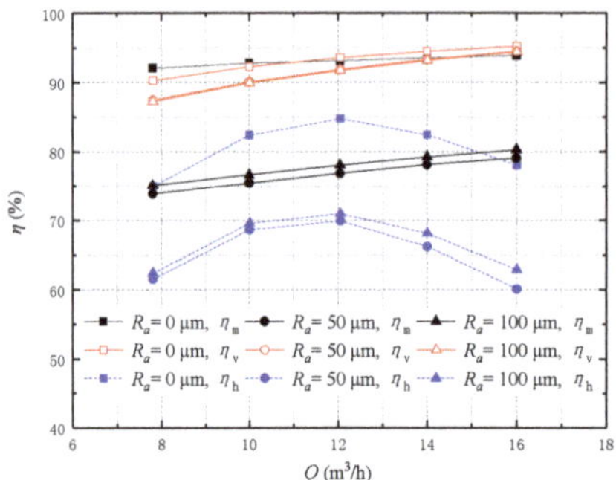

**Figure 13.** Three kinds of efficiency of the pump under five flow conditions.

### 4.5. Influence of Wall Roughness of Each Flow Part on the Pump Performance

To study the influence of wall roughness of each flow part on the pump performance, four flow parts of the pump, such as the impeller (including the blade and inner wall), shroud, pump cavity, and volute, were selected as research objects, and the wall roughness was set separately for numerical calculation. Figure 14 shows the pump head $H$ with different wall roughness $R_a$ of each flow part. It could be seen that with the increase of $R_a$ of the impeller and shroud, the working capacity of the pump was improved, and $H$ was gradually increased, but the increase rate was continuously decreased. The influence of $R_a$ of the impeller on $H$ was greater than that of the shroud. After considering the wall roughness of the pump cavity and volute, as $R_a$ increased, $H$ gradually decreased, and the decreasing rate continuously decreased. The influence of rough volute on the head was greater than that of all rough flow parts, indicating that the wall roughness of each flow part had no independent influence on the pump head, but interacted with each other.

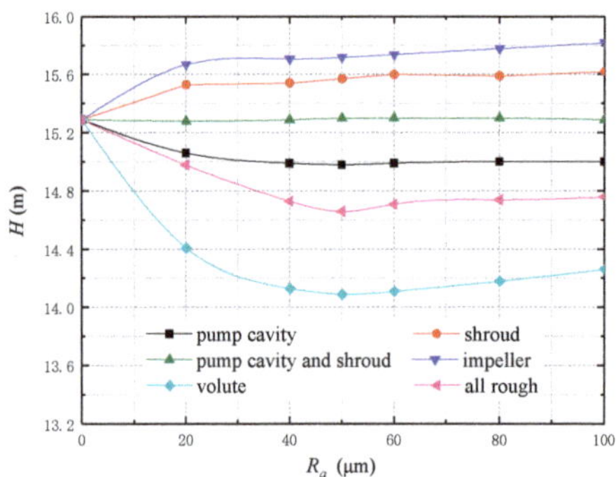

**Figure 14.** Pump's head with different wall roughness of each flow part under the rated flow condition.

Figure 15 shows the pump efficiency $\eta$ with different wall roughness $R_a$ of each flow part. It could be seen from Figure 15a that after considering the wall roughness of each flow part, $\eta$ decreased with the increase of $R_a$, and the decreasing rate gradually reduced. When $R_a$ reached the critical wall roughness ($R_a > 50$ μm), $\eta$ kept stable basically. The influence of $R_a$ of the impeller and shroud (pump cavity and volute) on $\eta$ was basically the same, while the influence of $R_a$ of the pump cavity and volute on $\eta$ was greater than that of the impeller and shroud. The comprehensive influence of wall roughness of all the flow parts on the pump efficiency was smaller than the sum influence of that of each flow part, indicating once again that the influence of wall roughness on the pump performance was interconnected. Figure 15b shows the mechanical efficiency $\eta_m$ with different wall roughness $R_a$ of each flow part. It could be seen that $R_a$ of the impeller and volute had little effect on $\eta_m$, while that of the shroud and pump cavity had a strong effect on $\eta_m$. Obviously, $R_a$ of the shroud was most sensitive to $\eta_m$ of the pump, indicating that the disk friction loss was closely related to the wall roughness of the shroud. Figure 15c shows the volumetric efficiency $\eta_v$ of the pump with different wall roughness $R_a$ of each flow part. It could be seen that $R_a$ of the impeller and shroud had a positive effect on $\eta_m$, while that of the volute and pump cavity had a negative effect. Moreover, $R_a$ of the shroud and pump cavity was rather sensitive to $\eta_v$, indicating the flow in the pump cavity was closely related to the volumetric leakage. Figure 15d shows the hydraulic efficiency $\eta_h$ of the pump with different wall roughness $R_a$ of each flow part. It could be seen that $R_a$ of the shroud had a slight positive effect on $\eta_m$, while that of the pump cavity, impeller, and volute had an obvious negative effect. The wall roughness of the volute was most sensitive to the hydraulic loss. In summary, the volumetric efficiency was affected by the wall roughness of the pump cavity and shroud, the mechanical efficiency was affected by the wall roughness of the shroud, and the hydraulic efficiency was affected by the wall roughness of the impeller and volute.

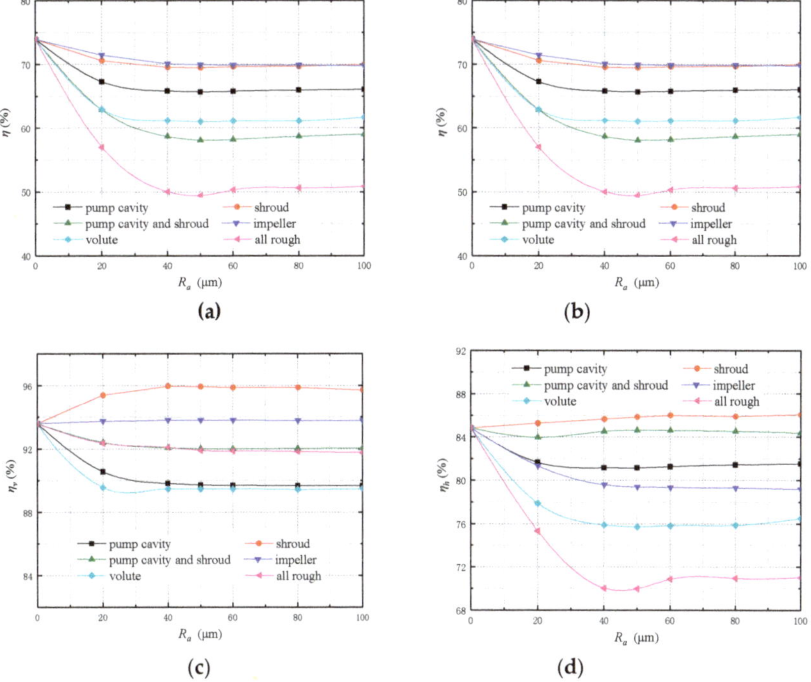

**Figure 15.** Pump efficiency $\eta$ with different wall roughness $R_a$ of each flow part under the rated flow condition: (**a**) total efficiency; (**b**) mechanical efficiency; (**c**) volumetric efficiency; (**d**) hydraulic efficiency.

Figure 16 shows the shaft power $P$ with different wall roughness $R_a$ of each flow part. As could be seen from Figure 16a, after considering the wall roughness of the volute, shroud, impeller, and pump cavity, respectively, $P$ increased with the increase of $R_a$, and the increasing rate gradually reduced to zero when $R_a$ reached to the critical value of the wall roughness. Figure 16b shows the disk friction loss power $P_m$. It could be seen that $R_a$ of the impeller had little effect on $P_m$, while $P_m$ increased with the $R_a$ of the volute, pump cavity, and shroud. Obviously, $R_a$ of the shroud had the most effects on the $P_m$ of the pump. Figure 16c shows the hydraulic power $P_h$ of the pump. It could be seen that $R_a$ of the shroud had a negative effect on $P_h$, while $P_h$ increased with the $R_a$ of the volute, pump cavity, and impeller. Obviously, $R_a$ of the impeller had the most effects on $P_h$. In conclusion, the disk friction loss power was affected by the wall roughness of the shroud, while the hydraulic power was affected by that of the impeller.

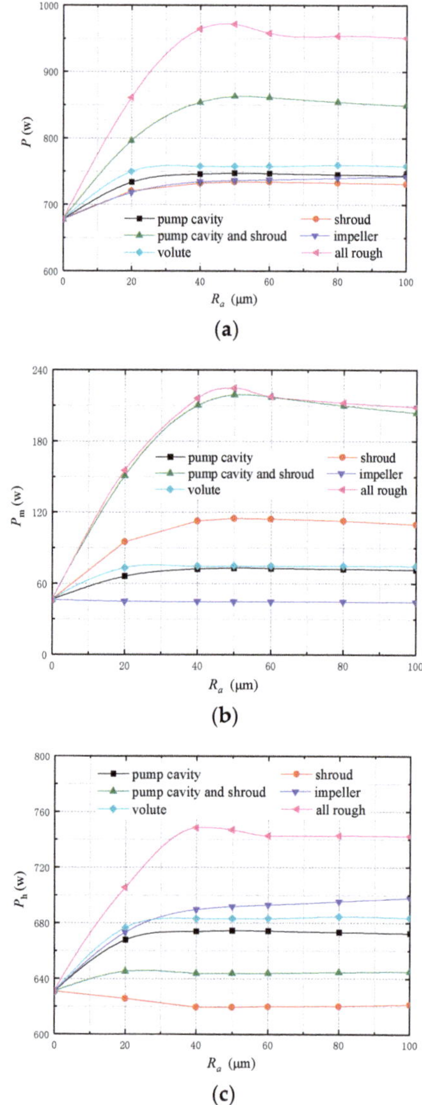

**Figure 16.** Shaft power $P$ of the pump with different wall roughness $R_a$ of each flow part: (**a**) total power; (**b**) disk friction loss power; (**c**) hydraulic power.

*4.6. Comparison Between the Numerical and Experimental Results*

The prototype of the pump was made, and the relevant experimental results could also be obtained. The whole experiment could be divided into four steps:

1. The original model was processed and then tested (All rough);

2. The impeller channel was polished and then tested (Smooth impeller channel);
3. The front and the rear shroud of the impeller were polished (Smooth shroud);
4. The volute channel and the inner wall of the pump cavity were polished (All smooth).

Every test was repeated to ensure the accuracy of the experimental data. After polishing the wall, the average wall roughness was about 5 μm. The schematic diagram of the test bench is shown in Figure 17. A turbine flowmeter was used to measure the flow Q, and the precision of the turbine flowmeter was ±0.3%. Speed n was measured by a tachometer (PROVA RM-1500, Taiwan). During the experiment, only one dynamic pressure transmitters (CYG1401) was used to measure the outlet pressure. The precision of CYG1401 was ±0.2%.

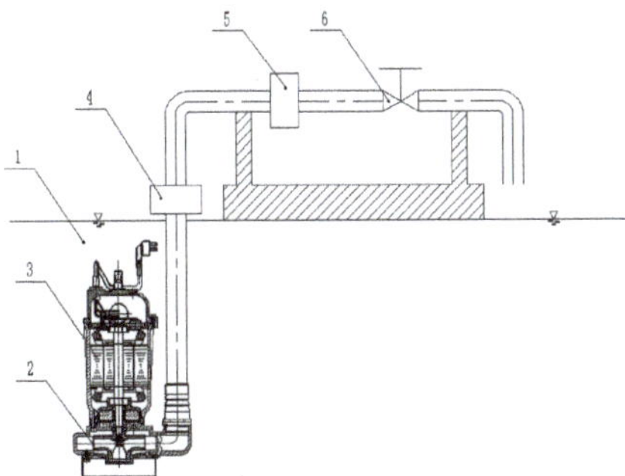

**Figure 17.** Schematic diagram of the test rig. 1. Pool; 2. Pump; 3. Motor; 4. Outlet pressure transmitters; 5. Turbine flowmeter; 6. Flow control valve.

Figure 18 illustrates the numerical and experimental results of the pump with four steps under five flow conditions ($Q$ = 7.8, 10, 12, 14, 16 m$^3$/h). As could be seen from Figure 18a, the experimental value of the head was consistent with the numerical value. When all the flow parts were smooth, the experimental and numerical head of the pump was largest. From Figure 18b, it could be seen that the numerical value of the efficiency was slightly higher than the experimental value because there is some backflow or deflow in the pump, and it's rather difficult to simulate the disordered flow by using CFD. Moreover, the numerical calculation did not consider the mechanical friction loss power at the bearing seal of the bearing shaft, which might account for 1% to 2% of the total power. However, the deviation between the numerical and experimental efficiency was only within 3%. Therefore, it's rather reliable to predict the pump's performance by using CFD.

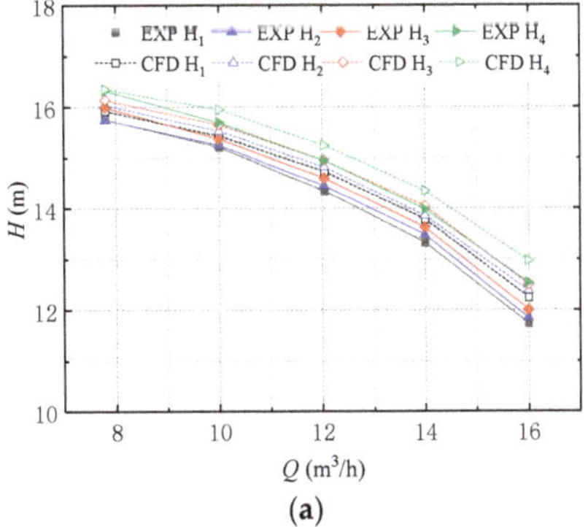


**Figure 18.** *Cont.*

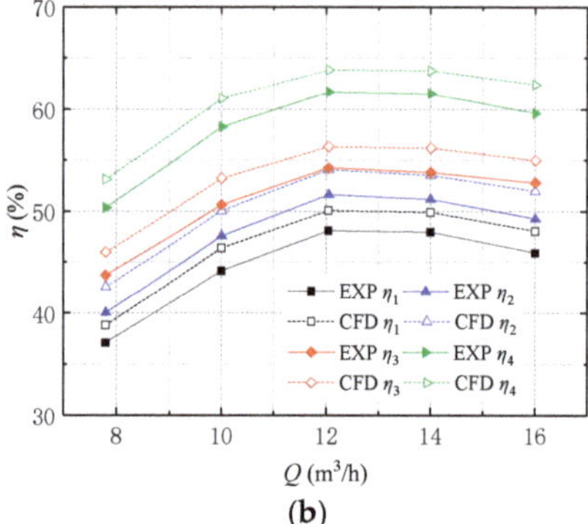



**(b)**

**Figure 18.** Numerical and experimental results of the pump with four steps under five flow conditions: (**a**) head; (**b**) efficiency.

## 5. Conclusions

(1) Wall roughness affected the performance of the double-channel swage pump, and there was also a critical wall roughness. When the wall roughness was less than the critical wall roughness, the wall roughness had a great effect on the performance of the pump. With increasing the wall roughness, the efficiency and head were reduced, and the shaft power was increased. The effect of wall roughness on the shaft power and efficiency was much greater than that on the head. When the roughness exceeded the critical wall roughness, the wall roughness had little effect on the performance of the pump.

(2) The volumetric efficiency was affected by the wall roughness of the pump cavity and shroud, the mechanical efficiency was affected by the wall roughness of the shroud, and the hydraulic efficiency was affected by the wall roughness of the impeller and volute. In addition, the effect of the wall roughness of different flow parts was interactive.

(3) For general centrifugal pumps, reducing the volumetric leakage loss was the most effective way to increase pump efficiency. Moreover, it was beneficial to improving the pump efficiency and reducing the pump shaft power by polishing the shroud and pump cavity.

(4) On the basis of the complete calculation model and appropriate numerical method, it was rather reliable to use CFD to predict the performance of the double-channel sewage pump.

**Author Contributions:** C.W. conceived and designed the experiments; Y.Z. and C.Z. performed the experiments and simulation; L.C. and K.C. analyzed the data; X.H. and B.H. wrote the paper; L.C. funding acquisition. All authors have read and agreed to the published version of the manuscript.

**Funding:** This work was supported by a project funded by the Priority Academic Program Development of Jiangsu Higher Education Institutions (PAPD), Nature Science Foundation of China (Grant No. 51609105 and No. 51979240).

**Conflicts of Interest:** The authors declare no conflicts of interest.

## References

1. Wang, C.; He, X.; Shi, W.; Wang, X.; Wang, X.; Qiu, N. Numerical study on pressure fluctuation of a multistage centrifugal pump based on whole flow field. *AIP Adv.* **2019**, *9*, 035118. [CrossRef]
2. Chang, H.; Shi, W.; Li, W.; Wang, C.; Zhou, L.; Liu, J.; Yang, Y. Agarwal Rameshe. Experimental optimization of jet self-priming centrifugal pump based on orthogonal design and grey-correlational method. *J. Therm. Sci.* **2019**. [CrossRef]
3. Lu, Y.; Zhu, R.; Wang, X.; Wang, Y.; Fu, Q.; Ye, D. Study on the complete rotational characteristic of coolant pump in the gas-liquid two-phase operating condition. *Ann. Nucl. Energy* **2019**, *123*, 180–189.

4.   Bai, L.; Zhou, L.; Jiang, X.; Pan, Q.; Ye, D. Vibration in a multistage centrifugal pump under varied conditions. *Shock Vib.* **2019**, *2019*, 2057031. [CrossRef]

5.   Wang, C.; Hu, B.; Zhu, Y.; Wang, X.; Luo, C.; Cheng, L. Numerical study on the gas-water two-phase flow in the self-priming process of self-priming centrifugal pump. *Processes* **2019**, *7*, 330. [CrossRef]

6.   Bhushan, S.; Ståhl, A.; Nilsson, S. Submersible wastewater pump association. *World Pumps* **2000**, *2000*, 44.

7.   Chen, B.; Zhang, K.W. Characteristics and significance of high head submersible sewage pump for Three Gorges Project. *Pump Technol.* **2002**, *2002*, 23–25.

8.   Zhao, B.; Hou, D.; Chen, H. Influence of impeller runner structure on performance of dual-channel pump. *J. Drain. Irrig. Mach. Eng.* **2013**, *31*, 294–299.

9.   Minggao, T.; Rong, D.; Houlin, L.; Liang, D.; Naichang, H. Analysis on pressure pulsation under different impeller diameters in double channel sewage pump. *Trans. Chin. Soc. Agric. Eng.* **2015**, *31*, 53–59.

10.  Tan, M.G.; Lian, Y.C.; Wu, X.F. Transient dynamics in a two-channel pump. *J. Drain. Irrig. Mach. Eng.* **2017**, *35*, 17–22.

11.  Wang, C.; He, X.; Zhang, D.; Hu, B.; Shi, W. Numerical and experimental study of the self-priming process of a multistage self-priming centrifugal pump. *Int. J. Energy Res.* **2019**, *43*, 4074–4092. [CrossRef]

12.  Wang, C.; Shi, W.; Wang, X.; Jiang, X.; Yang, Y.; Li, W.; Zhou, L. Optimal design of multistage centrifugal pump based on the combined energy loss model and computational fluid dynamics. *Appl. Energy* **2017**, *187*, 10–26. [CrossRef]

13.  Hu, B.; Li, X.; Fu, Y.; Zhang, F.; Gu, C.; Ren, X.; Wang, C. Experimental investigation on the flow and flow-rotor heat transfer in a rotor-stator spinning disk reactor. *Appl. Therm. Eng.* **2019**, *162*, 114316. [CrossRef]

14.  Wang, C.; Chen, X.X.; Qiu, N.; Zhu, Y.; Shi, W.D. Numerical and experimental study on the pressure fluctuation, vibration, and noise of multistage pump with radial diffuser. *J. Braz. Soc. Mech. Sci. Eng.* **2018**, *40*, 481. [CrossRef]

15.  Li, X.; Chen, B.; Luo, X.; Zhu, Z. Effects of flow pattern on hydraulic performance and energy conversion characterisation in a centrifugal pump. *Renew. Energy* **2020**. [CrossRef]

16.  Yang, H.; Zhang, W.; Zhu, Z. Unsteady mixed convection in a square enclosure with an inner cylinder rotating in a bi-directional and time-periodic mode. *Int. J. Heat Mass Transf.* **2019**, *136*, 563–580. [CrossRef]

17.  Hemeida, A.M. Effect of wall roughness in turbulent pipe flow of a pseudoplastic crude oil: An evaluation of pipeline field data. *J. Pet. Sci. Eng.* **1993**, *10*, 163–170. [CrossRef]

18.  Kandlikar, S.G. Roughness effects at microscale-reassessing Nikuradse's experiments on liquid flow in rough tubes. *Bull. Pol. Acad. Sci. Tech. Sci.* **2005**, *53*, 343–349.

19.  Li, D.; Kang, Y.; Wang, X.; Ding, X.; Fang, Z. Effects of nozzle inner surface roughness on the cavitation erosion characteristics of high speed submerged jets. *Exp. Therm. Fluid Sci.* **2016**, *74*, 444–452. [CrossRef]

20.  Tang, G.H.; Li, Z.; He, Y.L.; Tao, W.Q. Experimental study of compressibility, roughness and rarefaction influences on microchannel flow. *Int. J. Heat Mass Transf.* **2007**, *50*, 2282–2295. [CrossRef]

21.  Gamrat, G.; Favre-Marinet, M.; Le Person, S.; Baviere, R.; Ayela, F. An experimental study and modelling of roughness effects on laminar flow in microchannels. *J. Fluid Mech.* **2008**, *594*, 399–423. [CrossRef]

22.  Li, D.; Fan, Z.; Zhang, J.; Yang, X. Research on the impact of compressor blade roughness on its performance decline. *Aero Engine* **2009**, *5*, 36–39.

23.  Li, F.; Bu, Q. Effect of relative roughness of flow path of similar fans on its flow efficiency. *FAN Technol.* **2009**, *1*, 31–32.

24.  Li, C.; Liang, Q.; Zhao, Y. Effect of surface roughness on micro-gap leakage of oil-free scroll compressors. *Fluid Mach.* **2011**, *4*, 47–50.

25.  Gao, L. Experimental Research on the Effect of Surface Roughness on the Performance of Compressor Cascades. Ph.D. Thesis, University of Chinese Academy of Sciences, Beijing, China, 2015.

26.  Han, F.; Du, L.; Li, W.; Li, C. Effect of blade surface roughness on the aerodynamic performance of compressors in a class environment. *J. Dalian Jiaotong Univ.* **2015**, *2*, 50–54.

27.  Wang, C.; He, X.; Cheng, L.; Luo, C.; Xu, J.; Chen, K.; Jiao, W. Numerical Simulation on Hydraulic Characteristics of Nozzle in Waterjet Propulsion System. *Processes* **2019**, *7*, 915. [CrossRef]

28.  Jiao, W.; Cheng, L.; Zhang, D.; Zhang, B.; Su, Y.; Wang, C. Optimal design of inlet passage for waterjet propulsion system based on flow and geometric parameters. *Adv. Mater. Sci. Eng.* **2019**, *2019*, 2320981. [CrossRef]

29. Zhu, Y.; Tang, S.; Wang, C.; Jiang, W.; Yuan, X.; Lei, Y. Bifurcation Characteristic research on the load vertical vibration of a hydraulic automatic gauge control system. *Processes* **2019**, *7*, 718. [CrossRef]
30. Zhu, Y.; Tang, S.; Wang, C.; Jiang, W.; Zhao, J.; Li, G. Absolute stability condition derivation for position closed-loop system in hydraulic automatic gauge control. *Processes* **2019**, *7*, 766. [CrossRef]
31. Zhu, Y.; Tang, S.; Quan, L.; Jiang, W.; Zhou, L. Extraction method for signal effective component based on extreme-point symmetric mode decomposition and Kullback-Leibler divergence. *J. Braz. Soc. Mech. Sci. Eng.* **2019**, *41*, 100. [CrossRef]
32. Zhu, Y.; Qian, P.; Tang, S.; Jiang, W.; Li, W.; Zhao, J. Amplitude-frequency characteristics analysis for vertical vibration of hydraulic AGC system under nonlinear action. *AIP Adv.* **2019**, *9*, 035019. [CrossRef]
33. He, X.; Jiao, W.; Wang, C.; Cao, W. Influence of surface roughness on the pump performance based on Computational Fluid Dynamics. *IEEE Access* **2019**, *7*, 105331–105341. [CrossRef]
34. Guelich, J.F. Effect or Reynolds Number and Surface Roughness on the Efficiency of Centrifugal Pumps. *J. Fluids Eng.* **2003**, *125*, 670–679. [CrossRef]
35. Zhu, H.; Zhai, B.; Zhou, J. Study on the Influence of Wall Roughness on Hydraulic Performance of Axial Flow Pump. *J. Irrig. Drain.* **2006**, *25*, 87–90.
36. Yamazaki, Y.; Nakayama, T.; Narabayashi, T.; Kobayashi, H.; Shakouchi, T. Effect of Surface Roughness on Jet Pump Performance. *JSME Int. J. Ser. B* **2006**, *49*, 928–932. [CrossRef]
37. Aldas, K.; Yapici, R. Investigation of Effects of Scale and Surface Roughness on Efficiency of Water Jet Pumps Using CFD. *Eng. Appl. Comput. Fluid Mech.* **2014**, *8*, 14–25. [CrossRef]
38. Feng, J.J.; Zhu, G.J.; He, R. Effect of surface roughness on the performance of axial flow pump. *J. Northwest Univ. Nat. Sci. Ed.* **2016**, 196–202.
39. Pan, Z.Y.; Gu, L.Q.; Gu, M.F. Numerical simulation analysis of flow field characteristics of axial flow pump with wall roughness. *Water Technol. Econ.* **2018**, *24*, 43–46.
40. Deshmukh, D.; Siddique, M.H.; Kenyery, F.; Samad, A. Critical surface roughness for wall bounded flow of viscous fluids in an electric submersible pump. In Proceedings of the APS Meeting Abstracts, New Orleans, LA, USA, 13–17 March 2017.
41. Gu, M.F.; Yang, X.H.; Sun, F.M. Study on the influence of different roughness on the hydraulic performance of axial flow pump. *Tech. Superv. Water Resour.* **2018**, *6*, 146–148.
42. Lim, S.E.; Sohn, C.H. CFD analysis of performance change in accordance with inner surface roughness of a double-entry centrifugal pump. *J. Mech. Sci. Technol.* **2018**, *32*, 697–702. [CrossRef]
43. CFX-Solver Theory Guide, ANSYS 13.0 Help [EB/OL]. 2010. Available online: http://www.doc88.com/p-908539666887.html (accessed on 6 October 2012).
44. Brian Whalley, W.; Rea, B.R. A digital surface roughness meter. *Earth Surf. Processes Landf.* **1994**, *19*, 809–814. [CrossRef]
45. Johnson, F.; Brisco, B.; Brown, R.J. Evaluation of limits to the performance of the surface roughness meter. *Can. J. Remote Sens.* **1993**, *19*, 140–145. [CrossRef]
46. Yan, Z.; Yong, T. Uncertainty Analysis and Calculation of Surface Roughness Comparison Sample Measurement. *Mod. Meas. Lab. Manag.* **2009**, *4*, 30–31.
47. Zhao, B.; Wang, Y.; Chen, H. Impact of wall roughness on the flow law in chamber of a centrifugal pump at off-design operating condition. *J. Eng. Thermophys.* **2015**, *36*, 1927–1932.
48. Sun, H.; Ye, N.; Wang, S. Effect of Roughness on Boundary Layer Flow and Aerodynamic Performance of Compressor. *J. Harbin Eng. Univ.* **2017**, *38*, 554–560.
49. Colebrook, C.F.; White, C.M. Experiments with fluid friction in roughened pipes. *Proc. R. Soc. Lond. Ser. A Math. Phys. Sci.* **1937**, *161*, 367–381.
50. Adams, T.; Grant, C.; Watson, H. A simple algorithm to relate measured surface roughness to equivalent sand-grain roughness. *Int. J. Mech. Eng. Mechatron.* **2012**, *1*, 66–71. [CrossRef]
51. Deshmukh, D.; Samad, A. CFD-based analysis for finding critical wall roughness on centrifugal pump at design and off-design conditions. *J. Braz. Soc. Mech. Sci. Eng.* **2019**, *41*, 58. [CrossRef]

 © 2020 by the authors. Licensee MDPI, Basel, Switzerland. This article is an open access article distributed under the terms and conditions of the Creative Commons Attribution (CC BY) license (http://creativecommons.org/licenses/by/4.0/).

*Article*

# Multi-Disciplinary Optimization Design of Axial-Flow Pump Impellers Based on the Approximation Model

**Lijian Shi, Jun Zhu, Fangping Tang and Chuan Wang ***

College of Hydraulic Science and Engineering, Yangzhou University, Yangzhou 225000, China;
shilijian@yzu.edu.cn (L.S.); JJ1293247635@hotmail.com (J.Z.); tangfp@yzu.edu.cn (F.T.)
* Correspondence: wangchuan@ujs.edu.cn

Received: 1 January 2020; Accepted: 6 February 2020; Published: 11 February 2020

**Abstract:** This study adopts a multi-disciplinary optimization design method based on an approximation model to improve the comprehensive performance of axial-flow pump impellers and fully consider the interaction and mutual influences of the hydraulic and structural designs. The lightweight research on axial-flow pump impellers takes the blade mass and efficiency of the design condition as the objective functions and the head, efficiency, maximum stress value, and maximum deformation value under small flow condition as constraints. In the optimization process, the head of the design condition remains unchanged or varies in a small range. Results show that the mass of a single blade was reduced from 0.947 to 0.848 kg, reaching a decrease of 10.47%, and the efficiency of the design condition increased from 93.91% to 94.49%, with an increase rate of 0.61%. Accordingly, the optimization effect was evident. In addition, the error between the approximate model results and calculation results of each response was within 0.5%, except for the maximum stress value. This outcome shows that the accuracy of the approximate model was high, and the analysis result is reliable. The results provide guidance for the optimal design of axial-flow pump impellers.

**Keywords:** axial-flow pump; impeller; approximation model; optimization design; multi-disciplinary

---

## 1. Introduction

A multi-disciplinary optimization (MDO) design, which is among the latest and most active fields in the current research on complex system optimization design, is mainly used in specializations such as aerospace and torpedo missile design. The research on MDO in fluid machinery is mainly focused on the optimization design of turbine and wind turbine blades. However, pumps are classified as general machinery with varied applications [1–5]. The axial-flow pump impeller is the core and most important flow component of a pump device. The result of the design directly determines the comprehensive effects of the pump device and the entire pumping station. In recent years, approximately half of the axial-flow pump impellers produced annually have been used to replace scrapped products caused by blade problems [6]. To improve the comprehensive performance of axial-flow pump impellers, a multi-disciplinary optimization design of an axial-flow pump impeller should be implemented.

MDO is mainly used in some large-scale systems engineering. Sun et al. [7] used the MDO platform to focus on integral solid propellant ramjet supersonic cruise vehicles and constructed two MDO frameworks through discipline codes. Thus, the optimization of the detailed parameters and high fidelity were achieved. Chen X et al. [8] adopted MDO by introducing the multi-disciplinary feasible architecture and gradient-based optimization algorithm, which further improved the efficiency of gradient calculation. The optimized design was more energy-saving than the initial design. Xu HW et al. [9] dealt with time-varying uncertainty by proposing a multi-disciplinary robust design optimization method based on time-varying sensitivity analysis through a multi-disciplinary optimization design. MDO provides a good solution in the fields of launch vehicle design optimization, hard rock tunnel boring machine performance design optimization, and all-electric GEO satellite design optimization [10–14]. The MOD technology has incomparable advantages in solving the optimization design problem of large-scale complex systems engineering.

The optimal design of a pump is based on the design theory and method of pumps, which uses optimal design theory to achieve superior comprehensive performance [15]. At present, optimization methods for pumps mainly include those based on simplified model prediction and accurate model analysis. To improve the efficiency of a centrifugal pump, Wang WJ et al. [16] and Pei J et al. [17] optimized the impeller and guide vane, thereby obtaining a highly efficient hydraulic model. Miao F et al. [18] proposed a multi-objective optimization of the impeller shape of an axial-flow pump based on the modified particle swarm optimization (MPSO) algorithm. This novel algorithm was successfully applied for the optimization of axial-flow pump impeller shape designs by comparing the results of the MPSO and Computational Fluid Dynamics (CFD) simulation results. Yun JE [19] studied the impact loss at the inlet of rotor blades and obtained the optimal design of a blade shape by numerical analysis and optimization. Some studies [20,21] have adopted optimization design methods to achieve improved cavitation and good hydraulic performance of axial-flow pumps. Numerous studies have shown that the numerical simulation method has become the most important research method in the field of pumps. However, the majority of these studies have used a single subject optimization method to obtain the design result of pumps, in which the hydraulic performance and structural strength were optimized separately.

Only a few studies have been conducted on the fluid–structure coupling of pumps. These studies used the fluid–structure coupling analysis method based on the MDO platform to optimize the blades of pumps [22,23]. Tong YF et al. [24] implemented a multi-disciplinary energy-saving optimization design of bridge cranes and adopted the finite element analysis and multi-disciplinary optimization technology to reduce the total quality and energy-saving design of cranes. However, these studies have yet to achieve the degree of coordinated design optimization of hydraulic performance and structural strength.

The current study draws on the relevant research methods in the aforementioned fields to adopt an MDO design method for large axial-flow pump blades. At present, the rapid development of computational fluid and structural mechanics has resulted in numerous and successful research in the field of hydraulic performance and structural performance of axial-flow pump impellers. However, the traditional single-discipline optimization analysis method disregards the interaction and mutually affects between the hydraulic and structural designs. The method also fails to achieve a real coordinated design, which does not consider the reliability and accuracy of the optimized design results. To further improve the overall performance of pumps in axial-flow pump impeller optimization designs, the theory and method of multi-disciplinary design optimization of axial-flow pumps were applied in this research. This study focuses on the two disciplines of hydraulics and structure. Therefore, the MDO problem in this research can be presented as follows: considering the blade quality and efficiency of the design condition as objective functions; using the head of small flow (i.e., flow under the highest operating head), efficiency of small flow, maximum stress value, and maximum deformation value of small flow as constraints; and ensuring that the blade head of the design conditions remains unchanged or varies in a small range.

## 2. Numerical Calculation

### 2.1. Governing Equations

The solution of the fluid–structure interaction problem should consider the flow and structure fields and the data transfer between them. To considerably study this interaction, the current research used the calculation method of unidirectional fluid–solid coupling based on the system coupling module of the Ansys Workbench platform. First, the calculation of the fluid domain is performed. Second, information on the fluid domain is transmitted to the structural field by the fluid–solid coupling interface. Lastly, finite element structure analysis is performed.

In the flow field calculation, three-dimensional (3D) Reynolds-averaged N–S equations are used to solve the turbulent flow in the impeller of the axial-flow pump. By solving the equation, the fluid information of the discretized time and space flow field is obtained. The performance shows that the parameters can obtain the flow field flow characteristics. The basic control equations are as follows:

$$\frac{\partial v_j}{\partial x_j} = 0 \tag{1}$$

$$\frac{\partial(\rho v_i)}{\partial t} + \frac{\partial(\rho v_i v_j)}{\partial x_j} = -\frac{\partial p}{\partial x_i} + \mu \frac{\partial v_i}{\partial x_i \partial x_j} \tag{2}$$

where $v$ is the inflow velocity, $p$ is the flow field pressure, $\rho$ is the fluid density, $\mu$ is the fluid dynamic viscosity, and the subscripts $i$ and $j$ are the x- and y-coordinates, respectively.

The numerical simulation adopted the renormalization group(RNG) $k$-$\varepsilon$ turbulence model. When water flows through the blade passage of the axial-flow pump impeller, it will produce a higher strain rate and larger bending streamline. The RNG $k$-$\varepsilon$ model adopts a statistical technique called renormalization grouping to correct the turbulence viscosity by considering the condition of rotation and swirl. The RNG $k$-$\varepsilon$ turbulence model has better analysis ability than other turbulence models in the internal flow field calculation process of axial-flow pumps. The governing equations are as follows:

$$\frac{\partial}{\partial t}(\rho k) + \frac{\partial}{\partial x_i}(\rho k u_i) = \frac{\partial}{\partial x_j}(\alpha_k \mu_{eff} \frac{\partial k}{\partial x_j}) + G_k - \rho\varepsilon \tag{3}$$

$$\frac{\partial}{\partial t}(\rho\varepsilon) + \frac{\partial}{\partial x_i}(\rho\varepsilon u_i) = \frac{\partial}{\partial x_j}(\alpha_\varepsilon \mu_{eff} \frac{\partial\varepsilon}{\partial x_j}) + \frac{C_{1\varepsilon}\varepsilon}{k}G_k - C_{2\varepsilon}\rho\frac{\varepsilon^2}{k} \tag{4}$$

where $G_k$ represents the generation of turbulence kinetic energy caused by the mean velocity gradients. The quantities $\alpha_k$ and $\alpha_\varepsilon$ are the inverse effective Prandtl numbers for $k$ and $\varepsilon$. $C_{1\varepsilon} = 1.42$, $C_{2\varepsilon} = 1.68$. $\mu_{eff}$ is the effective viscosity accounting for turbulence. $k$ is turbulence energy. $\varepsilon$ is dissipation rate of turbulent kinetic energy.

The structural calculation used the finite element method to analyze the structural field of the axial-flow pump blade. The dynamic equation of the axial-flow pump blade under hydrodynamic action is defined as follows:

$$[M](\ddot{x}) + [C](\dot{x}) + [K](x) = \{F\} \tag{5}$$

where $[M]$ is the structural mass matrix, $[C]$ is the structural damping matrix, $[K]$ is the structural stiffness matrix, $(x)$ is the structural displacement, $(\dot{x})$ is the structural velocity, $(\ddot{x})$ is the structural acceleration, and $\{F\}$ represents the flow field force of the structure under a fluid–solid coupling.

When the static calculation of axial-flow pumps blade is performed through fluid–solid coupling, the fluid and solid systems are included in such a calculation. On the coupling surface, the velocity and stress continuous equations should be satisfied. The water pressure on the blade surface under a certain operating condition of pumps can be approximated and is constant with time. That is, the time-related quantity is disregarded. Thus, the relationship can be simplified as follows:

$$[K](x) = \{F\} \tag{6}$$

where $K = \sum \iiint_V B^T CB dV$ and $[B]$ is the strain matrix.

The stress requirement for the MDO design of axial-flow pumps is the most important. The stress distribution of blades under the design conditions is unnecessary because the general pumping station project focuses on the highest head. The most important aspect in terms of hydraulic performance is the efficiency of the design conditions. The highest operating head conditions focus on safe operations. The practical application requirements of pump stations and the corresponding flow given under the highest operation head of these stations indicate the calculation of stress distribution by using the head as a constraint under this flow. In energy saving and emission-reduction environments, blade quality is regarded as the goal of structural optimization design. Therefore, the MDO problem in this research can be presented as follows: considering the blade quality and efficiency of the design condition as objective functions; using the head of small flow (i.e., flow under the highest operating head), efficiency of small flow, and maximum stress and maximum deformation values of small flow as constraints; and ensuring that the blade head of the design conditions remains unchanged or varies in a small range. Under these conditions, the impeller of the axial-flow pump undergoes multi-constraint, multi-objective, multi-modality, and multi-disciplinary optimization.

### 2.2. Parameterized Model

A blade modeling program written in MATLAB based on the Zhukovsky airfoil can generate blade curve files for Turbo-Grid and UG. In this manner, the 3D model of the fluid and 3D model of the blade structure can be established conveniently. The flow–solid coupling calculation analysis of axial-flow pumps is only for the impeller. Thus, the analysis only involves the calculation of the single channel flow field to save calculation time and improve calculation efficiency in the flow field analysis. Structural calculations consider the single blade in the corresponding coordinate quadrant. The majority of the axial-flow pumps in actual pumping stations are fully regulated vane pumps. To fit the actual situation, a 3D modeling of the model impeller round shank is performed instead of a 3D hub shape, with the inner surface of the round shank being a fixed end. The 3D model of the blade structure is shown in Figure 1.

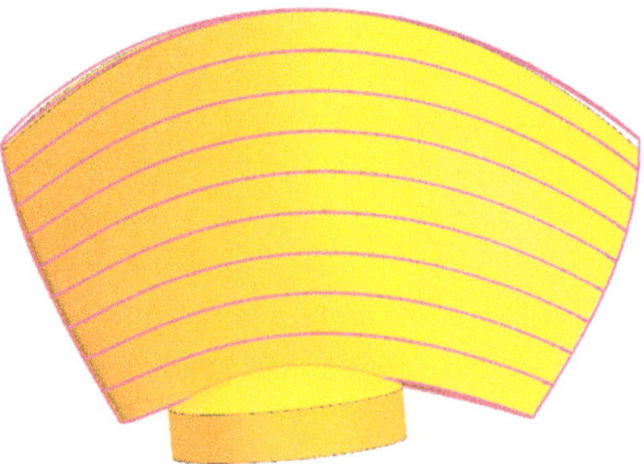

**Figure 1.** Three-dimensional shape of blade.

### 2.3. Grid and Load

The fluid domain calculation model is mainly for the impeller of axial-flow pumps. In this study, the nominal specific speed of the axial-flow impeller $n_s$ is 750, design flow $Q$ is 360 L/s, design head $H$ is 7.0 m, rotating speed $n$ is 1450 r/min, and the blade tip unilateral gap is 0.15 mm. The diameter of the impeller is 300 mm. The number of impeller blades is four. The impeller is modeled by using Turbo-Grid according to the 3D coordinate data points. The axial-flow pump impeller is the computational domain. The inlet and outlet of the computational domain are respectively set as the total pressure inlet and the flow outlet. No-slip conditions are used for all the walls.

The structural grid of the axial-flow impeller is meshed by the software of Turbo-Grid. The quality of the grid is good and can meet orthogonal requirement. The grid of the impeller is shown in Figure 2.

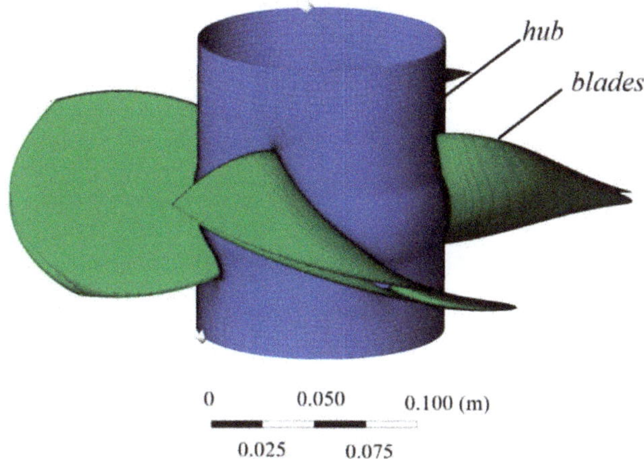


**Figure 2.** Fluid grid of the impeller.

In the mesh elements ranging from $8.0 \times 10^4$ to $1.6 \times 10^5$, six sets of single channel grids of the impeller are used to do the grid independence analysis. Numerical calculation results of the head and efficiency of the impeller reveal a small difference when the grid elements are up to $1.21 \times 10^5$ (Figure 3). Therefore, 121,527 is selected as the number of single channel computational mesh elements for numerical simulation in this study.

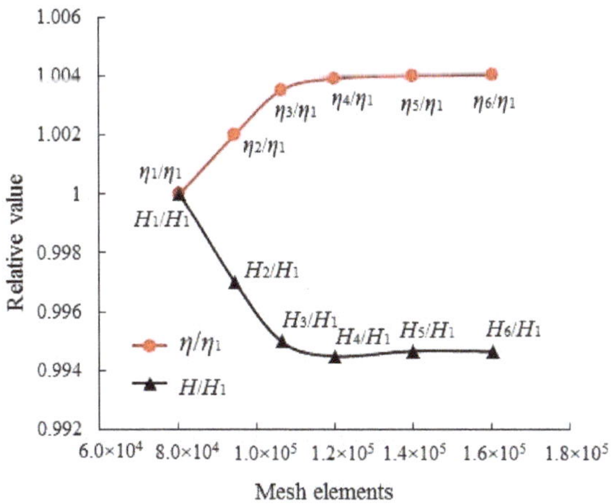


**Figure 3.** Mesh independence of the impeller.

Bernoulli energy equations were adopted to calculate the head of axial-flow pump. The head of the pump is shown as follows:

$$H_{net} = \frac{\int\limits_{s_2} P_2 u_{t2} ds}{\rho Q g} + H_2 + \frac{\int\limits_{s_2} u_2^2 u_{t2} ds}{2 Q g} - \left( \frac{\int\limits_{s_1} P_1 u_{t2} ds}{\rho Q g} + H_1 + \frac{\int\limits_{s_1} u_1^2 u_{t1} ds}{2 Q g} \right) \tag{7}$$

where $H_1$ is the inlet water level, $H_2$ is the outlet water level (m), $s_1$ is the inlet section area, $s_2$ is the outlet section area (m$^2$), $u_1$ is flow velocity of inlet section, $u_2$ is flow velocity of outlet section (m/s), $u_{t1}$ is the normal component velocity of inlet section, $u_{t2}$ is the normal component velocity of outlet section (m/s), $P_1$ is static pressure of inlet section, and $P_2$ is static pressure of outlet section (Pa).

The efficiency of the axial-flow pump is shown as follows:

$$\eta = \frac{\rho g Q H_{net}}{T_p \omega} \tag{8}$$

where $T_p$ is torque of the blade (N·m), and $\omega$ is the angular velocity of the impeller (rad/s).

According to the actual situation of the blade structure, the 3D solid model is imported into the Design Modeler module of the Workbench platform to set the properties of the model material to structural steel.

The material property parameters of the axial-flow pump blade are as follows: modulus of elasticity E = 2.0 × 10$^{11}$ Pa, Poisson's ratio $\mu$ = 0.3, and density $\rho$ = 7850 kg/m$^3$. The diameter and thickness of the round shank are 60 mm and 12.5 mm, respectively. The solid grid model is divided in the mesh module of a model that uses unstructured grids. The number of grids of a single blade is around approximately 120,000. The grid model is shown in Figure 4.

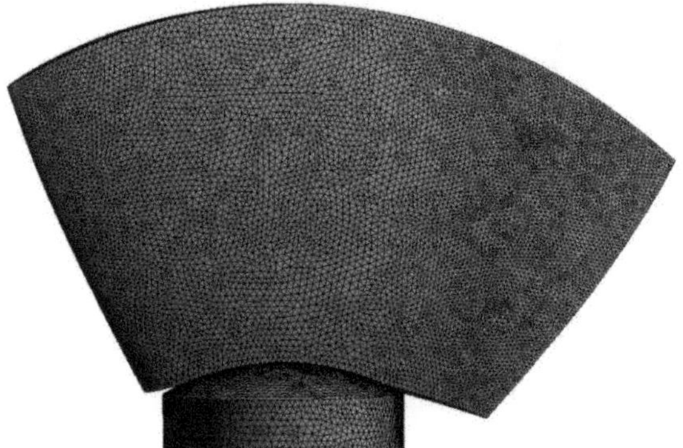

**Figure 4.** Solid grid model.

This study calculated the stress distribution of a pump blade under inertial and pressure loads. Inertial load is the centrifugal force received by the pump. The pump's rate of speed is $n$ = 1450 r/min. The pressure load is the water pressure. In the structural analysis, the influence of bearing load and bolt load on the stress distribution is disregarded for convenience of calculation.

### 2.4. Model Test Verification

A field performance test of the initial scheme was carried out on an axial-flow pump model test bench in China. The impeller of the axial-flow pump was machined with copper material according to design results of the initial scheme. The rotating speed of the impeller was 1450 r/min, and the diameter of the impeller was 300 mm. The number of the blades was four. The impeller of the axial-flow pump is shown in Figure 5a. According to the hydraulic industry standard of the pump model test of China, the model test of the axial-flow pump section includes impeller, guide vane, inlet pipe and outlet pipe, etc. The guide vane adopts the matching design guide vane. The pump installation is shown in Figure 5b. The comparison of numerical simulation and model test is shown in Figure 6. The predicted performance curve of numerical simulation is consistent with the experimental curve. The curves have a good agreement, and the error of each point is less than 3%. The results show that the numerical simulation of the axial-flow pump is accurate and reliable.

(a) Impeller

(b) The figure of the model test

**Figure 5.** Model test physical map.

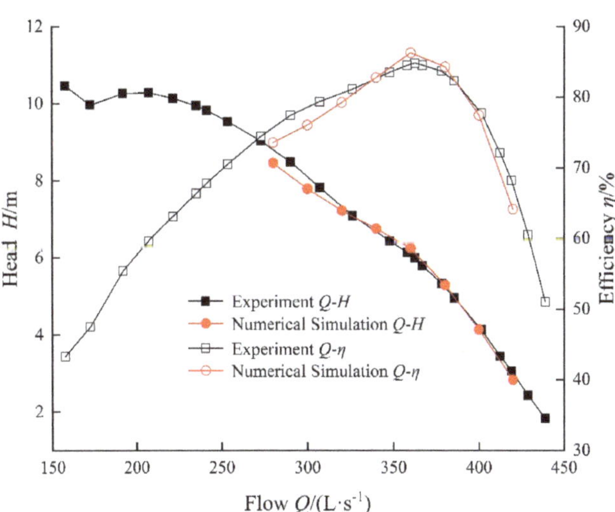


**Figure 6.** Comparison of numerical simulation and model test.

## 3. Analysis of Design of Experiment

The optimal Latin hypercube method was selected for the calculation of the sample points in the flow–solid coupling calculation of the axial-flow pump. This method allows many more points and more combinations to be studied for each factor. Experiment points are spread evenly, allowing higher order effects to be captured. The engineer has total freedom in selecting the number of designs to run as long as it is greater than the number of factors. The optimal Latin hypercube method is based on the random Latin hypercube method to improve the uniformity. The level selection of the factors is regulated by the corresponding algorithm, so that the selected sample points have better space filling and more uniform distribution in the whole design space.

The design variables included four design parameters: cascade density, airfoil placement angle, airfoil arch ratio, and airfoil thickness ratio. Airfoil selects the Zhukovsky airfoil structure used in the flushing angle research. All parametric modeling works were implemented in MATLAB. The 3D coordinate data of the blade section were generated using MATLAB, while the 3D solid models were built in Turbo-Grid and UG. To save calculation time and improve efficiency, the 3D shape of the entire blade was controlled using the minimum design variables. By using the equal strength design method, the cascade density of 10 sections can be controlled by two design variables: cascade density multiple of the blade tip and cascade density multiple of the blade root. The third-order B-spline curve control can reduce the airfoil angle data of 10 sections to only three variables: airfoil angle of the hub, rim, and middle section. The maximum camber and maximum thickness of the 10 sections are linear from hub to rim. Therefore, each parameter is only controlled by two variables of the hub and the rim, which successfully reduces 40 design variables to nine design variables. The initial value and range of the nine design variables are as follows:

$$\text{Design variables} \begin{cases} x_1 = 0.8 \in [0.7, 0.9] \\ x_2 = 1.2 \in [1.15, 1.45] \\ x_3 = 15.6 \in [13.6, 17.6] \\ x_4 = 32 \in [30, 34] \\ x_5 = 48 \in [46, 50] \\ x_6 = 4.2 \in [3.5, 4.9] \\ x_7 = 6.2 \in [5.5, 6.9] \\ x_8 = 6 \in [5, 7] \\ x_6 = 12 \in [10, 14] \end{cases} \tag{9}$$

where $x_1$ is the cascade density of the blade tips, $x_2$ is the cascade density multiple of the blade root, $x_3$ is the airfoil angle of the hub (°), $x_4$ is the airfoil angle of the rim (°), $x_5$ is the airfoil angle of the middle section (°), $x_6$ is the maximum camber of the hub (mm), $x_7$ is the maximum camber of the rim (mm), $x_8$ is the maximum thickness of the hub (mm), and $x_9$ is the maximum thickness of the rim (mm).

Within the range of the design variables, 80 sample points were generated using an optimized Latin hypercube method. A complete fluid–solid coupling calculation is required for each set of sample points. Each fluid–solid coupling calculation requires a complete set of processes including 3D modeling of fluid and structure, mesh division, and flow field and finite element calculation. Each flow–solid coupling calculation process is shown in Figure 7.

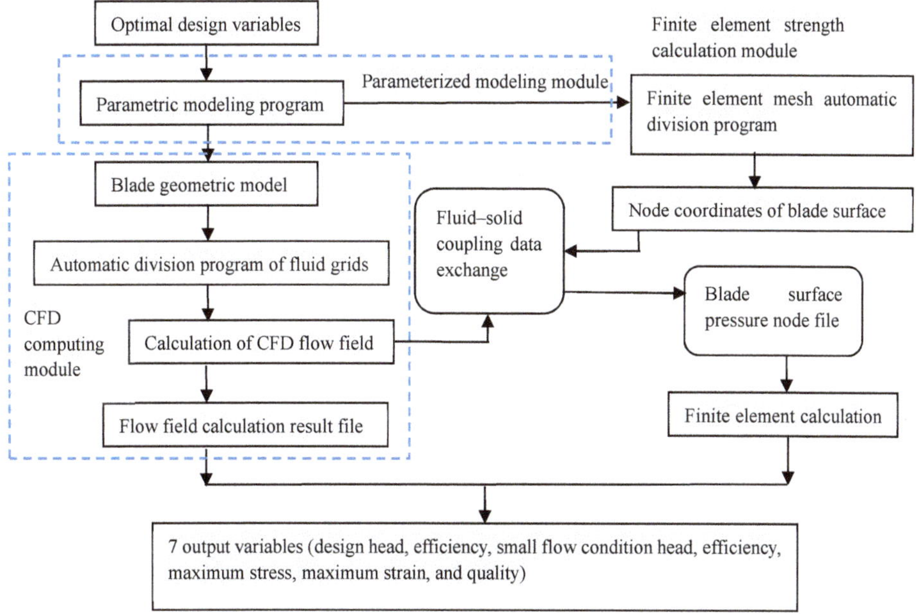



**Figure 7.** Flow chart of the fluid–solid coupling calculation.

According to the optimization requirements, the structural parameters, including mass, maximum stress, and maximum strain, should be calculated. Hydraulic parameters include the head and the efficiency under design conditions and small flow conditions. The design and small flow conditions are 360 L/s and 240 L/s, respectively. Table 1 shows the calculation results of the 80 sample points.

**Table 1.** Calculation results of the sample points. $y_1$ is the head of the small flow condition (m), $y_2$ is the efficiency of small flow condition (%), $y_3$ is the maximum deformation value of small flow condition (m), $y_4$ is the maximum stress value of small flow condition (Pa), $y_5$ is the mass of blade (kg), $y_6$ is the head of design condition (m), $y_7$ is the efficiency of design condition (%).

| Serial Number | $y_1$ | $y_2$ | $y_3$ | $y_4$ | $y_5$ | $y_6$ | $y_7$ |
|---|---|---|---|---|---|---|---|
| Initial value | 11.44 | 90.2147 | 0.00017979 | 103,620,000 | 0.94769 | 6.95309 | 93.9141 |
| 1 | 11.7532 | 90.1304 | 0.00018335 | 107,910,000 | 0.93693 | 6.71377 | 93.6598 |
| 2 | 10.9753 | 91.8065 | 0.00017092 | 89,098,000 | 0.90572 | 5.96531 | 93.3522 |
| 3 | 11.6557 | 91.557 | 0.00016117 | 71,270,000 | 0.99557 | 6.50123 | 93.5828 |
| 4 | 12.8308 | 89.2097 | 0.00017691 | 92,624,000 | 0.9928 | 7.85457 | 94.291 |
| 5 | 11.9646 | 90.8667 | 0.00015263 | 92,273,000 | 0.95464 | 7.40514 | 94.233 |
| 6 | 12.4638 | 90.1697 | 0.00015064 | 76,457,000 | 1.0196 | 7.57474 | 94.0972 |
| 7 | 10.7539 | 90.9872 | 0.00016243 | 79,116,000 | 0.86208 | 6.48433 | 93.861 |
| 8 | 11.4467 | 89.449 | 0.00016261 | 71,086,000 | 0.86371 | 7.30957 | 94.0902 |
| 9 | 10.9185 | 90.9655 | 0.00012786 | 65,613,000 | 0.89861 | 6.58435 | 93.679 |
| 10 | 10.5674 | 91.0425 | 0.00013858 | 72,479,000 | 0.87519 | 6.00038 | 93.4396 |
| ... | ... | ... | ... | ... | ... | ... | ... |
| 71 | 11.6549 | 88.9273 | 0.00025333 | 111,840,000 | 0.81188 | 7.55323 | 94.1173 |
| 72 | 11.3588 | 90.6635 | 0.00014668 | 81,217,000 | 0.85662 | 7.2884 | 94.2028 |
| 73 | 11.6204 | 91.7638 | 0.00012582 | 66,575,000 | 0.95936 | 6.81371 | 94.0114 |
| 74 | 12.3619 | 89.0101 | 0.00017739 | 89,281,000 | 0.90464 | 7.85347 | 94.1984 |
| 75 | 11.1076 | 91.171 | 0.0002191 | 109,260,000 | 0.87431 | 6.60246 | 93.8747 |
| 76 | 12.0918 | 89.3128 | 0.00014278 | 76,217,000 | 0.951 | 7.51066 | 94.0422 |
| 77 | 11.1648 | 90.4989 | 0.00020433 | 107,720,000 | 0.86179 | 6.84219 | 93.794 |
| 78 | 11.0768 | 90.8673 | 0.00013344 | 74,040,000 | 0.94497 | 6.37273 | 93.4434 |
| 79 | 11.707 | 90.0439 | 0.00021392 | 120,720,000 | 0.92188 | 6.89957 | 93.7827 |
| 80 | 11.3428 | 89.1842 | 0.00019157 | 90,251,000 | 0.82919 | 7.34594 | 94.1117 |

The calculation results of hydraulic performance and structural performance of the axial-flow pump were obtained according to 80 different blade design schemes. In the calculation results of 80 groups of samples, the minimum value of the head was 10.38 m and the maximum value was 12.83 m under the small flow condition (Figure 8a). The lowest value of the pump efficiency was 87.41%, and the highest value was 92.36% under the small flow condition (Figure 8b). Under the small flow condition, the maximum deformation of the pump blade was in the range of 0.126~0.287 mm (Figure 8c). Under the small flow condition, the maximum stress of the pump blade was in the range of 65.6~152.9 MPa (Figure 8d). When the axial-flow pump was operated under small flow conditions, the pump head was relatively high, and the blades in this operating area were most vulnerable to structural damage. Therefore, it is necessary to consider structural deformation and structural stress under small flow condition when doing structural optimization design. The minimum mass value of the single axial-flow pump blade under different design schemes was 0.79 kg, and the maximum value was 1.04 kg (Figure 8e). Under the design condition, the minimum value of the head was 5.60 m, and the maximum value was 8.19 m (Figure 8f). Under design condition, the minimum value of the pump efficiency was 92.97%, and the highest value was 94.43% (Figure 8g). The head and efficiency of the axial-flow pump under design condition are very important evaluation parameters for the design of the axial-flow pump. In order to ensure that the specific speed of the pump is constant, the head under the design condition should be kept basically unchanged, and at the same time, the efficiency is expected to be higher. The difference between the calculation results of the different sample points is relatively large, particularly the structural performance (Figure 8). The maximum calculation results of the maximum stress and deformation value were more than twice the minimum value. This result shows that the change of the design parameters has immense influence on the structural performance of pump.

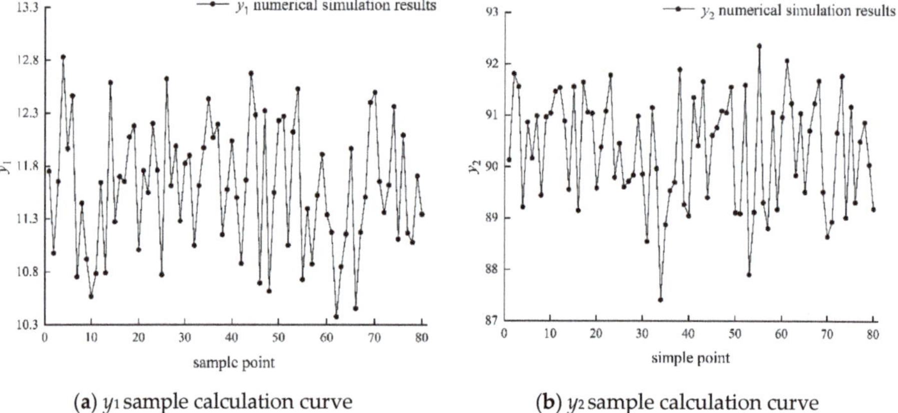


(**a**) $y_1$ sample calculation curve　　　　(**b**) $y_2$ sample calculation curve

**Figure 8.** *Cont.*

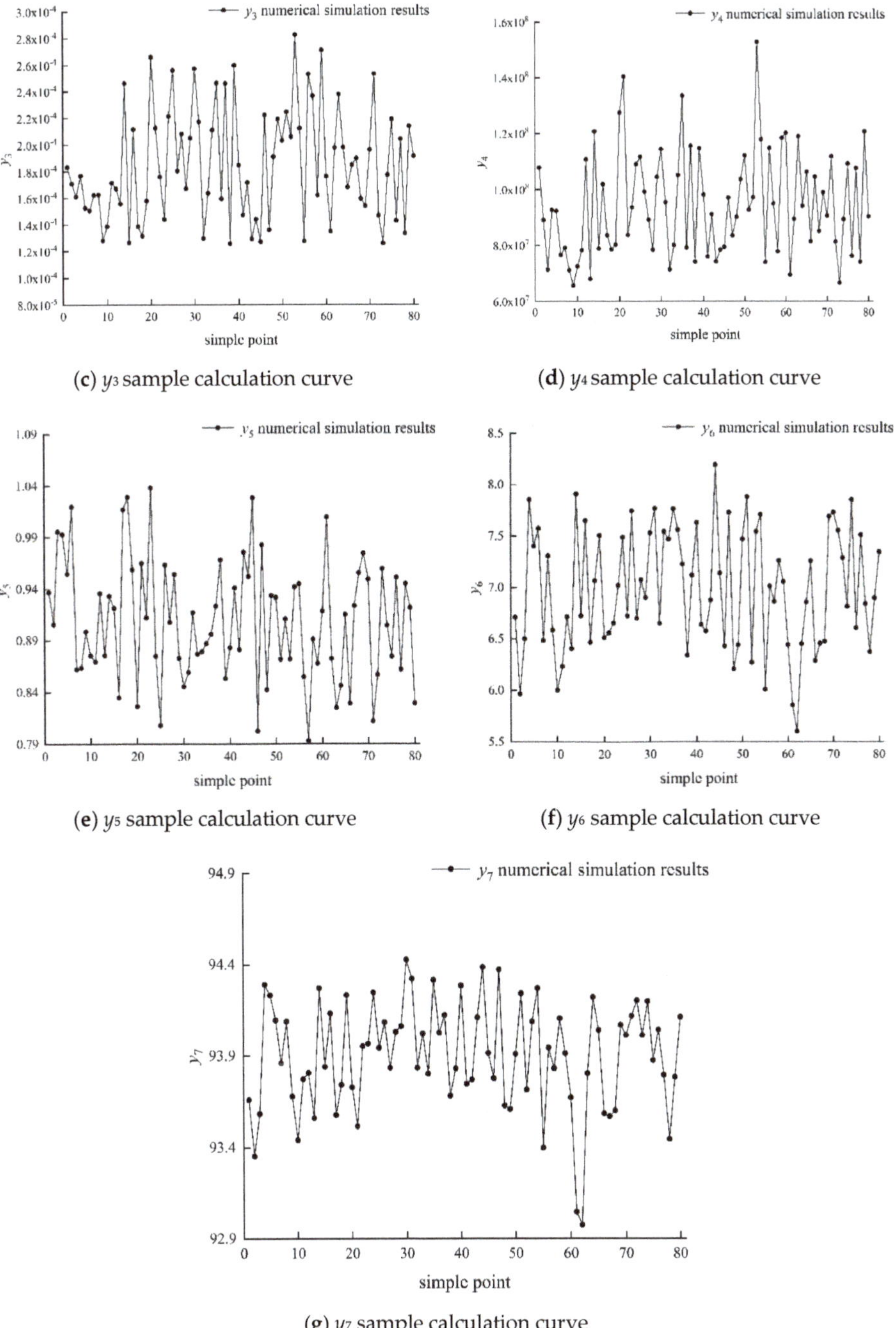

(**c**) $y_3$ sample calculation curve

(**d**) $y_4$ sample calculation curve

(**e**) $y_5$ sample calculation curve

(**f**) $y_6$ sample calculation curve

(**g**) $y_7$ sample calculation curve

**Figure 8.** Calculation results of the sample points.

The three typical conditions of stress and strain distribution are shown in Figures 9 and 10.

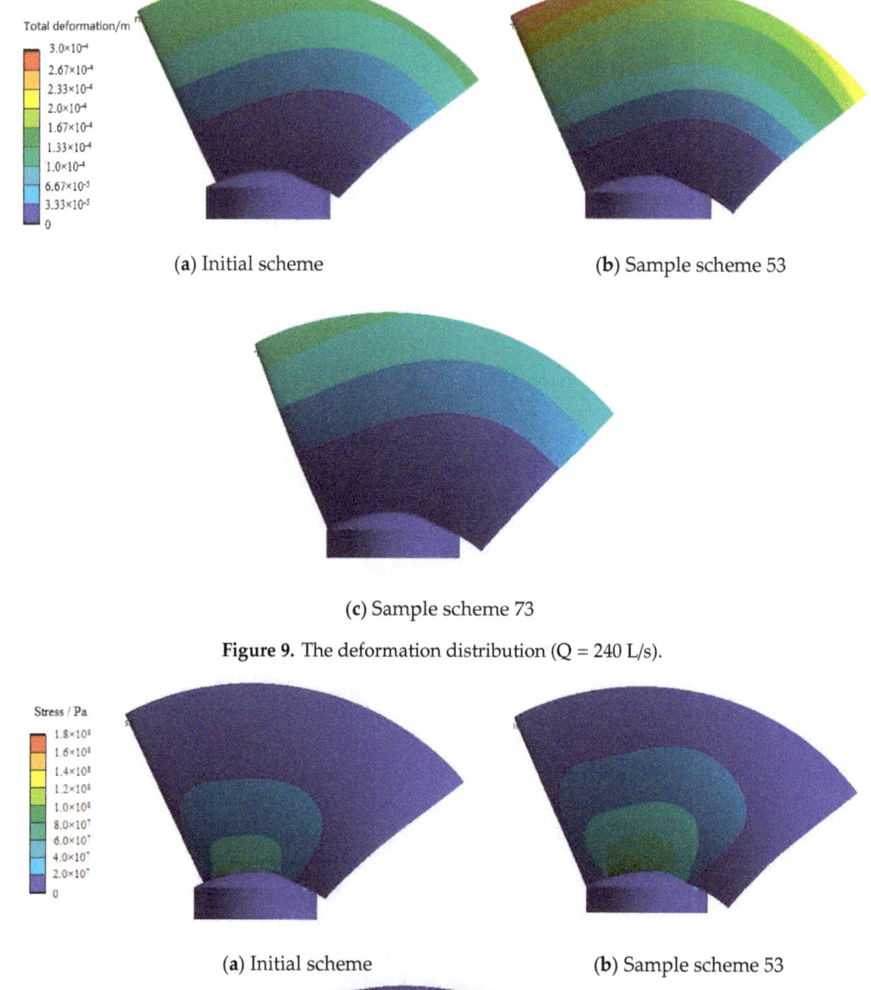

(**a**) Initial scheme

(**b**) Sample scheme 53

(**c**) Sample scheme 73

**Figure 9.** The deformation distribution (Q = 240 L/s).

(**a**) Initial scheme

(**b**) Sample scheme 53

(**c**) Sample scheme 73

**Figure 10.** Stress distribution (Q = 240 L/s).

The structural performance calculation results of sample points 73 and 53 were taken out to analyze their structural characteristics because the structural performance of sample points 73 and 53 were the maximum and minimum values respectively in the whole sample. Evidently, the maximum deformation position of the blade of the axial-flow pump was near the edge of the blade inlet section. The maximum stress position was near the hub of the blade. The maximum stress and maximum deformation position and distribution trend corresponding to the different design parameters were consistent. However, the maximum deformation and maximum stress values differed substantially. The solid line in the graph indicates the position of the blade when it was not subjected to external stress. The cloud image shows the position after deformation. The deformation direction of the blade was the same. The deformation was found toward the inlet direction of the impeller; the closer the rim, the more substantial the deformation was.

## 4. Approximation Models

Approximation models are methods of approximating a set of input variables (i.e., independent variables) and output variables (i.e., response variables) through a mathematical model. The approximate model uses the following equation to describe the relationship between the input variable and output response:

$$y(x) = \widetilde{y}(x) + \varepsilon, \tag{10}$$

where $y(x)$ is the response to the actual value, which is an unknown function; $\widetilde{y}(x)$ is a response to the approximation, which is a known polynomial; and $\varepsilon$ is the random error between the approximation and actual value, which typically follows the standard normal distribution of $(0, \sigma^2)$.

An approximate model structure was constructed and aimed at the 80 sample points of the axial-flow pump multi-disciplinary optimization design. $R^2$ is a regression coefficient that characterizes the correlation between the predicted and actual values. When $R^2$ is equal to 1, the predicted value of each sample point is the best fit with the actual value; the closer the $R^2$ coefficient of the approximate model to 1, the closer the approximate model to the real solution. The $R^2$ coefficients of the different approximation models are listed as follows.

R-squared can be used to analyze the degree of consistency between the approximate model and the sample points. An R-squared value of 1.00 indicates that the approximate model has high reliability. The credibility is low when the value of R-squared is less than the set acceptance level of 0.8. The approximate models were established by means of RSM method, neural network method, and Kriging method. The R-squared value of Table 2 displayed in red is closest to 1. It shows that the approximate model displayed in red is the most ideal scheme, and it has high reliability. Table 2 shows that the Kriging model cannot fit the approximate model data of the two responses—the largest deformation and maximum stress values and the effect is not good when fitting other responses. Hence, the Kriging model was not used in this study. When using neural network models to construct approximate models, the radial and elliptical basis models have minimal effect on the fitting results. These models only have a good fitting effect on part of the responses. The response surface model has a superior fitting effect. Table 2 shows that the more complex the response of the maximum stress value, the higher the order and the better the fitting effect, which was not the case in the other responses, particularly in the quality response. The fitting quality of the fourth-order response surface model decreases substantially. This study used the second-order response surface method to construct an approximate model of the small flow rate efficiency and quality, in accordance with the comparative results of the fitting effect. The third-order response surface method was used to construct an approximate model of the small flow head and the design condition head. The fourth-order response surface method was used to construct an approximate model of the small flow maximum deformation, maximum stress, and design efficiency.

**Table 2.** Analysis of the $R^2$ correlation coefficient of different approximate models.

| Evaluation Index | Response | RSM | | | | Neural Network | | Kriging |
|---|---|---|---|---|---|---|---|---|
| | | First Order | Second Order | Third Order | Fourth Order | RBF | EBF | |
| $R^2$ | $y_1$ | 0.94685 | 0.99669 | 0.99693 | 0.98667 | 0.99343 | 0.99282 | 0.96766 |
| | $y_2$ | 0.88756 | 0.98814 | 0.98739 | 0.98634 | 0.97069 | 0.9654 | 0.95398 |
| | $y_3$ | 0.96856 | 0.99898 | 0.99293 | 0.99949 | 0.98135 | 0.99294 | 0 |
| | $y_4$ | 0.70425 | 0.8275 | 0.87918 | 0.88162 | 0.67383 | 0.68307 | 0 |
| | $y_5$ | 0.99692 | 0.99991 | 0.99869 | 0.8498 | 0.9984 | 0.99933 | 0.95499 |
| | $y_6$ | 0.96582 | 0.99926 | 0.99929 | 0.9915 | 0.99727 | 0.99651 | 0.95914 |
| | $y_7$ | 0.8369 | 0.99039 | 0.99135 | 0.99155 | 0.9524 | 0.96781 | 0.90637 |

## 5. Optimal Design

### 5.1. The Optimization Model

The multi-disciplinary optimization design of the axial-flow pump blades adopted a discipline analysis method to construct an approximate model, reduce calculation time, and save on calculation cost. Optimization aimed to make the axial-flow pump blades meet the operational requirements of the highest head of the pump station under the small flow condition. Within the constraints of a few changes in the head of the design conditions, and satisfying the condition that the maximum blade stress was lower than the yield strength of the material, the highest efficiency and lightest quality or design conditions of the highest efficiency and minimum deformation optimization goals under axial-flow pump blade design conditions were achieved.

The optimization model is as follows:

Objective function:

$$\text{Max} y_7 \text{ and min} y_5 \tag{11}$$

$$\text{Constraints}: \begin{cases} y_1 \geq 11.5 \\ y_2 \geq 90 \\ y_3 \leq 3.0 \times 10^{-4} \\ y_4 \leq 1.1 \times 10^8 \\ 6.8 \leq y_6 \leq 7.2 \end{cases} \tag{12}$$

The relevant literature has indicated that the yield strength of steel is $\sigma > 207$ MPa, and the maximum stress of the blade in the sample point was 150 MPa, which is considerably below the yield strength of the material. Given that the head of the low-flow condition in this research did not reach the maximum operational head that the pump can run, the maximum stress should have a higher margin to ensure that the pump station operates under the high head. Therefore, the maximum stress constraints $\leq 110$ MPa were combined with the maximum stress analysis of the sample points. Given that the deformation was relatively small, the constraint range can be relaxed in the optimization process. The head of the design condition point is an important design parameter of the pump station operation. Moreover, the requirements are high. To ensure that the specific speed of the optimized impeller does not change substantially, the head of the design condition must be constrained to change within a small range.

### 5.2. Optimization Results and Analysis

In the optimization process, three multi-objective optimization methods can be realized because of the approximate model method, short optimization time, low computational cost, and minimally complex optimization objectives. This study selected the AMGA optimization algorithm. According to the optimization algorithm and approximation model, 15,000 iterations were carried out using the Isight 5.8 software. The optimization model and all Pareto solution sets of feasibility are shown in Figures 11 and 12, respectively.

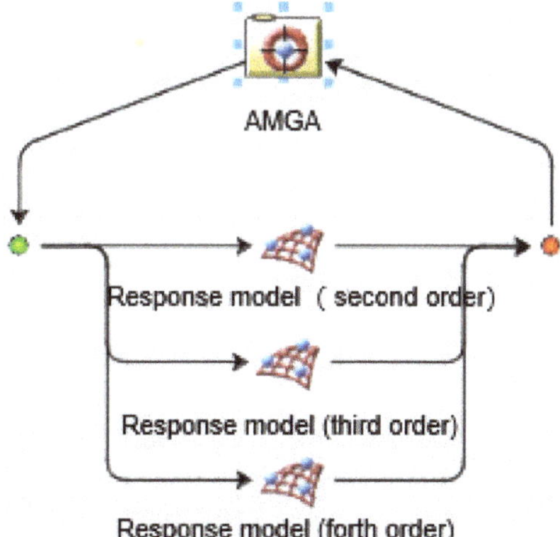



**Figure 11.** Optimization model.

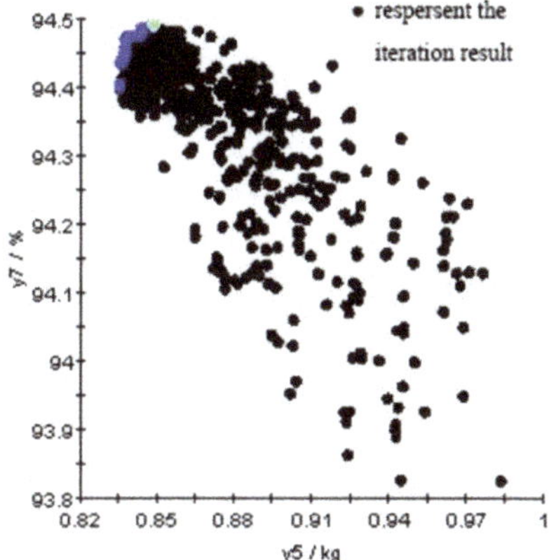



**Figure 12.** Pareto solution set.

All points in the graph are feasible solutions. The combination of the color points is the Pareto frontier, and the optimal solution can be selected flexibly according to different weight values. The green solution is the final result in this study. The optimized results were compared with the initial results. Table 3 shows the comparison of the design variables before and after optimization. Table 4 presents the comparison of the response values before and after optimization. The comparison of the model before and after optimization is shown in Figure 13. The cloud chart of the maximum stress and deformation distribution is shown in Figure 14.

**Table 3.** Comparison of the design variables before and after optimization.

| Design Variables | $x_1$ | $x_2$ | $x_3$ | $x_4$ | $x_5$ | $x_6$ | $x_7$ | $x_8$ | $x_9$ |
|---|---|---|---|---|---|---|---|---|---|
| Initial value | 0.8 | 1.2 | 15.6 | 32 | 48 | 4.2 | 6.2 | 6 | 12 |
| Optimal value | 0.745 | 1.448 | 17.408 | 31.269 | 49.803 | 4.538 | 5.984 | 5.057 | 11.360 |

**Table 4.** Comparison of the response values before and after optimization.

| Response | $y_1$ | $y_2$ | $y_3$ | $y_4$ | $y_5$ | $y_6$ | $y_7$ |
|---|---|---|---|---|---|---|---|
| Initial value | 11.44 | 90.2147 | 0.00017979 | 103,620,000 | 0.94769 | 6.95309 | 93.9141 |
| Optimal value | 11.5174 | 90.2637 | 0.00021611 | 98,923,000 | 0.84843 | 7.1999 | 94.4913 |
| Effect | 0.67% | 0.054% | 20.20% | −4.53% | −10.47% | 3.54% | 0.61% |
| Computation value | 11.5564 | 90.2068 | 0.00021412 | 102,600,000 | 0.84704 | 7.22086 | 94.4188 |
| Error | 0.33% | 0.063% | 0.92% | 3.71% | 0.16% | 0.29% | 0.076% |

Note: Minus sign represents a decrease.

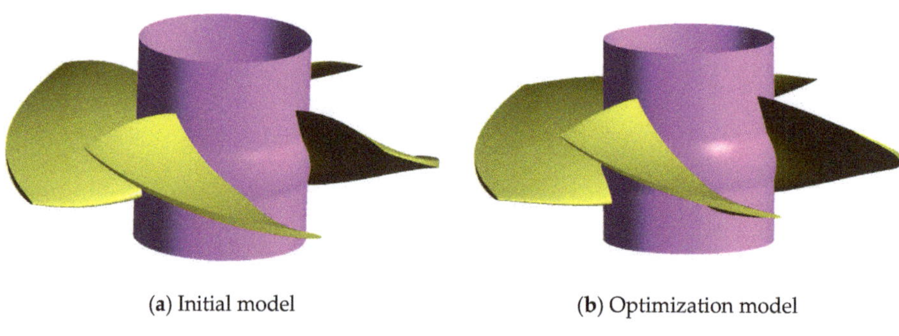

(**a**) Initial model      (**b**) Optimization model

**Figure 13.** Comparison of the model before and after optimization.

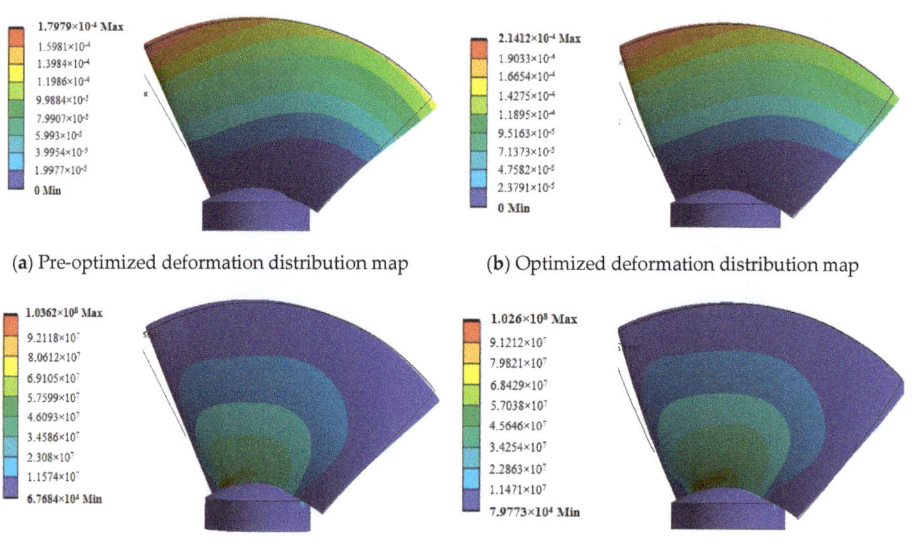

(**a**) Pre-optimized deformation distribution map      (**b**) Optimized deformation distribution map

(**c**) Pre-optimized stress distribution map      (**d**) Optimized stress distribution map

**Figure 14.** Cloud chart of the maximum stress and deformation distribution.

Table 3 and the Figure 13 show that the cascade density of the blade tip decreased after optimization, whereas the cascade density of the blade root increased. $x_1$ is the cascade density of blade tip, and $x_2$ is the cascade density multiple of blade root. The most direct effect of the change of the cascade density is the change of the shape and the mass of the blade. At the same time, it can reduce the length difference between the hub and rim airfoils and increase the efficiency of the axial-flow pump when reducing the cascade density of the blade tip and the cascade density of the blade hub. From the change trend of design parameters of $x_3$, $x_4$, and $x_5$ in Table 3, it can be found that the airfoil angle of the rim and hub were increased. Hence, the attack angle of the blade was increased. Consequently, the airfoil work ability was improved, and the airfoil angle of the middle section was reduced, thereby making the distribution regular considerably reasonable. From the change trend of design parameters of $x_6$ and $x_7$ in Table 3, the camber of the airfoil on the flange side increased, while that on the hub side decreased. That means it can reduce the distortion of the blade and increase the work capacity of the blade. The airfoil thickness decreased from the change trend of the design parameter of $x_8$ and $x_9$ in Table 3. The reduction of the airfoil thickness had little effect on the hydraulic performance of axial-flow pump, but it can affect the mass of the axial-flow pump blade. The 3D blade models were obtained as shown in Figure 13 according to the design parameters before and after optimization in Table 3.

Table 4 shows that all constraints of the optimization result meet the requirements. The 'optimal value' in Table 4 is the theoretical calculation result by the approximation model, which is corresponding to the highlighted scheme in Figure 10. The 'computation value' in Table 4 is the numerical calculation result by the CFD software. The 'error' in Table 4 is the difference between numerical simulation results and theoretical calculation results. From the value of the error, it can be found that the numerical simulation results by CFD verify the theoretical calculation results by the approximation model well.

From Table 4, the head of the small flow condition point was above 11.5 m. Accordingly, the efficiency increased, but the increase was small. The constraint conditions were satisfied. From the optimization target, the quality decreased from 0.947 to 0.848 kg, the decline reached 10.47%, and design efficiency increased from 93.91% to 94.49%. Given an increase of 0.61%, the optimization effect is evident. Apart from the maximum deformation, other responses were optimized. According to the optimized design variables, the results can be obtained using a fluid–structure interaction calculation. Compared with the approximate model optimization results, the errors of other responses were below 0.5%, except for the maximum stress value error. The 3.71% error is acceptable because maximum stress varies widely, and the constraint value is less than half of the yield strength of the material. According to Figure 14, the maximum stress and maximum deformation distribution trends were consistent before and after optimization. The maximum stress and maximum deformation position remained the same, but the maximum value changed. The maximum deformation of the initial scheme was 0.18 mm, and the maximum deformation of the optimized scheme was 0.21 mm (Figure 14a,b), but all of them were within 0.3 mm of the constraint condition, which was caused by the increase of the working capacity and the decrease of the thickness of the blades. The maximum stress of the initial scheme was 103.6 MPa, and the maximum stress of the optimized scheme was 102.6 MPa (Figure 14c,d). After optimization, the structural strength increased due to the shorter blade length.

## 6. Conclusions

This study applied the theory and method of multi-disciplinary design optimization of the axial-flow pump to further improve the overall performance of the axial-flow pump impeller optimization design in this research. The MDO problem in this research was addressed can be concluded as follows: considering the blade quality and efficiency of design condition as an objective function; using the head of the small flow, efficiency of the small flow, maximum stress value, and maximum deformation value of the small flow as constraints; and ensuring that the blade head of the design conditions remains unchanged or varies in a small range. According to the range of the design variables, 80 sample points were generated using an optimized Latin hypercube method. Thereafter, the response surface method was used to build the approximate model according to the sample points. Lastly, the optimal design of axial-flow pump impeller was carried out by the approximate model, and the impeller satisfying the constraint conditions was obtained. The following results were obtained through calculation and analysis.

(1)   From the 80 sample points, the difference between the calculation results of the different sample points was relatively large, particularly for the structural performance. The maximum calculation results of the maximum stress and deformation value were over twice the minimum value,

which shows that the change of design parameters had a substantial influence on the structural performance of the pump. The response surface model had a superior fitting effect. This study used the second-order response surface method to construct the approximate model of the small flow rate efficiency and quality, in accordance with the comparative results of the fitting effect. The third-order response surface method was used to construct the approximate model of the small flow and design condition head. The fourth-order response surface method was used to construct the approximate model of the small flow maximum deformation, maximum stress, and design efficiency.

(2)     For the construction of the approximate model of the MDO design, the fitting effect of the response surface approximate model was better than that of the other approximate models. The results of the MDO design of the impeller are as follows: the quality decreased from 0.947 kg to 0.848 kg; the decline reached 10.47%, whereas the design efficiency increased from 93.91% to 94.49%; the increase was 0.61%; and the optimization effect was obvious. In addition, the errors of the other responses were below 0.5%, except for the maximum stress value error. The approximate model has high accuracy and the analysis results are reliable.

As we all know, the impeller of the axial-flow pump is the core and most important flow part in the pump station. The design quality of the impeller directly determines the overall efficiency of the pump device and even the pumping station. With the increasingly complex operational requirements, the pump design has to meet higher design requirements [25–27]. A multi-disciplinary design optimization [28] method can combine hydraulic design and structural design, so that it can get the optimized impeller to match the complex operating conditions. This present study solved some key technical problems in the MDO process of the axial-flow impeller. However, in addition to the mass of the blade, how to combine vibration characteristics, hydraulic noise, and materials to carry out multi-disciplinary optimization design is the main content of future studies.

**Author Contributions:** Data curation, J.Z.; Formal analysis, F.T.; Writing—original draft, L.S.; Writing—review and editing, C.W. All authors have read and agreed to the published version of the manuscript.

**Funding:** This study was supported by the National Natural Science Foundation of China (Grant Number 51376155), the Natural Science Foundation of Jiangsu Province of China (Grant Number BK20190914), China Postdoctoral Science Foundation Project (2019M661946); the Natural Science Foundation of Jiangsu Higher Education Institutions of China (Grant Number 19KJB570002), the National Science Foundation of Yangzhou of China(Grant Number YZ2018103), and Priority Academic Program Development of Jiangsu Higher Education Institutions (Grant Number PAPD).

**Conflicts of Interest:** The authors declare no conflict of interest.

## References

1.     Li, X.; Chen, B.; Luo, X.; Zhu, Z. Effects of flow pattern on hydraulic performance and energy conversion characterisation in a centrifugal pump. *Renew. Energy* 2020. [CrossRef]

2.     Wang, C.; Chen, X.; Qiu, N.; Zhu, Y.; Shi, W. Numerical and experimental study on the pressure fluctuation, vibration, and noise of multistage pump with radial diffuser. *J. Braz. Soc. Mech. Sci. Eng.* **2018**, *40*, 481. [CrossRef]

3.     Lu, Y.; Zhu, R.; Wang, X.; Wang, Y.; Fu, Q.; Ye, D. Study on the complete rotational characteristic of coolant pump in the gas-liquid two-phase operating condition. *Ann. Nucl. Energy* **2019**, *123*, 180–189.

4.     Wang, C.; He, X.; Cheng, L.; Luo, C.; Xu, J.; Chen, K.; Jiao, W. Numerical simulation on hydraulic characteristics of nozzle in waterjet propulsion system. *Processes* **2019**, *7*, 915. [CrossRef]

5.     He, X.; Zhang, Y.; Wang, C.; Zhang, C.; Cheng, L.; Chen, K.; Hu, B. Influence of critical wall roughness on the performance of double-channel sewage pump. *Energies* **2020**, *13*, 464. [CrossRef]

6.     Zhu, D.; Tao, R.; Xiao, R.F.; Yang, W.; Liu, W.C.; Wang, F.J. Optimization design of hydraulic performance in vaned mixed-flow pump. *Proc. Inst. Mech. Eng. Part A J. Power Energy* **2019**, 1–13. [CrossRef]

7.     Sun, X.J.; Ge, J.Q.; Yang, T.; Xu, Q.Q.; Zhang, B. Multifidelity multidisciplinary design optimization of integral solid propellant ramjet supersonic cruise vehicles. *Int. J. Aerosp. Eng.* 2019. [CrossRef]

8.  Chen, X.; Wang, P.; Zhang, D.Y.; Dong, H.C. Gradient-based multidisciplinary design optimization of an autonomous underwater vehicle. *Appl. Ocean Res.* **2018**, *80*, 101–111. [CrossRef]

9.  Xu, H.W.; Li, W.; Li, M.F.; Hu, C.; Zhang, S.C.; Wang, X. Multidisciplinary robust design optimization based on time-varying sensitivity analysis. *J. Mech. Sci. Technol.* **2018**, *32*, 1195–1207. [CrossRef]

10. Brevault, L.; Balesdent, M.; Defoort, S. Preliminary study on launch vehicle design: Applications of multidisciplinary design optimization methodologies. *Concurr. Eng. Res. Appl.* **2018**, *26*, 93–103. [CrossRef]

11. Sun, W.; Wang, X.B.; Shi, M.L.; Wang, Z.Q.; Song, X.G. Multidisciplinary design optimization of hard rock tunnel boring machine using collaborative optimization. *Adv. Mech. Eng.* **2018**, *10*, 1–12. [CrossRef]

12. Pavese, C.; Tibaldi, C.; Zahle, F.; Kim, T. Aeroelastic multidisciplinary design optimization of a swept wind turbine blade. *Wind Energy* **2017**, *20*, 1941–1953. [CrossRef]

13. Shi, R.H.; Liu, L.; Long, T.; Liu, J.; Yuan, B. Surrogate assisted multidisciplinary design optimization for an all-electric GEO satellite. *Acta Astronaut.* **2017**, *138*, 301–317. [CrossRef]

14. Hosseini, M.; Nosratollahi, M.; Sadati, H. Multidisciplinary design optimization of uav under uncertainty. *J. Aerosp. Technol. Manag.* **2017**, *9*, 160–169. [CrossRef]

15. Wang, C.; Shi, W.; Wang, X.; Jiang, X.; Yang, Y.; Li, W.; Zhou, L. Optimal design of multistage centrifugal pump based on the combined energy loss model and computational fluid dynamics. *Appl. Energy* **2017**, *187*, 10–26. [CrossRef]

16. Wang, W.J.; Yuan, S.Q.; Pei, J.; Zhang, J.F. Optimization of the diffuser in a centrifugal pump by combining response surface method with multi-island genetic algorithm. *Proc. Inst. Mech. Eng. Part E J. Process. Mech. Eng.* **2017**, *231*, 191–201. [CrossRef]

17. Pei, J.; Wang, W.J.; Yuan, S.Q.; Zhang, J.F. Optimization the impeller of a low-specific-speed centrifugal pump for hydraulic performance improvement. *Chin. J. Mech. Eng.* **2016**, *29*, 992–1002. [CrossRef]

18. Miao, F.; Park, H.S.; Kim, C.; Ahn, S. Swarm intelligence based on modified PSO algorithm for the optimization of axial-flow pump impeller. *J. Mech. Sci. Technol.* **2015**, *29*, 4867–4876. [CrossRef]

19. Yun, J.E. CFD analysis for optimization of guide vane of axial-flow pump. *Trans. Korean Soc. Mech. Eng. B* **2016**, *40*, 519–525. [CrossRef]

20. Cao, L.L.; Watanabe, S.; Imanishi, T.; Yoshimura, H.; Furukawa, A. Experimental analysis of flow structure in contra-rotating axial flow pump designed with different rotational speed concept. *J. Therm. Sci.* **2013**, *22*, 345–351. [CrossRef]

21. Li, W.G. Verifying performance of axial-flow pump impeller with low NPSHr by using CFD. *Eng. Comput.* **2011**, *28*, 557–577. [CrossRef]

22. Meng, D.; Liu, M.; Yang, S.Q.; Zhang, H.; Ding, R. A fluid-structure analysis approach and its application in the uncertainty-based multidisciplinary design and optimization for blades. *Adv. Mech. Eng.* **2018**, *10*, 1–7. [CrossRef]

23. Huang, H.; An, H.C.; Wu, W.R.; Zhang, L.Y.; Wu, B.B.; Li, W.P. Multidisciplinary design modeling and optimization for satellite with maneuver capability. *Struct. Multidiscip. Optim.* **2014**, *50*, 883–898. [CrossRef]

24. Tong, Y.F.; Ye, W.; Yang, Z.; Li, D.B.; Li, X.D. Research on multidisciplinary optimization design of bridge crane. *Math. Probl. Eng.* 2013. [CrossRef]

25. Wu, X.F.; Tian, X.; Tan, M.G.; Liu, H.L. Multi-parameter optimization and analysis on performance of a mixed flow pump. *J. Appl. Fluid Mech.* **2020**, *13*, 199–209. [CrossRef]

26. Derakhshan, S.; Pourmahdavi, M.; Abdolahnejad, E.; Reihani, A.; Ojaghi, A. Numerical shape optimization of a centrifugal pump impeller using artificial bee colony algorithm. *Comput. Fluids* **2013**, *81*, 145–151. [CrossRef]

27. Xie, Y.; Hu, P.; Zhu, N.; Lei, F.; Xing, L.; Xu, L. Collaborative optimization of ground source heat pump-radiantceiling air conditioning system based on response surface method andNSGA-II. *Renew. Energy* **2020**, *147*, 249–264. [CrossRef]

28. Wei, Z.; Long, T.; Shi, R.; Tang, Y.; Li, H. Multidisciplinary design optimization of long-range slender guided rockets considering aeroelasticity and subsidiary loads. *Aerosp. Sci. Technol.* **2019**, *92*, 790–805. [CrossRef]

© 2020 by the authors. Licensee MDPI, Basel, Switzerland. This article is an open access article distributed under the terms and conditions of the Creative Commons Attribution (CC BY) license (http://creativecommons.org/licenses/by/4.0/).

*Article*

# Effects of the Impeller Blade with a Slot Structure on the Centrifugal Pump Performance

**Hongliang Wang [1], Bing Long [1], Chuan Wang [2,*], Chen Han [3] and Linjian Li [3]**

[1] School of Aerospace and Mechanical Engineering/Flight College, Changzhou Institute of Technology, Changzhou 213032, China; wanghl@czu.cn (H.W.); longb@czu.cn (B.L.)
[2] College of Hydraulic Science and Engineering, Yangzhou University, Yangzhou 225009, China
[3] National Research Center of Pumps, Jiangsu University, Zhenjiang, Jiangsu 212013, China; hanchen0622@outlook.com (C.H.); cllgeee@126.com (L.L.)
* Correspondence: wangchuan@ujs.edu.cn

Received: 7 February 2020; Accepted: 24 March 2020; Published: 2 April 2020

**Abstract:** An impeller blade with a slot structure can affect the velocity distribution in the impeller flow passage of the centrifugal pump, thus affecting the pump's performance. Various slot structure geometric parameter combinations were tested in this study to explore this relationship: slot position $p$, slot width $b_1$, slot deflection angle $\beta$, and slot depth $h$ with (3–4) levels were selected for each factor on an $L_{16}$ orthogonal test table. The results show that $b_1$ and $h$ are the major factors influencing pump performance under low and rated flow conditions, while $p$ is the major influencing factor under the large flow condition. The slot structure close to the front edge of the impeller blade can change the low-pressure region of the suction inlet of the impeller flow passage, thus improving the fluid velocity distribution in the impeller. Optimal slot parameter combinations according to the actual machining precision may include a small slot width $b_1$, slot depth $h$ of $\frac{1}{4}$ $b$, slot deflection angle $\beta$ of 45°–60°, and slot position $p$ close to the front edge of the blade at 20–40%.

**Keywords:** blade slot; orthogonal test; numerical simulation; centrifugal pump

---

## 1. Introduction

The pumps are classified as general machinery with varied applications [1–5]. Blade slotting was first used in the aviation industry to improve the separation of airflow over the wings [6]. Blade slotting technologies work by allowing the pressure difference to be adjusted between two sides of an airfoil so that gas on the high-pressure side flows through the slot, thus forming a jet upon reaching the low-pressure side. This jet effectively delays the separation of the boundary layer of the airfoil surface and increases the airfoil head coefficient to improve overall airfoil properties [7–9]. The effects of various slot parameters on the performance of blade centrifugal pump represent significant information for engineers.

Many scholars have investigated the effects of slotting technology on the mechanical performance of impellers. Tang et al. [10], for example, carried out a numerical simulation using the three-dimensional (3D) uncompressible Navier–Stokes equation to determine the influence of the slot position and slot width on the centrifugal fan performance. They found that the efficiency and total pressure of the slotted impeller improved under the design conditions; noise was also reduced as the performance under non-design conditions improved. Huang et al. [11] optimized slotting long and short blade impeller parameters based 3D, incompressible Navier–Stokes equations and a $k$-$\omega$ turbulence model followed by a prototype test; they concluded that the slotting technique improves blower performance. Wang et al. [12] used the RNG $k$-$\varepsilon$ model to simulate the multiphase flow in a centrifugal pump with a slotted blade structure. They found that an opening near the blade inlet improves the cavitation performance of the medium-low specific speed centrifugal pump. Ye [13] studied the effects of slotted blades on the centrifugal efficiency, head, and internal flow field of a pump; the blade slot was found to increase the head and flow of the pump under high flow conditions. Gao et al. [14] determined the loss of pressure and coefficient of heat transfer in the slotted turbine blade of a trailing edge by numerical simulation and experimental verification; they also analyzed the effects of slotting on the flow and heat transfer characteristics of the surface. Xing [15] studied the effects of long and short blades on the internal flow field and impeller performance of pumps using the numerical simulation method. Yuan [16] found that the use of splitter blades effectively improves the performance of low specific speed centrifugal pumps. Kergourlay et al. [17] used the unsteady Reynolds average Navier–Stokes (URANS) method to study the effect of blade slotting on the flow field in a centrifugal pump. An impeller with a slotted structure was found to make the circumferential speed and pressure distribution more uniform while slightly improving the pump performance. Gölcü et al. [18] found that a slotted-blade impeller is less loaded than the impeller without a slotted blade.

There has been a great deal of research on slotting the blades of low specific speed centrifugal pumps [12,13,19], but previously published techniques have certain limitations due to the sole analysis of a single condition. In an effort to systematically explore the effects of slotted blades on the performance of centrifugal pumps, the present study was conducted to test four parameters: slotted blade position, slotting width, slotting angle, and slotting depth. The effects of various combinations of different parameters on the performance of medium specific speed centrifugal pumps were explored accordingly via orthogonal design.

## 2. Numerical Simulation Method

### 2.1. Computation Model

A medium specific speed centrifugal pump with a speed ratio of $n_s = 85$ was selected to study the effects of blade slotting on the pump performance. The specific speed is formulated as follows:

$$n_s = \frac{3.65n \sqrt{Q}}{H^{3/4}} \tag{1}$$

where, $n$ is the rotating speed (r/min), $Q$ is the rated point flow rate (m$^3$/s), and $H$ is the rated point head (m).

The main design parameters of the pump are shown in Table 1.

**Table 1.** Main design parameters of centrifugal pump.

| Design Parameter | Value |
| --- | --- |
| Rated point head | 35 m |
| Rated point flow | 50 m$^3$/h |
| Impeller speed | 2850 r/min |
| Impeller inlet diameter | 74 mm |
| Impeller outlet diameter | 174 mm |
| Impeller outlet width | 12 mm |
| Blade wrap angle | 108° |
| Blade exit angle | 31° |
| Blade inlet placement angle | 37.5° |
| Volute base diameter | 184 mm |
| Spiral inlet width | 20 mm |

The total flow field was computed to consider the effects of ring clearance leakage on the pump's performance. As shown in Figure 1, the computational domains in this case include the inlet section, impeller, pump cavity, volute, and outlet section, wherein the pump cavity portion includes the front and ring cavities. The extension lengths of the inlet and outlet sections in the computational domain were set to four times the diameters of the inlet and outlet pipes, respectively, to ensure a sufficient flow development.

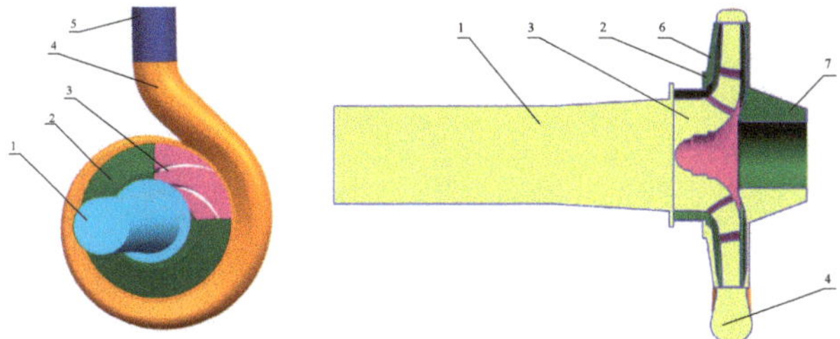

**Figure 1.** Simulation pump model. 1. Inlet section; 2. pump cavity; 3. impeller; 4. volute; 5. outlet section; 6. front cavity and 7. rear cavity.

## 2.2. Meshing

Commercial ANSYS-ICEM17.0 software was utilized to mesh various computational domains. As the structured meshes only contained quadrilaterals or hexahedrons in this case, their topological structure was equivalent to a uniformly orthogonal mesh within a rectangular domain. Accordingly, the nodes on each layer of the mesh lines can be effectively adjusted to ensure a high quality [20–24]. The overall computational domain was structured and meshed and the boundary layer of the meshes in the vicinity of the near wall of the blade was refined. The quality of meshes within all computational domains was above 0.35 (Figure 2).

(a) Mesh of the impeller.

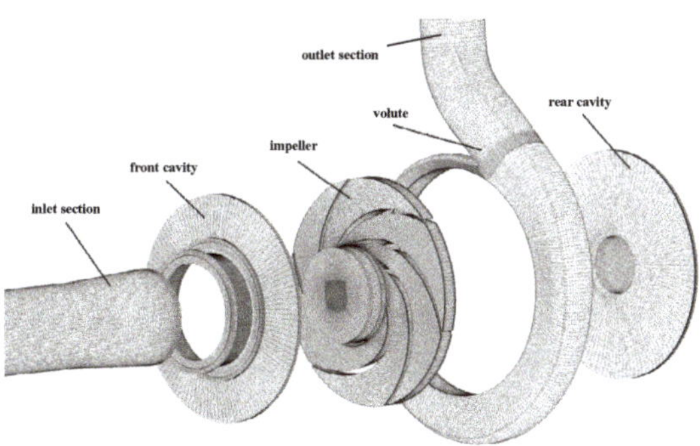



(b) Computational domain and structure grid.

**Figure 2.** Mesh of the pump.

The efficiency, head, and power of the rated point of the pump were considered indexes for mesh independence verification of the unslotted centrifugal pump model. The global maximum mesh size was used to control the mesh density of each computational domain. Local meshes within each computational domain were specifically refined to ensure the desired mesh quality. The meshing results are shown in Table 2.

**Table 2.** Analysis of the grid independent.

| Case | Global Maximum Mesh Size/mm | Total Number of Grids | Efficiency/% | Head/m | Power/kW |
|------|------|------|------|------|------|
| 1 | 3 | 1,269,483 | 78.5 | 36.22 | 6.25 |
| 2 | 2 | 1,688,511 | 78.93 | 36.18 | 6.24 |
| 3 | 1 | 4,760,103 | 79.31 | 36.33 | 6.24 |

Figure 3 shows the effects of the number of meshes in different cases on the head, power, and efficiency of the simulated pump. A numerical calculation on a mesh with a global mesh size of 2 mm was conducted with both the computational cycle and computational accuracy taken into account.

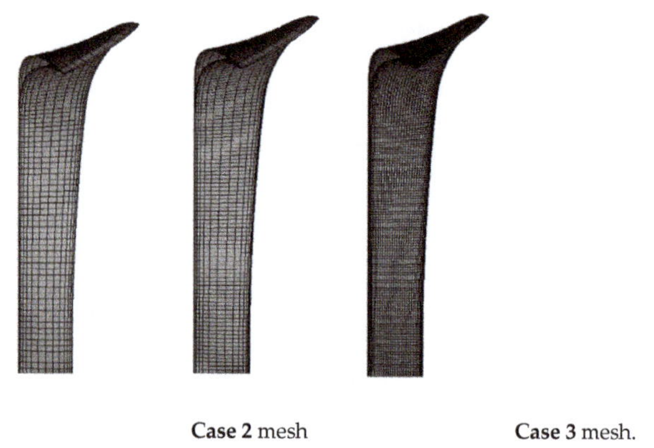

**Case 1** mesh       **Case 2** mesh       **Case 3** mesh.

**Figure 3.** Schematic diagram of the blade structure grid.

### 2.3. Computational Cases and Boundary Conditions

Numerical calculations were performed in ANSYS-CFX19.2 software (ANSYS CFX. 19.2" ANSYS CFX 19.2 Documentation. 2019). The turbulence model in this case was a standard $k$-$\omega$ turbulence model. The computational impeller domain was set to a rotational domain and all other computational domains were set to static. Data transfer at the interface between the static domains and the rotational domains was achieved by the frozen rotor method.

To consider the effects of the impeller cover plate on the flow, all other inner wall surfaces within the pump cavity excluding those in contact with the impeller outlet surface were set to a rotational wall surface. The roughness of each computational domain surface was set to 10 μm to observe the effects of the material on the internal flow characteristics of the pump. The boundary conditions were set to pressure inlet and mass outflow. The reference pressure was set to a standard atmospheric pressure, the wall surface was placed under a non-slip boundary condition, and a standard wall surface function was used with the convergence accuracy set to $10^{-4}$.

### 2.4. Orthogonal Design of Blade Slots

To explore the effects of blade-slotting on the medium specific speed pump systematically, four factors including the slotting position, slotting width, slotting depth, and slotting angles of the blades were studied via the orthogonal design method. The orthogonal design method is a scientific design technique wherein test plans are reasonably arranged to determine the main factors that influence certain indexes within a brief testing time [25]. Many researchers have used orthogonal designs to study the performance of centrifugal pumps [26–29]. Considering the time and cost burdens of the test, the geometrical factors of the slots as they affect pump performance were observed in this study by combining CFD technology with an orthogonal design.

### 2.5. Determining the Test Factors

As discussed above, four sets of geometric parameters were taken as test factors: slot position $p$, slot width $b_1$, slot deflection angle $\beta$ and slot depth $h$ (Figure 4). Slot position $p$ is the position of the slot on the blade. Based on the arc length of the blade profile, four uniform levels were taken from the inlet edge of the blade to the outlet edge of the blade: 20%$p$, 40%$p$, 60%$p$, and 80%$p$. Three levels of the slot width $b_1$ were selected: 0.5 mm, 1 mm, and 1.5 mm. As shown in Figure 4, the angle $\beta$ is the included angle between the slot and the tangent line of the blade in the slot position. The deflection angle of the slot relates to the effects of the slot jet on the liquid flow in the flow passage. Three angles were tested: 30°, 45°, and 60°. The slot depth $h$ is the axial distance from the inner surface of the front cover plate, relative to the blade outlet width $b$. Three depths were selected: 1/4 $b$, 2/4 $b$, and 3/4 $b$.

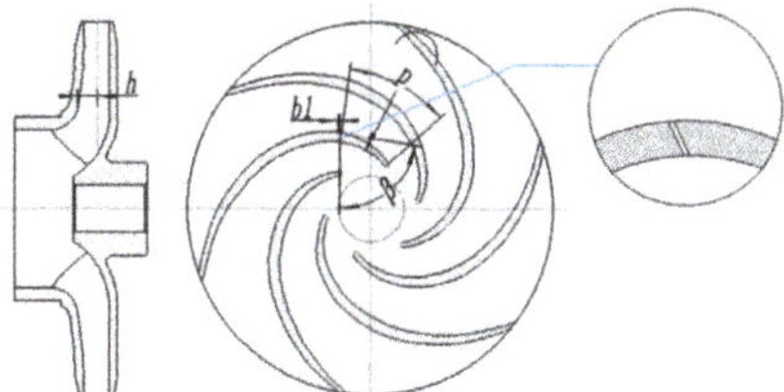

**Figure 4.** The schematic diagram of the gap geometry parameters.

As shown in Table 3, the $L_{16}$ orthogonal table was selected for these four factors.

**Table 3.** Factor level table.

| Orthogonal Design Case | A | B | C | D |
|---|---|---|---|---|
| | $p$ | $b_1$/mm | $\beta$/° | $h$/mm |
| 1 | 20% | 0.5 | 30 | 1/4b |
| 2 | 40% | 1 | 30 | 2/4b |
| 3 | 60% | 1.5 | 30 | 3/4b |
| 4 | 80% | 0.5 | 30 | 1/4b |
| 5 | 60% | 1 | 45 | 1/4b |
| 6 | 80% | 0.5 | 45 | 2/4b |
| 7 | 20% | 0.5 | 45 | 3/4b |
| 8 | 40% | 1.5 | 45 | 1/4b |
| 9 | 80% | 1.5 | 60 | 1/4b |
| 10 | 60% | 0.5 | 60 | 2/4b |
| 11 | 40% | 0.5 | 60 | 3/4b |
| 12 | 20% | 1 | 60 | 1/4b |
| 13 | 40% | 0.5 | 30 | 1/4b |
| 14 | 20% | 1.5 | 30 | 2/4b |
| 15 | 80% | 1 | 30 | 3/4b |
| 16 | 60% | 0.5 | 30 | 1/4b |

## 3. Analysis of Numerical Simulation Results

### 3.1. Test Verification

To validate the numerical simulation method used in this study, the original model was tested. As shown in Figure 5, the test rig is an open-type system, which is composed of two parts, namely, the data acquisition system and the water circulation system. The DN100 electromagnetic flowmeter whose maximum allowable error is ±0.5% was used to measure the flow rate $Q$. The valve of the pump inlet pipeline was fully opened during the test and the flow condition points were collected through the pump outlet pipeline valve. To secure a smooth external characteristic curve during the collection process, recording was performed at an interval of 5 m³/h from the shutoff point to the large flow condition point for a total of 17 operating points.

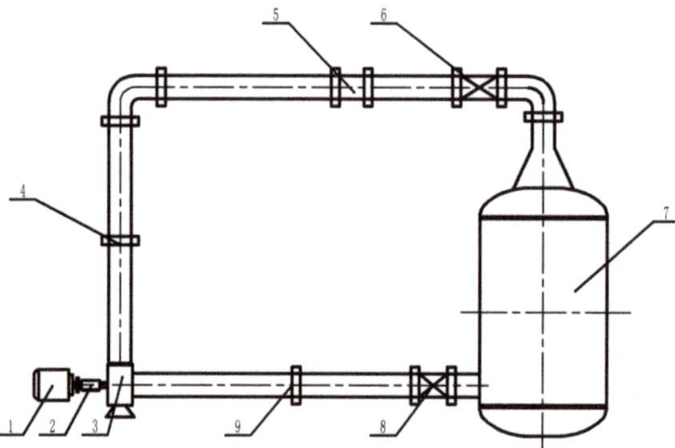

**Figure 5.** Schematic diagram of the test rig. 1. Motor; 2. torque meter; 3. centrifugal pump; 4. outlet pipeline pressure section; 5. DN100 electromagnetic flowmeter; 6. outlet pipeline valve; 7. water tank; 8. inlet pipeline valve and 9. inlet pipeline pressure section.

Table 4 shows the pump performance test results. As the rotational speed of the pump was not constant at 2850 r/min during actual operation, for an effective comparison against the numerical calculation results, the external characteristic data of the pump was converted to a rated speed of 2850 r/min according to the rules of similarity theory.

**Table 4.** Pump performance test results.

| $Q$/(m³/h) | Inlet Pressure/kPa | Outlet Pressure/kPa | $h$/m | $P$/kW | $\eta$/% |
|---|---|---|---|---|---|
| 0.11 | 113.05 | 505.17 | 36.62 | 2.42 | 0.46 |
| 5.04 | 109.31 | 505.44 | 37.05 | 2.68 | 18.98 |
| 9.64 | 106.77 | 506.24 | 37.47 | 3.04 | 32.32 |
| 15.06 | 105.82 | 506.46 | 37.67 | 3.51 | 44.01 |
| 19.55 | 104.97 | 504.31 | 37.75 | 3.87 | 51.93 |
| 23.95 | 103.92 | 501.57 | 37.79 | 4.26 | 57.87 |
| 28.54 | 102.59 | 496.04 | 37.60 | 4.56 | 64.10 |
| 33.59 | 100.92 | 487.67 | 37.25 | 4.97 | 68.46 |
| 38.46 | 98.98 | 475.52 | 36.61 | 5.27 | 72.68 |
| 43.13 | 97.04 | 459.79 | 35.73 | 5.64 | 74.34 |
| 48.05 | 94.37 | 441.22 | 34.57 | 5.99 | 76.60 |
| 52.94 | 91.84 | 417.51 | 33.06 | 6.22 | 76.56 |
| 57.80 | 89.16 | 391.54 | 31.29 | 6.43 | 76.54 |
| 62.41 | 85.86 | 364.43 | 29.53 | 6.67 | 75.20 |
| 67.05 | 82.86 | 330.38 | 27.12 | 6.87 | 72.02 |
| 69.51 | 80.99 | 276.46 | 22.51 | 6.81 | 62.51 |

Figure 6 shows a comparison between the test-based and simulated pump performance indicators. To completely reflect the external characteristic variation curve from the shutoff point to the maximum flow condition during the pump test, eight flow condition points were simulated from 0.1$Q$ to 1.4$Q$. The numerical simulation results accurately predicted the external characteristic curve of the pump within the whole range of operating conditions as observed in the test. The relative errors in the head, efficiency, and power of the rated operating points were 4.4%, 2.95%, and 4%, respectively. All were smaller than 5%, which indicates that the numerical simulation method was accurate. Further, these results suggest that the orthogonal design case accurately reflects slotting effects numerically.

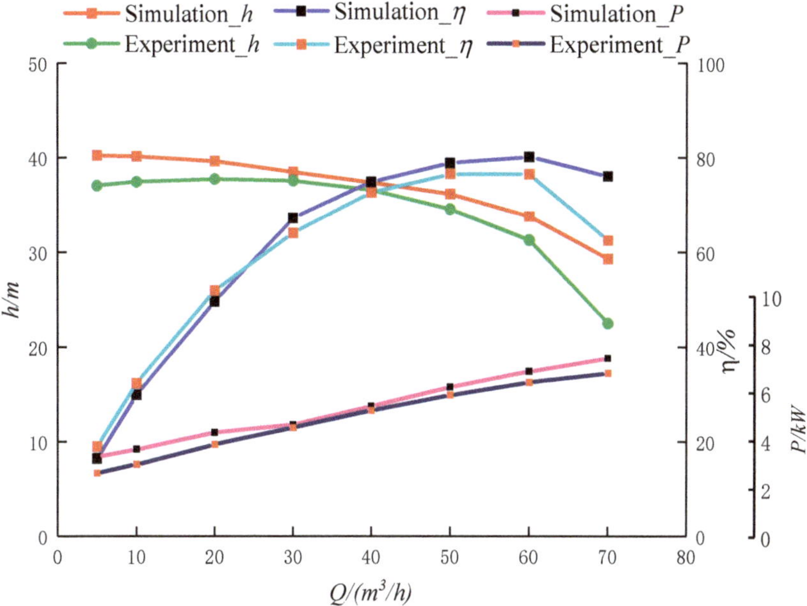



**Figure 6.** Comparison of test and numerical results.

### 3.2. Direct Analysis of Orthogonal Design Case

Sixteen sets of orthogonal design cases were used in this study. Prototyping all of them would be costly and time-consuming, so considering the accuracy of the numerical calculation method, the full flow field numerical simulation method was selected as the research tool for this orthogonal design. To observe the effects of slotting on the performance of the medium specific speed centrifugal pump, full flow field numerical simulations were conducted at four operating condition points: 0.6$Q$, 0.8$Q$, 1.0$Q$, 1.2$Q$, and 1.4$Q$ for 16 sets of slotted impellers in conjunction with the volutes.

This study centers on the effects of different slotting cases on pump performance. Tables 5–7 show the numerical simulation results of 0.6$Q$, 1.0$Q$, and 1.4$Q$, respectively. The orthogonal test data was processed to assess the main factors influencing the pump head and efficiency in the slotted blade case [18]. The range analysis method was used to observe the effects of the levels of the factors at 0.6$Q$, 1.0$Q$, and 1.4$Q$ operating conditions on the pump's performance.

**Table 5.** 0.6$Q$ numerical simulation results.

| Case | 1 | 2 | 3 | 4 | 5 | 6 | 7 | 8 |
|---|---|---|---|---|---|---|---|---|
| $h$/m | 39.3 | 38.85 | 38.59 | 38.89 | 38.69 | 39.19 | 38.57 | 38.43 |
| $\eta$/% | 65.98 | 65.57 | 62.76 | 66.39 | 66.48 | 66.05 | 66.15 | 66.07 |
| $P$/kW | 4.87 | 4.84 | 5.03 | 4.79 | 4.76 | 4.85 | 4.77 | 4.76 |
| Case | 9 | 10 | 11 | 12 | 13 | 14 | 15 | 16 |
| $h$/m | 39.12 | 38.95 | 38.81 | 39.29 | 39.23 | 38.48 | 38.85 | 39.15 |
| $\eta$/% | 66.17 | 65.32 | 65.42 | 65.46 | 66.14 | 64.79 | 64.1 | 66.15 |
| $P$/kW | 4.83 | 4.87 | 4.85 | 4.91 | 4.85 | 4.86 | 4.95 | 4.84 |
| Case | Original Model | | | | | | | |
| $h$/m | 38.5 | | | | | | | |
| $\eta$/% | 67.28 | | | | | | | |
| $P$/kW | 4.68 | | | | | | | |

**Table 6.** 1.0$Q$ numerical simulation results.

| Case | 1 | 2 | 3 | 4 | 5 | 6 | 7 | 8 |
|---|---|---|---|---|---|---|---|---|
| $h$/m | 35.9 | 35.1 | 32.8 | 35.9 | 35.3 | 35.9 | 35.6 | 35.2 |
| $\eta$/% | 78.6 | 77.9 | 74.5 | 78.5 | 78 | 78.3 | 78.4 | 77.8 |
| $P$/kW | 6.2 | 6.1 | 5.9 | 6.2 | 6.2 | 6.2 | 6.2 | 6.2 |
| Case | 9 | 10 | 11 | 12 | 13 | 14 | 15 | 16 |
| $h$/m | 35.4 | 35.6 | 35.3 | 35.8 | 35.7 | 35.6 | 35.2 | 35.8 |
| $\eta$/% | 78.3 | 78.2 | 78.2 | 78.4 | 78.5 | 78.2 | 77 | 78.4 |
| $P$/kW | 6.2 | 6.2 | 6.2 | 6.2 | 6.2 | 6.2 | 6.2 | 6.2 |
| Case | Original Model | | | | | | | |
| $h$/m | 36.18 | | | | | | | |
| $\eta$/% | 78.93 | | | | | | | |
| $P$/kW | 6.24 | | | | | | | |

**Table 7.** 1.4$Q$ numerical simulation results.

| Case | 1 | 2 | 3 | 4 | 5 | 6 | 7 | 8 |
|---|---|---|---|---|---|---|---|---|
| $h$/m | 30.15 | 28.85 | 25.61 | 30.05 | 28.52 | 29.76 | 30.12 | 28.49 |
| $\eta$/% | 77.86 | 76.23 | 70.49 | 77.72 | 76.13 | 76.82 | 77.83 | 75.93 |
| $P$/kW | 7.39 | 7.21 | 6.93 | 7.37 | 7.15 | 7.38 | 7.38 | 7.17 |
| Case | 9 | 10 | 11 | 12 | 13 | 14 | 15 | 16 |
| $h$/m | 28.72 | 29.44 | 29.22 | 30.04 | 29.99 | 30.05 | 27.34 | 29.65 |
| $\eta$/% | 76.26 | 76.63 | 76.78 | 77.79 | 77.75 | 77.69 | 73.99 | 77.29 |
| $P$/kW | 7.18 | 7.33 | 7.26 | 7.37 | 7.36 | 7.37 | 7.04 | 7.32 |
| Case | Original model | | | | | | | |
| $h$/m | 30.05 | | | | | | | |
| $\eta$/% | 77.09 | | | | | | | |
| $P$/kW | 7.43 | | | | | | | |

In the case of a greater range, different levels of a given factor lead to a larger amplitude of variations in the test indicators. To this effect, the factor corresponding to the maximum range was the most important factor. $K_i(i = 1,2,3,4)$ denotes the sum of the tests of the same level in any of the columns in Table 3, where $i$ corresponds to different levels of the same factor, $k_i = K_i/n$ denotes the arithmetic mean value of different levels of the same factor, $n$ denotes the number of occurrence of the same level in any of the columns in the table, and $R = max(k_1, k_2, k_3, k_4) - min(k_1, k_2, k_3, k_4)$ denotes the range. A range analysis of 0.6Q are shown in Tables 8 and 9.

**Table 8.** $0.6Q$ head analysis.

| $h/m$ | $A$ | $B$ | $C$ | $D$ |
|---|---|---|---|---|
| $K_1$ | 155.64 | 312.09 | 311.34 | 312.1 |
| $K_2$ | 155.32 | 155.68 | 154.88 | 155.47 |
| $K_3$ | 155.38 | 154.62 | 156.17 | 154.82 |
| $K_4$ | 156.05 | - | - | - |
| $k_1$ | 38.91 | 39.01 | 38.92 | 39.01 |
| $k_2$ | 38.83 | 38.92 | 38.72 | 38.87 |
| $k_3$ | 38.85 | 38.66 | 39.04 | 38.71 |
| $k_4$ | 39.01 | - | - | - |
| $R$ | 0.18 | 0.36 | 0.32 | 0.31 |

**Table 9.** $0.6Q$ efficiency analysis.

| $h/m$ | $A$ | $B$ | $C$ | $D$ |
|---|---|---|---|---|
| $K_1$ | 262.38 | 527.6 | 521.88 | 528.84 |
| $K_2$ | 263.2 | 261.61 | 264.75 | 261.73 |
| $K_3$ | 260.71 | 259.79 | 262.37 | 258.43 |
| $K_4$ | 262.71 | - | - | - |
| $k_1$ | 65.59 | 65.95 | 65.24 | 66.105 |
| $k_2$ | 65.8 | 65.40 | 66.19 | 65.43 |
| $k_3$ | 65.18 | 64.95 | 65.59 | 64.61 |
| $k_4$ | 65.68 | - | - | - |
| $R$ | 0.62 | 1.00 | 0.95 | 1.50 |

The primary and secondary geometric parameters of slots influencing the pump performance at $0.6Q$ were obtained as shown in Table 10.

**Table 10.** The order of influence of the gap geometry parameters on pump performance at $0.6Q$.

| Index | Major Factor | $\longrightarrow$ | Secondary Factor | |
|---|---|---|---|---|
| $h/m$ | $b_1$ | $\beta$ | $h$ | $p$ |
| $\eta/\%$ | $h$ | $b_1$ | $\beta$ | $p$ |

A range analysis of $1.0Q$ are shown in Tables 11 and 12.

**Table 11.** $1.0Q$ head analysis.

| $h/m$ | $A$ | $B$ | $C$ | $D$ |
|---|---|---|---|---|
| $K_1$ | 142.9 | 285.7 | 282 | 285 |
| $K_2$ | 141.3 | 141.4 | 142 | 142.2 |
| $K_3$ | 139.5 | 139 | 142.1 | 138.9 |
| $K_4$ | 142.4 | - | - | - |
| $k_1$ | 35.73 | 35.71 | 35.25 | 35.63 |
| $k_2$ | 35.34 | 35.35 | 35.5 | 35.55 |
| $k_3$ | 34.89 | 34.75 | 35.53 | 34.73 |
| $k_4$ | 35.6 | - | - | - |
| $R$ | 0.85 | 0.96 | 0.28 | 0.9 |

**Table 12.** $1.0Q$ efficiency analysis.

| $h/m$ | $A$ | $B$ | $C$ | $D$ |
|---|---|---|---|---|
| $K_1$ | 313.6 | 627.1 | 621.6 | 626.5 |
| $K_2$ | 312.4 | 311.3 | 312.5 | 312.6 |
| $K_3$ | 309.1 | 308.8 | 313.1 | 308.1 |
| $K_4$ | 312.1 | - | - | - |
| $k_1$ | 78.4 | 78.39 | 77.7 | 78.313 |
| $k_2$ | 78.1 | 77.83 | 78.13 | 78.15 |
| $k_3$ | 77.28 | 77.2 | 78.28 | 77.03 |
| $k_4$ | 78.03 | - | - | - |
| $R$ | 1.13 | 1.19 | 0.58 | 1.29 |

The primary and secondary geometric parameters of slots influencing the pump performance at $1.0Q$ were obtained as shown in Table 13.

**Table 13.** The order of influence of the gap geometry parameters on pump performance at $1.0Q$.

| Index | Major Factor | $\longrightarrow$ | | Secondary Factor |
|---|---|---|---|---|
| $h/m$ | $b_1$ | $h$ | $p$ | $\beta$ |
| $\eta/\%$ | $h$ | $b_1$ | $p$ | $\beta$ |

A range analysis of $1.4Q$ are shown in Tables 14 and 15.

**Table 14.** $1.4Q$ head analysis.

| $h/m$ | $A$ | $B$ | $C$ | $D$ |
|---|---|---|---|---|
| $K_1$ | 120.36 | 238.38 | 231.69 | 235.61 |
| $K_2$ | 116.55 | 114.75 | 116.89 | 118.1 |
| $K_3$ | 113.22 | 112.87 | 117.42 | 112.29 |
| $K_4$ | 115.87 | - | - | - |
| $k_1$ | 30.09 | 29.79 | 28.96 | 29.45 |
| $k_2$ | 29.14 | 28.69 | 29.22 | 29.53 |
| $k_3$ | 28.31 | 28.22 | 29.36 | 28.07 |
| $k_4$ | 28.97 | - | - | - |
| $R$ | 1.79 | 1.58 | 0.39 | 1.45 |

**Table 15.** $1.4Q$ efficiency analysis.

| $h/m$ | $A$ | $B$ | $C$ | $D$ |
|---|---|---|---|---|
| $K_1$ | 311.17 | 618.68 | 609.02 | 616.73 |
| $K_2$ | 306.69 | 304.14 | 306.71 | 307.37 |
| $K_3$ | 300.54 | 300.37 | 307.46 | 299.09 |
| $K_4$ | 304.79 | - | - | - |
| $k_1$ | 77.79 | 77.3 | 76.12 | 77.09 |
| $k_2$ | 76.67 | 76.03 | 76.67 | 76.84 |
| $k_3$ | 75.13 | 75.09 | 76.86 | 74.77 |
| $k_4$ | 76.19 | - | - | - |
| $R$ | 2.66 | 2.24 | 0.74 | 2.32 |

The primary and secondary geometric parameters of slots influencing the pump performance at $1.4Q$ were obtained as shown in Table 16.

**Table 16.** The order of influence of the gap geometry parameters on pump performance at 1.4*Q*.

| Index | Major Factor | $\longrightarrow$ | Secondary Factor | |
|---|---|---|---|---|
| $h$/m | $p$ | $b_1$ | $h$ | $\beta$ |
| $\eta$/% | $p$ | $h$ | $b_1$ | $\beta$ |

Range analyses of the operating condition points, 0.6*Q*, 1.0*Q*, and 1.4*Q* indicated that the slot width and depth under the small flow conditions and rated conditions have the greatest effects on the pump head and efficiency among the parameters tested. The slot position appeared to have little effect on the performance of the pump under small flow conditions. In the case of large flow conditions, however, the slot position had a greater effect on pump performance than any other parameter.

To analyze the effects of the changes in factor levels on the pump performance more intuitively, a trend variation chart was plotted with the head and efficiency of the pump as indicators. As shown in Figure 7, the head $h_{0.6Q}$ was the largest when the blade slotting position $p$ was close to the outlet side and the slot deflection angle $\beta$ was the smallest under the small flow condition 0.6*Q*. The head $h_{0.6Q}$ decreased progressively as slot width $b_1$ and the slot depth $h$ increased, and an inflection point emerged on the curve of the head $h_{0.6Q}$ as slot position $p$ varied. Based on the steepness of the curve variation trend, the primary and secondary factors influencing the head $h_{0.6Q}$ were slot width $b_1$, slot deflection angle $\beta$, slot depth $h$, and slot position $p$, respectively. This result was consistent with the range analysis results. The efficiency $\eta_{0.6Q}$ also increased as $b_1$ and $h$ decreased. The efficiency $\eta_{0.6Q}$ curve trend also presented an inflection point with the changes in the slot position. Based on the trend graph, in order of intensity, the factors influencing the efficiency $\eta_{0.6Q}$ were slot depth $h$, slot width $b_1$, slot deflection angle $\beta$, and slot position $p$.

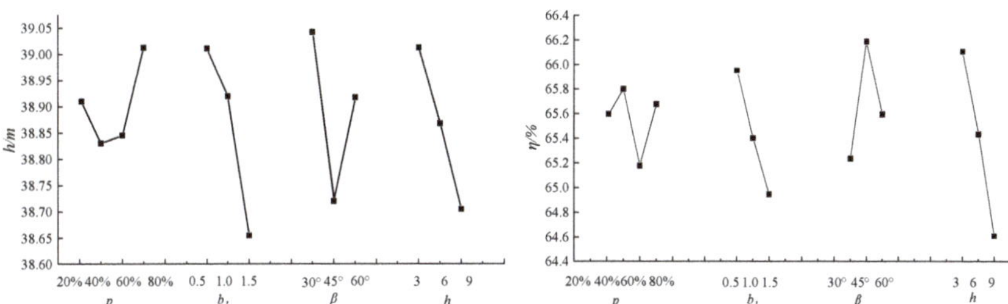

**Figure 7.** 0.6*Q* head (**left**) and efficiency (**right**)-factor relationship.

As shown in Figures 8 and 9, under the conditions of 1.0*Q* and 1.4*Q*, the heads of $h_{1.0Q}$ and $h_{1.4Q}$ were the largest when the blade slot position $p$ was in the vicinity of the inlet edge of the blade. Like 0.6*Q*, the heads of $h_{1.0Q}$ and $h_{1.4Q}$ and the efficiencies of $\eta_{1.0Q}$ and $\eta_{1.4Q}$ decreased progressively as slot width $b_1$ and the slot depth $h$ increased. $h_{1.0Q}$, $h_{1.4Q}$, $\eta_{1.0Q}$, and $\eta_{1.4Q}$ also increased progressively as slot deflection angle $\beta$ increased.

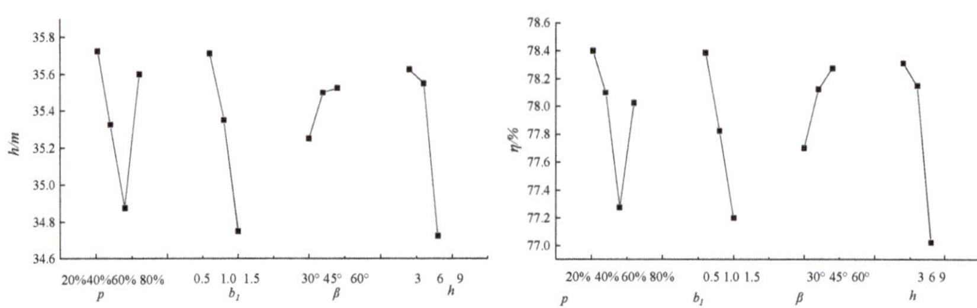

**Figure 8.** 1.0*Q* head (**left**) and efficiency (**right**)-factor relationship.

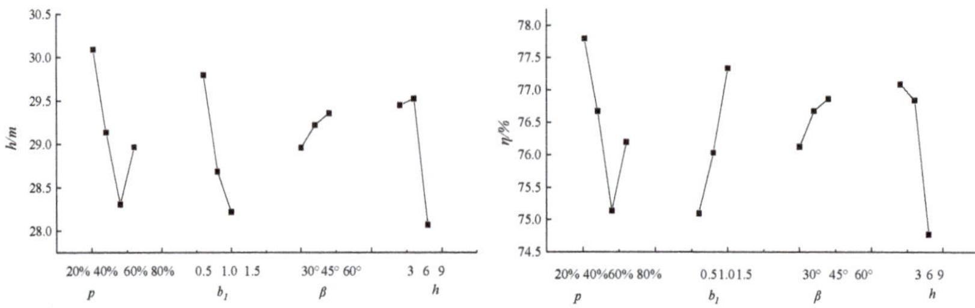

**Figure 9.** 1.4*Q* head (**left**) and efficiency (**right**)-factor relationship.

### 3.3. Analysis of Internal Flow Field

The orthogonal test results suggest that blade slotting improved the head at small flow condition points and the efficiency at large flow condition points, which is consistent with previously published results. Under the working condition of 0.6*Q*, the head of Case 1 was 39.3 m; in the original case the head was 38.5 m. The head and efficiency in Case 1 for the 1.4*Q* condition were 30.15 m and 77.86%, respectively, and in the original case were 30.05 m and 77.09%. To further explore the effects of the geometric slot parameters on pump performance, the distributions of performance curves of the original model and Case 1 were compared as shown in Figure 10.

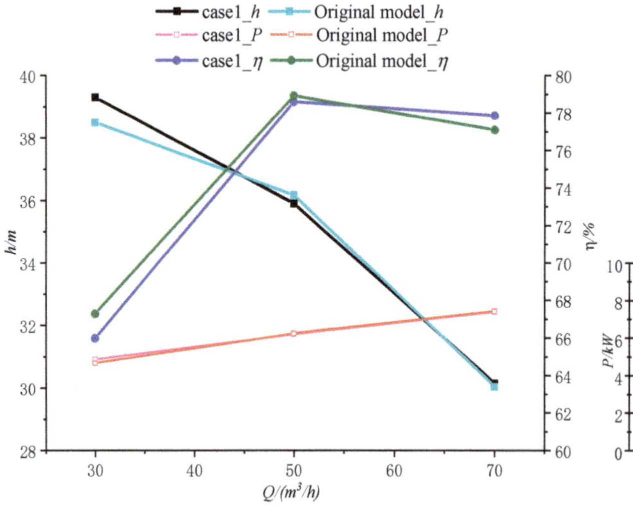

**Figure 10.** Comparison of case 1 and original model.

Figures 11 and 12 show cloud diagrams of the static pressure distribution of the blade unfolding at the section of the pump impeller flow passage (the section value Span was 0.9) in the original model and Case 1 of slotted blades under the conditions of 0.6Q and 1.4Q, respectively. As shown in Figure 11, under the 0.6Q condition, the static pressure distributions of the blade unfolding in Case 1 and the original model differed significantly. The distribution of pressure in the impeller flow passage from the blade inlet to the outlet was characterized by a low-pressure region in the first half-section of the impeller flow passage and a high-pressure region in the second half-section of the passage. The pressure gradient in the second half-section of the impeller flow passage was large because the flow passage diffusion was severe, which might have created a secondary back flow at the outlet of the impeller under small flow conditions. This is also likely a cause of the low efficiency of the medium-low specific speed centrifugal pump at the small flow condition point.

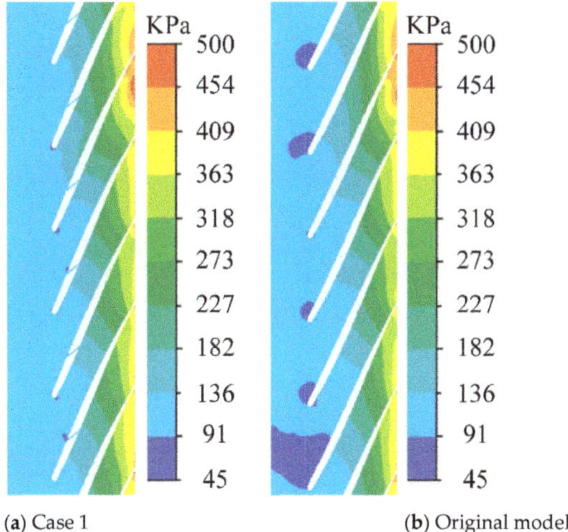

(**a**) Case 1        (**b**) Original model

**Figure 11.** Static pressure distributions on the cross section of the impeller channel at 0.6Q.

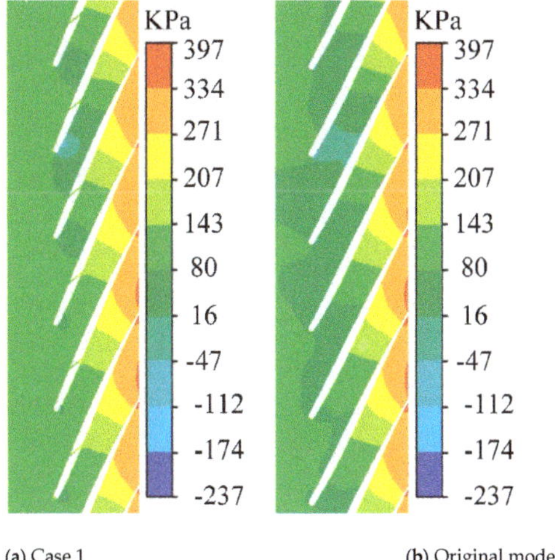

(**a**) Case 1        (**b**) Original model

**Figure 12.** Static pressure distributions on the cross section of the impeller channel at 1.4Q.

A significant low-pressure region was also observed in the position close to the inlet edge of the original model. Due to slotting in the position close to the inlet edge of the impeller, the distribution of blade unfolding static pressure disappears in the low-pressure region close to the inlet position of the blade in Case 1 and the pressure distribution is significantly more uniform than that in the original model. Vortexes and back flows are unlikely to form in the inlet position of the blade in this case, which is also one of the reasons why the head of the model in Case 1 is larger than that of the original model under the 0.6Q condition.

Under the 1.4Q large-flow condition, the diagram for the blade unfolding static pressure distribution in the case of the original model was similar to that in Case 1, however, the original model had a significant low-pressure region with considerable variations in the pressure gradient in the first half-section of the impeller flow passage inlet. This is mainly because the fluid flow angle of the incoming liquid increases with the flow rate while the inlet setting angle of the blade remains unchanged. As a result, the inlet setting angle is smaller than the liquid flow angle; a flow cutoff forms at the working surface in the position of the blade inlet creating a low-pressure region. Similarly, the changes in pressure gradient in the static pressure distribution diagram of blade unfolding in Case 1 are smaller than those of the original model due to the fact that the blade is slotted near the inlet.

As shown in Figure 13, the pressure distribution is shown on the blade surface at the mean circumferential flow surface. Under 0.6Q, the pressure distribution of the original model and the Case 1 model were quite different (Figure 13). At the position near the inlet side, the pressure of the pressure surface and the suction surface of the Case 1 model were larger than the original model. This further illustrates that the Case 1 model improved the pressure distribution at the inlet edge of the blade due to slotting. The pressure near the inlet edge of the blade was higher than that of the original model under the 1.4Q large-flow condition.

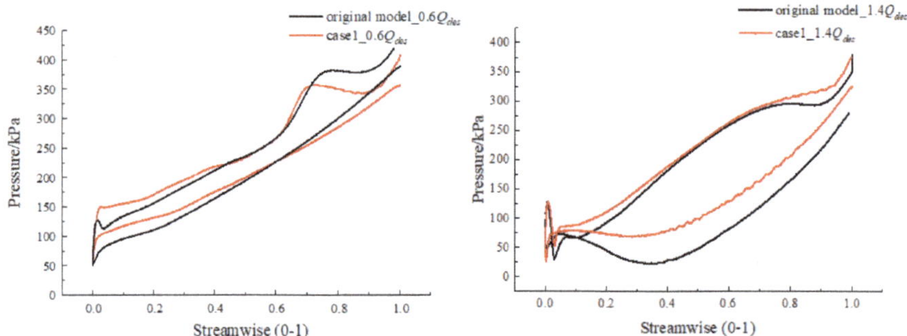

**Figure 13.** Variation of the blade load with streamwise at 0.6Q (**left**) and at 1.4Q (**right**).

Figures 14 and 15 show the distribution of pressure clouds in the middle plane of the blade flow channel. The pressure gradient distribution of the Case 1 model is more uniform than the original model under the 0.6Q condition (Figure 14); the original model shows a lower pressure than the Case 1 model near the blade inlet as well, which is consistent with the findings shown in Figures 11 and 12. The enlarged view in the figure shows where, due to the existence of a gap, the local low-pressure gradient distribution was more uniform in the original model. This gap jet made the streamline in the inlet low-pressure area closer to the profile of the blade airfoil, thereby improving the local flow field. Figure 15 shows that under the large flow rate of 1.4Q, the impact of the gap on the local area was relatively small. The local enlarged view did not show similar phenomena to the small flow conditions. Generally speaking, the gap improved the local flow field under small flow conditions.

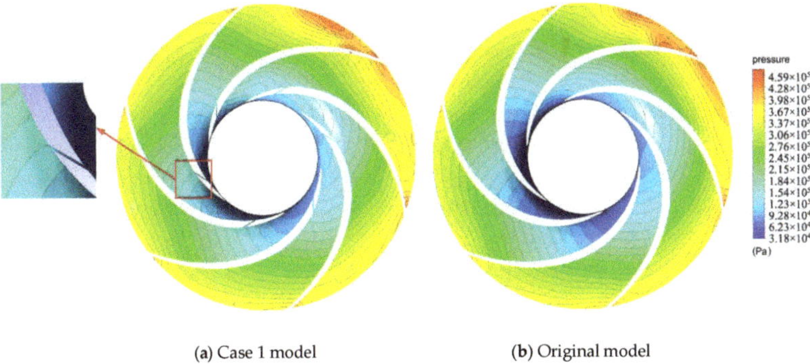

(**a**) Case 1 model     (**b**) Original model

**Figure 14.** Static pressure distribution in the middle plane of the blade flow channel at 0.6*Q*.

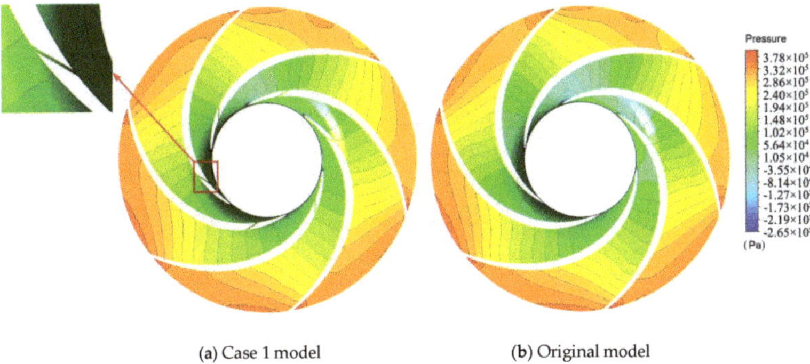

(**a**) Case 1 model     (**b**) Original model

**Figure 15.** Static pressure distribution in the middle plane of the blade flow channel at 1.4*Q*.

Figure 16 shows a diagram of the relative velocity distribution under the 1.4*Q* operating condition in the impeller calculation domain. This distribution was normal; the average velocity of the impeller calculation domain was basically 11 m/s. The velocity distribution amplitude of velocity in Case 1 was larger than that of the original model, which indicates that the velocity within the impeller was concentrated near the desired value and was uniform throughout the impeller calculation domain. This was also one of the reasons why the efficiency and head of Case 1 were larger than those of the original model.

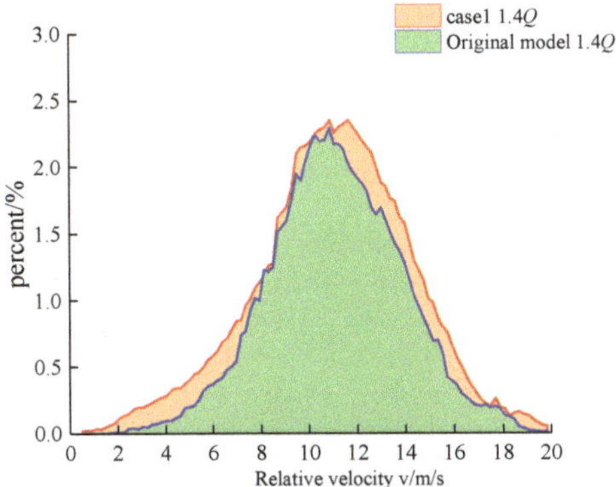


**Figure 16.** Velocity distribution in the impeller calculation domain at 1.4*Q*.

## 4. Conclusions

(1) Orthogonal test results show that various geometrical slot parameters affected pump performance. Under small flow conditions, the factors most intensely influencing the pump head were slot width $b_1$ > slot deflection angle $\beta$ > slot depth $h$ > slot position $p$. The factors most intensely influencing the pump efficiency were slot depth $h$ > slot width $b_1$ > slot deflection angle $\beta$ > slot position $p$. The main factors influencing the pump head and efficiency under rated flow conditions were slot width $b_1$ and slot depth $h$, whereas the main factor influencing the pump head and efficiency under large flow conditions was slot position $p$.

(2) The orthogonal test results also indicate that under low flow conditions and rated flow conditions, the head and efficiency of the pump decreased as blade slot width increased. These effects were linear.

(3) Different combinations of slot geometric parameters in Case 1 were found to increase the pump head at the small flow condition point as well as the efficiency and head of the pump at the large flow condition point compared to the original model without slots. The internal flow field shows that slotting near the front edge of the blade improved the low-pressure region of the impeller inlet flow passage and brought the flow velocity distribution in the impeller field under the large flow conditions closer to the desired value as the flow velocity distribution grew more uniform.

(4) To improve the performance of the pump, optimal slot parameter combinations according to the actual machining precision might include a small slot width $b_1$, slot depth $h$ of $\frac{1}{4} b$, slot deflection angle $\beta$ of 45–60°, and slot position $p$ close to the front edge of the blade at 20–40%.

**Author Contributions:** C.W. performed the writing—reviewing and editing; H.W. made the investing; B.L. performed the data curation; C.H. made the software; L.L. made the methodology. All authors have read and agreed to the published version of the manuscript.

**Funding:** This work was supported by the National Natural Science Foundation of China (Grant No. 51609105), Changzhou Sci&Tech Program (Grant No. CJ20190048), Jiangsu Province Graduate Practice Innovation Project (Grant No. SJCX18_0743).

**Acknowledgments:** In this section you can acknowledge any support given that is not covered by the author contribution or funding sections. This may include administrative and technical support, or donations in kind (e.g., materials used for experiments).

**Conflicts of Interest:** The authors declare no conflict of interest.

## References

1. Li, X.; Chen, B.; Luo, X.; Zhu, Z. Effects of flow pattern on hydraulic performance and energy conversion characterisation in a centrifugal pump. *Renew. Energy* **2020**. [CrossRef]
2. Li, X.; Shen, T.; Li, P.; Guo, X.; Zhu, Z. Extended compressible thermal cavitation model for the numerical simulation of cryogenic cavitating flow. *Int. J. Hydrog. Energy* **2020**. [CrossRef]
3. Shi, L.; Zhang, W.; Jiao, H.; Tang, F.; Wang, L.; Sun, D.; Shi, W. Numerical simulation and experimental study on the comparison of the hydraulic characteristics of an axial-flow pump and a full tubular pump. *Renew. Energy* **2020**, *153*, 1455–1464. [CrossRef]
4. Wang, C.; Chen, X.; Qiu, N.; Zhu, Y.; Shi, W. Numerical and experimental study on the pressure fluctuation, vibration, and noise of multistage pump with radial diffuser. *J. Braz. Soc. Mech. Sci. Eng.* **2018**, *40*, 481. [CrossRef]
5. Shi, L.; Zhu, J.; Tang, F.; Wang, C. Multi-Disciplinary Optimization Design of Axial-Flow Pump Impellers Based on the Approximation Model. *Energies* **2020**, *13*, 779. [CrossRef]
6. Ni, P.; Wu, G.; Wang, Y.; Gao, Y. The application of slotted blade. *Mech. Des. Manuf.* **2014**, *3*, 90–92.
7. Li, J.; Tian, H.; Niu, Z. Study on the flow fields in a centrifugal fan with slots along the blade ends. *J. Eng. Thermophys.* **2009**, *30*, 2028–2030.
8. Wang, Y.; Mei, Y.; Liu, B.; Cao, Z. Numerical approach for turbine blade with trailing edge ejection. *J. Propuls. Technol.* **2002**, *23*, 315–317.
9. Wang, Y.; Liu, B.; Jiang, J.; Chen, Y. Experiment and numerical simulation investigation of turbine blade with trailing edge ejection. *J. Aerosp. Power* **2006**, *21*, 474–479.

10. Tang, X.; Huang, D.; Zhu, Z. Application of boundary layer control technology in centrifugal impeller. *Fluid Mach.* **1998**, *9*, 15–19.
11. Huang, D.; Bian, X.; Tang, X.; Lu, Y. Application of slotted technique on splitter blade in centrifugal fan. *J. Tsinghua Univ. (Sci. Technol.)* **1999**, *39*, 6–9.
12. Wang, Y.; Xie, S.; Wang, W. Numerical simulation of cavitation performance of low specific speed centrifugal pump with slotted blades. *J. Drain. Irrig. Mach. Eng.* **2016**, *34*, 210–215.
13. Ye, D. *Research on the Performance of Centrifugal Pump with Slotted Blades*; Jiangsu University: Zhenjiang, China, 2013.
14. Gao, Y.; Yan, X.; Li, J. Investigation on characteristics of flow and heat transfer in a turbine blade with trailing-Edge cutback. *J. Xi'an Jiaotong Univ.* **2018**, *52*, 31–40.
15. Xing, G. *Numerical Study on the Flow Field inside a Centrifugal Impeller with Slotted Blade*; Chongqing University: Chongqing, China, 2008.
16. Yuan, S. Advances in hydraulic design of centrifugal pumps. In Proceedings of the 1997 ASME Fluids Engineering Division Summer Meeting, Vancouver, BC, Canada, 22–26 June 1997; pp. 1–15.
17. Kergourlay, G.; Younsi, M.; Bakir, F.; Rey, R. Influence of splitter blades on the flow field of a centrifugal pump: Test-analysis comparison. *Int. J. Rotating Mach.* **2007**, *2007*, 1–13. [CrossRef]
18. Gölcü, M.; Pancar, Y. Investigation of performance characteristics in a pump impeller with low blade discharge angle. *World Pumps* **2005**, *468*, 32–40. [CrossRef]
19. Yuan, J.; Li, S.; Fu, Y. Splitter blades' effect on characteristics of centrifugal pump by orthogonal experiment. *Drain. Irrig. Mach.* **2009**, *27*, 306–309.
20. Uzol, O.; Camci, C. Aerodynamic loss characteristics of a turbine blade with trailing edge coolant ejection: Part2 External aerodynamics, Total pressure losses and predictions. *ASME J. Turbomach.* **2001**, *123*, 249–257. [CrossRef]
21. Yousefi, H.; Noorollahi, Y.; Tahani, M.; Fahimi, R.; Saremian, S. Numerical simulation for obtaining optimal impeller's blade parameters of a centrifugal pump for high-viscosity fluid pumping. *Sustain. Energy Technol. Assess.* **2019**, *34*, 16–26. [CrossRef]
22. He, X.; Zhang, Y.; Wang, C.; Zhang, C.; Cheng, L.; Chen, K.; Hu, B. Influence of critical wall roughness on the performance of double-channel sewage pump. *Energies* **2020**, *13*, 464. [CrossRef]
23. Wang, H.; Long, B.; Yang, Y.; Xiao, Y.; Wang, C. Modelling the influence of inlet angle change on the performance of submersible well pumps. *Int. J. Simul. Model.* **2020**, *19*, 100–111. [CrossRef]
24. Wang, C.; Shi, W.; Wang, X.; Jiang, X.; Yang, Y.; Li, W.; Zhou, L. Optimal design of multistage centrifugal pump based on the combined energy loss model and computational fluid dynamics. *Appl. Energy* **2017**, *187*, 10–26. [CrossRef]
25. Li, Y.; Hu, C. *Experimental Design and Data Processing*; Chemical Industry Press: Beijing, China, 2008.
26. Shi, W.; Zhou, L.; Lu, W.; Zhang, L.; Wang, C. Orthogonal test and optimization design of high-head deep-well centrifugal pump. *J. Jiangsu Univ. (Nat. Sci. Ed.)* **2011**, *32*, 400–404.
27. Wang, H.; Shi, W.; Lu, W.; Zhou, L.; Wang, C. Optimization design of deep well pump based on latin square test. *Trans. Chin. Soc. Agric. Eng.* **2010**, *41*, 56–63.
28. Yuan, S.; Zhang, J.; Yuan, J.; FU, Y. Orthogonal experimental study effect of main geometry factors of splitter blades on pump performance. *Drain. Irrig. Mach.* **2008**, *26*, 1–5.
29. Zhang, J.; Zhu, H.; Li, Y.; Yang, C. Optimization design of hybrid pump impeller based on orthogonal design method. *J. China Univ. Pet. (Ed. Nat. Sci.)* **2009**, *33*, 105–110.

© 2020 by the authors. Licensee MDPI, Basel, Switzerland. This article is an open access article distributed under the terms and conditions of the Creative Commons Attribution (CC BY) license (http://creativecommons.org/licenses/by/4.0/).

*Article*

# CFD-DEM Simulation for the Distribution and Motion Feature of Solid Particles in Single-Channel Pump

**Cheng Tang [1] and Youn-Jea Kim [2],***

[1]   Graduate School of Mechanical Engineering, Sungkyunkwan University, Suwon 16419, Korea;
     tangcheng@skku.edu
[2]   School of Mechanical Engineering, Sungkyunkwan University, Suwon 16419, Korea
*   Correspondence: yjkim@skku.edu

Received: 14 August 2020; Accepted: 18 September 2020; Published: 23 September 2020

**Abstract:** Since various foreign bodies can cause clogging and wear in single-channel pumps, considerable attention has been focused on the numerical study of solid-liquid flows in the single-channel pump. However, conventional numerical simulation cannot responsibly assess the significant effect of the particle material properties, inter-particle collision, and size on the pump. In consideration of the particle features and behaviors, the Computational Fluid Dynamics (CFD)-Discrete Element Method (DEM) coupling method was applied for the first time to simulate the solid-liquid flows in a single-channel pump. The results showed that the smaller particles possessed a wider velocity distribution range and velocity peak, while the larger particles exerted a greater contact force. Additionally, the pie-shaped particles had the most severe collisions, and spherical particles had the least in total. Furthermore, the hub and shroud wall suffered a minor contact force, but the blade and volute wall both sustained a considerable contact force. This paper could present some supply data for future research on the optimization of a single-channel pump.

**Keywords:** single-channel pump; CFD-DEM coupling method; particle features and behaviors; solid-liquid two-phase flows

---

## 1. Introduction

A single-channel pump is one type of sewage pump with a specially designed impeller. For the transport of various foreign objects in sewage, such as solids and fibers, it is intended that there is only one blade in the impeller. With the advantage of anti-clogging, a single-channel pump has been widely used for domestic and industrial sewage transport systems. However, compared with conventional centrifugal pumps, this particular design also brings some problems that cannot be ignored, such as hydrodynamic unbalance and relatively lower efficiency [1]. In order to improve the hydraulic performance of a single-channel pump, many numerical and experimental studies have focused on the inner flow field in the pump. Benra et al. [2,3] used PIV (Particle Image Velocimetry) technology to measure the periodic unsteady flow field in the impeller of a single-channel pump and compared the experimental results with numerical ones. The velocity field inside the pump investigated by PIV showed a good agreement with numerical one, which validated that using commercial Navier–Stokes solvers appears to be reliable. Nishi et al. [4] adopted LDV (Laser Doppler Velocimetry) and CFD (Computational Fluid Dynamics) methods to study the internal flow field of a single-channel pump. The results showed that the flow stagnation point on the work surface of the blade is far away from the inlet edge of the blade, resulting in a flow separation zone near the inlet edge. Auvinen et al. [5] used the OpenFOAM software to research the velocity field in a single-channel pump, and the results indicated that on the premise of high grid quality, the grid resolution had a slight influence on the pump performance prediction. Generally, these CFD methods above treated the fluid in a single-channel pump as a single-phase flow and obtained the flow characteristics of a liquid. However, those solid particles and fibers in sewage systems in fact have an essential effect on the pump performance. The particle features, material properties, shapes, and sizes can not get an accurate assessment in this way. Moreover, these CFD methods cannot provide a reliable estimate for the behaviors of particle-particle and particle-wall, such as collision, which may damage the pump.

Alternatively, the Discrete Element Method (DEM), a professional numerical method for processing discrete phases, has been widely applied in the calculation of particle systems. DEM can provide an accurate description of the particle property and calculate its dynamic motions in accordance with Newton's second law. Given this, the DEM coupled with the CFD method was proposed to improve the computational efficiency and numerical accuracy in two-phase flow by Tsuji et al. [6] and Kafui et al. [7]. On the other hand, with an increasing realization of the relevance of particle shapes in soil behavior and flow patterns [8–10], the modeling of particle shapes has been developed into various constructs instead of a single sphere in the DEM. Cleary et al. [8] compared the effect of particle shapes on granular flows in hopper discharge. Santamarina et al. [9] concluded that particle shape emerges as a significant parameter needing to be properly characterized and documented as part of every soil characterization exercise. Pena et al. [10] studied the influence of particle shape on the global mechanical behavior of dense granular media.

Although the CFD-DEM coupling method has incomparable superiority, it was seldom applied to study solid-liquid two-phase flow in a centrifugal pump. Taking specific particle features, including size and material property into consideration, Huang et al. [11] used the CFD-DEM coupling method to calculate the unsteady solid-liquid two-phase flow in a centrifugal pump and obtained the solid phase distribution. Liu et al. [12] studied the crystal particles' behavior in a centrifugal pump with the CFD-DEM coupling method, and the results showed that some crystal particles begin to get aggregation near the inlet edges of blades, forming the larger particles. Li et al. [13] used this method to simulate the reflux of different-diameter particles with the same volume concentration in a mixed pump. The results indicated that particles with a diameter of 50 mm cannot be refluxed, and they are likely to accumulate at the junction of the impeller and vanes.

Based on the above, in consideration of the particle features, the CFD-DEM coupling method was employed for the first time to study the distribution and motion of solid particles in a single-channel pump. The distribution and motion of the particles were analyzed by a DEM code inside the commercial CFD tool STAR-CCM+. The calculation factors in the shape and size with the material properties of the particles and the interactions of particle–particle, particle-wall, and particle-liquid. These results could be used to highlight some design guidance for designing a single-channel pump by obtaining the flow characteristics of solid particles.

## 2. Methodology

### 2.1. Computational Domain and Meshing

In order to verify the accuracy of the numerical method compared with the experimental data [14], a single-channel pump consisting of the inlet pipe, impeller, and volute was selected in this work to study the inner flow field. This single-channel pump model was kept the same as that used in experiments, and the basic parameters of the centrifugal pump are as follows: design discharge $Q = 210\ \text{m}^3/\text{h}$, head $H = 20$ m, rotational speed $n = 1800$ rpm, inlet diameter $D_1 = 50$ mm, impeller diameter $D_2 = 138$ mm, pump outlet diameter $D_3 = 110$ mm. The polyhedron meshes are generated in the entire computational domain, as shown in Figure 1a. In addition, for the turbulent flow simulation, an appropriate resolution of the near-wall region is needed, and 5 prism layers are created next to all the wall surfaces (see Figure 1b) to improve the accuracy of the flow solution. Table 1 shows the results of the mesh dependency test for the head of the pump. It shows that the pump head remains steady when the grid number exceeds 535,665. Given the great amount of CFD-DEM coupling calculations, relatively few meshes are significant for simulation efficiency in the subsequent optimization process. Hence, the optimal number of mesh cells was determined as 535,665. As a kind of real-mass particle, the acceleration of gravity ($g = 9.81$ m/s$^2$) is also taken into consideration, with its direction opposite to the $y$-axis.

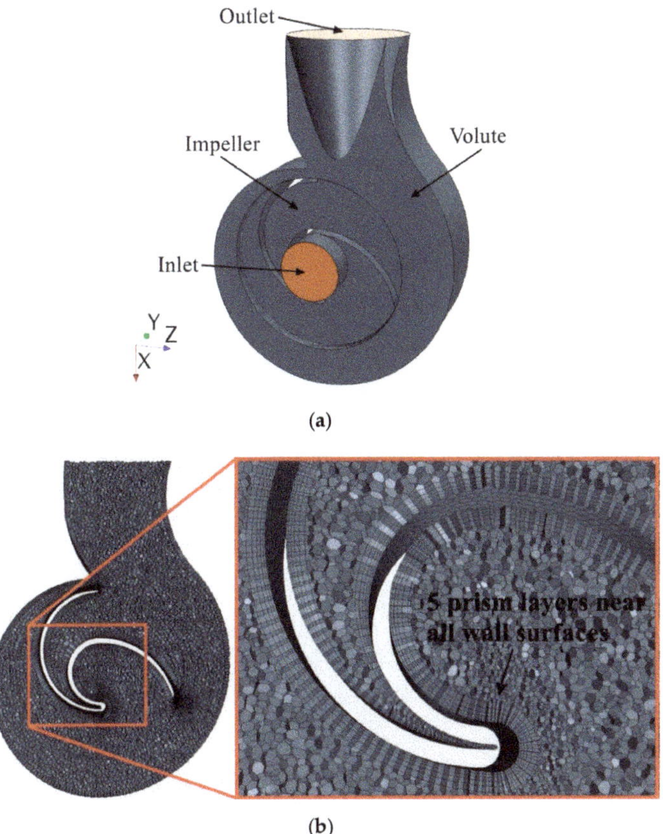


**Figure 1.** (**a**) Computational domain; (**b**) mesh cells.

**Table 1.** Mesh dependency test.

| Grid Number | Head [m] | Deviation [%] |
| --- | --- | --- |
| 250,493 | 18.09 | |
| 331,976 | 18.26 | 0.93 |
| 427,510 | 18.37 | 0.60 |
| 535,665 | 18.41 | 0.22 |
| 689,142 | 18.43 | 0.11 |

## 2.2. Governing Equations

The equations of continuity and momentum (Navier–Stokes equation) that govern the liquid phase are as follows in the tensor form:

$$\frac{\partial}{\partial t}\left(\alpha_f \rho_f\right) + \frac{\partial}{\partial x_j}\left(\alpha_f \rho_f u_j\right) = 0, \tag{1}$$

$$\frac{\partial}{\partial t}\left(\alpha_f \rho_f u_i\right) + \frac{\partial}{\partial x_j}\left(\alpha_f \rho_f u_i u_j\right) = -\frac{\partial p}{\partial x_i} + \frac{\partial}{\partial x_j}\left[\alpha_f \mu_{eff}\left(\frac{\partial u_i}{\partial x_j} + \frac{\partial u_j}{\partial x_i}\right)\right] + \alpha_f \rho_f \mathbf{g} + \mathbf{F}_s, \tag{2}$$

where $\rho_f$ is the fluid density, $u$ is the fluid velocity, $p$ is the pressure of the fluid, $\mu_{eff}$ is the effective viscosity, $x$ is the coordinates, $\mathbf{g}$ is the acceleration of gravity, and $\mathbf{F}_s$ is the drag force between the particles and the liquid. $\alpha_f$ represents the porosity around the particle, which can be calculated as:

$$\alpha_f = 1 - \sum_{i=1}^{n} V_{p,i}/V_{cell}, \tag{3}$$

where $V_{p,i}$ represents the volume of particle $i$ in a CFD cell, $n$ represents the number of particles inside the cell, $V_{cell}$ represents the volume of the cell.

Particle trajectory and particle behavior are controlled by the combined influence of hydrodynamic interactions and external force, which are described by Newton's laws of motion. According to DEM, the translational and rotational movements of a particle can be calculated with Newton's kinetic equation:

$$m\frac{d\mathbf{v}}{dt} = m\mathbf{g} + \sum \mathbf{F}_c + \mathbf{F}_{drag}, \tag{4}$$

$$\mathbf{I}\frac{d\boldsymbol{\omega}}{dt} = \sum \mathbf{T}_c + \mathbf{T}_f \tag{5}$$

where $\mathbf{F}_c$ represents the contact force, $\mathbf{F}_{drag}$ represents fluid drag force, m is the particle mass, and $\mathbf{I}$ is the moment of inertia of the particle. $d\mathbf{v}/dt$ is the translational acceleration of the particle, $d\boldsymbol{\omega}/dt$ is the angular acceleration of the particle, $\mathbf{T}_c$ is the contact torque, and $\mathbf{T}_f$ is the torque caused by the fluid.

### 2.3. CFD-DEM Coupling

During the two-way coupling adopted in this study, the modeling of the particle by DEM code is at the individual particle level, while the liquid flow by the CFD solver is at the computational cell level [15]. The coupling is initiated by analyzing the liquid flow based on CFD simulation. Once the iterative calculation in the CFD simulation is converged within a given time step, the DEM simulation then starts to calculate the instantaneous fluid–particle interaction force exerted on the individual particles from the CFD results. Thereafter, the new position and velocities of all the particles in the next fluid time step are determined. Inputting the updated particle information into the CFD solver, the forces on the fluid from particles are subsequently introduced into the liquid for the next loop through a series of momentum source terms [16].

### 2.4. Particle Model

In order to factor the size and shapes of particles, the simulation was divided into two groups. One group (mixed-size particles group) represents spherical particles with a certain range of diameters, and another (different-shaped particles group) represents three types of particles with different shapes. For the mixed-size particles group, the size of the particles varied from 1.0 to 5.0 mm in diameter in a random manner. For the different-shaped particles group, the shapes of particles in the simulation were set as a cylinder, pie, and sphere (see Figure 2a–c, respectively), which are the most three representative shapes of foreign matter in sewage. In addition, the cylindrical and pie-shaped particles were formed from rigid sphere clusters, where the adhesion force between spheres was set to infinity, and they would not separate during the simulation. The density of the particles was taken as 2000 kg/m$^3$, and the particles were selected in a diameter of 3.0 mm, while the particle flow rate was set to 4000 particles/s at the pump inlet. When the particles contact each other or solid boundaries, their momentum and energy are exchanged. This necessitates a DEM phase interaction model to describe the particle–particle and particle–wall interactions. In this model, the Hertz–Mindlin contact model [17] was adopted to solve the contact forces of particles, which were described by a soft-sphere model [18]. The Hertz–Mindlin contact model is the standard model that was used for describing the particle–particle and particle–wall interactions. The collision parameters used in the models are summarized in Table 2.

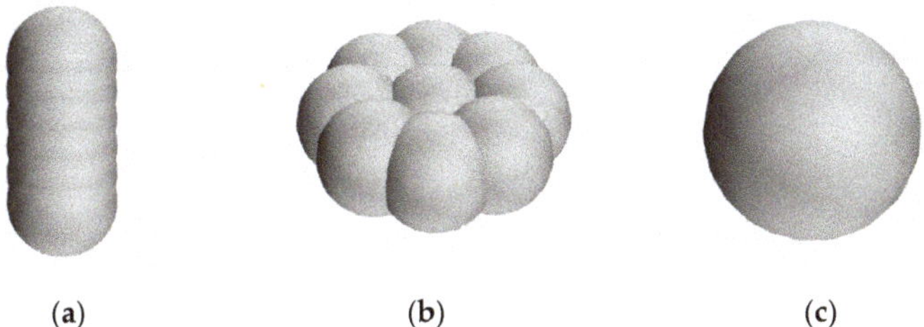

(a)        (b)        (c)

**Figure 2.** Particle shape models: (**a**) cylinder particle, (**b**) pie-shaped particle, (**c**) spherical particle.

**Table 2.** Collision parameters in the Hertz–Mindlin contact model.

| Collision Coefficient | Particle-Particle | Particle-Wall |
|---|---|---|
| Restitution | 0.5 | 0.5 |
| Static friction | 0.61 | 0.8 |
| Rolling friction | 0.01 | 0.01 |

### 2.5. Fluid Phase Setup

In this work, the transient fluid phase flow was analyzed through CFD simulations, based on solutions of the transient Reynolds averaged Navier–Stokes (RANS) equation. The realized two-layer k-ε model and "high-y+ wall treatment" in the STAR-CCM+ platform were employed for turbulence modeling. This wall treatment assumes that the near-wall cell lies within the region of the boundary layer. In the present simulations, the values of y+ ranged from 0.31 to 259.74, and the average y+ values for the impeller and volute surfaces are 80.38 and 109.64, respectively. The distribution of y+ for the impeller and volute is shown in Figure 3. The flow was assumed to be isothermal and incompressible with the properties of water ($\mu = 8.887 \times 10^{-3}$ Pa·s, $\rho = 998$ kg/m$^3$). The pump walls as well as particle surfaces were defined as no-slip walls, and the pump outlet was defined as a pressure outlet with $p = 1.0$bar. At the velocity inlet, a constant profile was specified. The absolute convergence criterion for the calculated residuals was set as $10^{-4}$ by default, together with monitoring the variations in average pressure at the pump inlet during the computations, as shown in Figure 4.

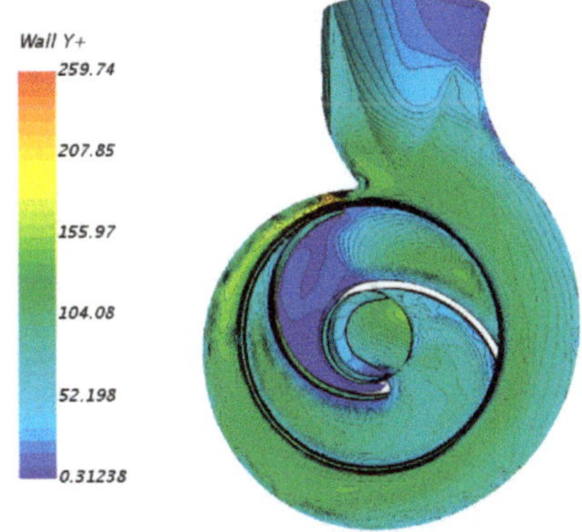


**Figure 3.** Distribution of y+ for the impeller and volute.

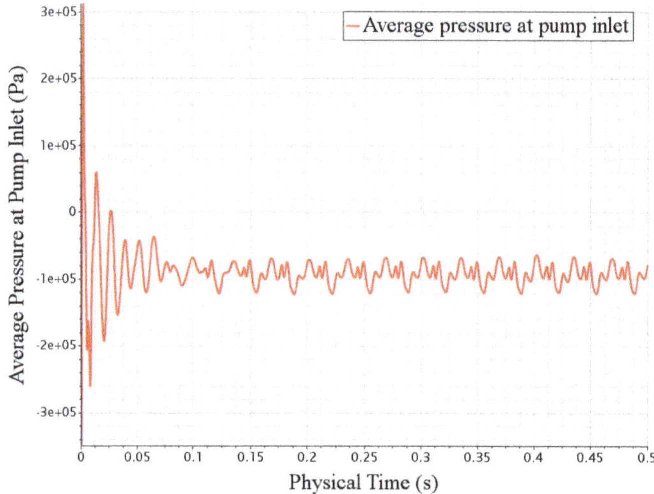



**Figure 4.** Variations in the average pressure at the pump inlet.

## 3. Results and Discussion

### 3.1. Validation

For verifying the reliability of the CFD-DEM coupling method, the numerical results of the total head were compared with the existing experimental data [14] for the single-channel pump, as shown in Figure 5. The flow rate, size, and rotational speed of the computational pump model are the same as those of the pump model 22 used in the experiment. The total computational head of the pump is 18.41m, showing a good agreement with that of the experimental one—19.85 m. Additionally, this indicates the feasibility of the CFD-DEM coupling method. Meanwhile, the head predicted by means of CFD-DEM was a little bit lower than the experimental one. This discrepancy could result from the clean water used in the experiment compared to the solid-liquid two-phase flow in the simulation.

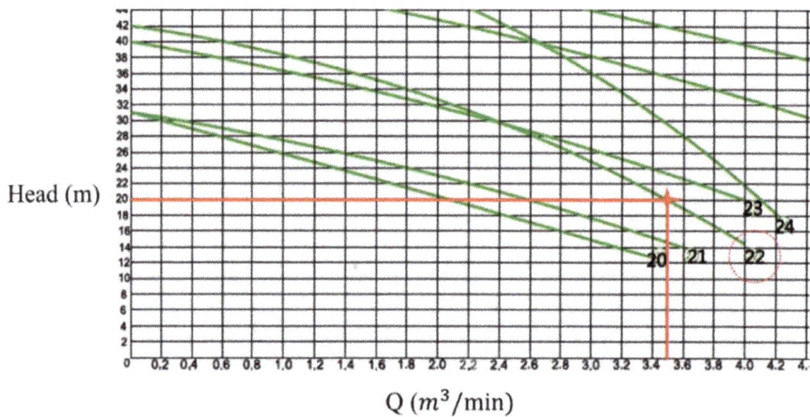



**Figure 5.** Comparison of the total head between the present simulation and the experimental data [12].

### 3.2. Mixed-Sized Particle

#### 3.2.1. Particle Trajectory and Distribution

Figure 6 shows the trajectory and size distribution of spherical particles whose diameter varied from 1.0 to 5.0 mm in a random manner (see Figure 6a). Remarkably, to reduce the impeller weight and unbalanced mechanical force, the blade is normally hollow (see Figure 6b). In this case, the hollow area is large enough that the flow status and particle motion in it cannot be ignored. In the impeller passage, the particle number density is different in different regions, with the characteristic of larger density in the vicinity of the inlet section and the blade pressure side.

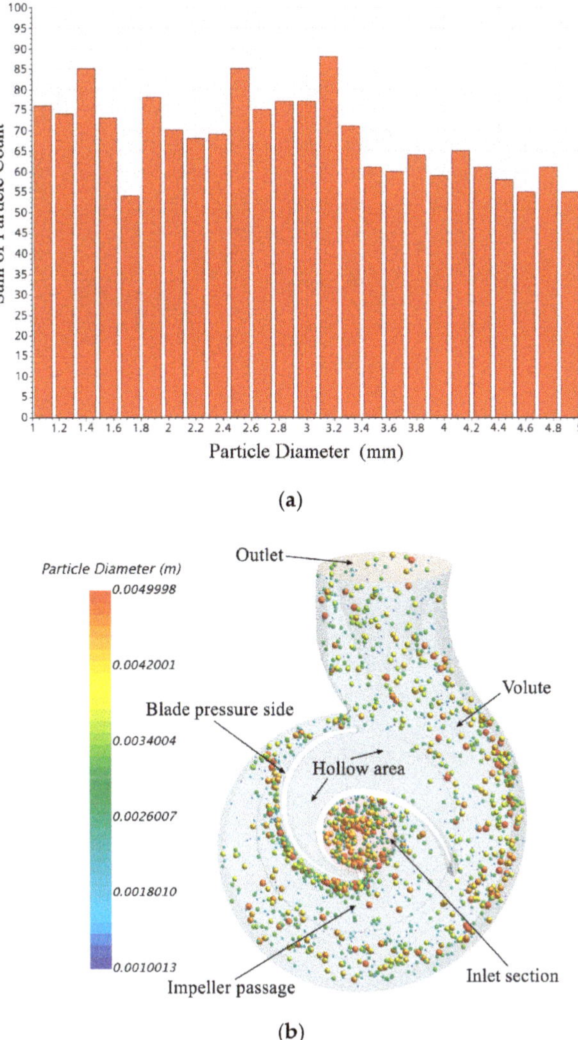


**Figure 6.** Particle size distribution (**a**) and trajectory (**b**) mixed-size particles.

Besides this, to distinguish the trajectories of different-sized particles, the larger particles are presented in warmer colors, while smaller particles are shown in cooler colors. In general, the particles tended to maintain a steady trajectory towards the volute that corresponds with the shape of the impeller blade. After entering the volute, the particles tended to cluster along the volute outside wall and move downstream towards the outlet. However, the smaller particles had a much more uniform distribution in the passage of the impeller and volute than the larger ones. The most likely cause of this was the lower inertia of the smaller particles.

### 3.2.2. Particle Velocity Distribution

Figure 7 shows the scatter plot of the particle velocity and diameter. It can be seen that the smaller size the particles possess, the greater the velocity distribution range and velocity peak they get. Generally, the majority of the particle velocity would be between 1 and 7.5 m/s. In this case, when the particle size is tiny enough, the maximum speed reaches 14.7 m/s. With the increase in particle size, the velocity distribution range becomes smaller. This can be manifested by the decrease in the maximum speed—that is, the smaller particles are more likely to acquire a higher velocity.

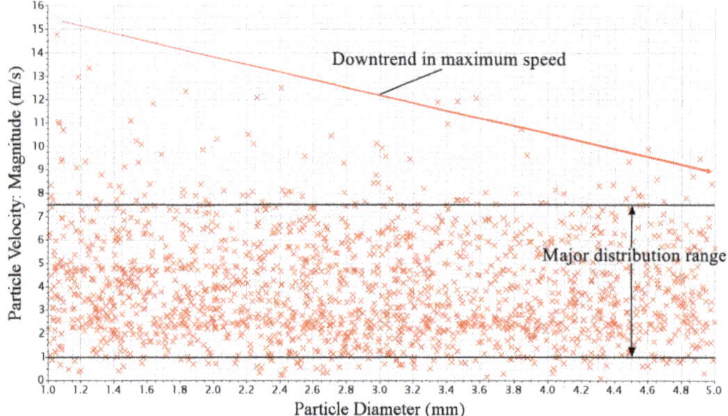


**Figure 7.** Scatter plot of particle velocity and diameter.

Figure 8 presents the variations in the volume-average velocities of both the liquid phase and solid phase in the pump domains. The results show that the average velocity of the liquid phase approached a nearly steady value of 4.7 m/s, while that of the solid phase reached a considerably lower value of 4.1 m/s. There are apparently slip velocities between the liquid and solid phase flows inside the pump. Furthermore, the average velocity of the liquid phase approached a steady value after t = 0.15 s, while the average velocity of particles spent 0.24 s attaining a stable value. Preliminary analysis suggests that the dominant influence on particles could be exerted by the liquid phase. Coupled with the inertia of the particles, the average velocity of the particles would take longer to reach a plateau.

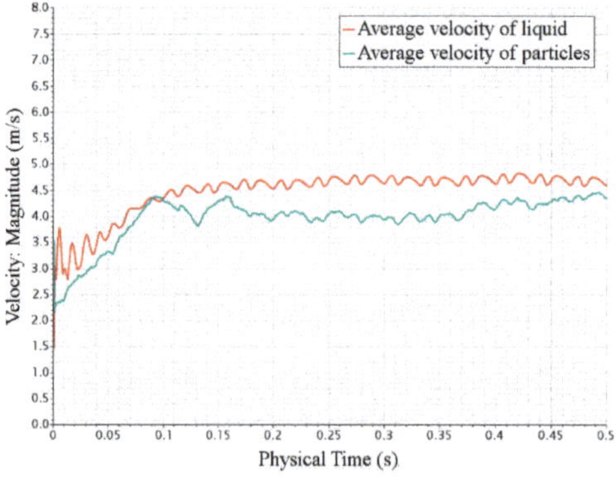


**Figure 8.** Variations in the volume-average velocities of the liquid and particles in flow domains.

### 3.2.3. Particle Contact Force Distribution

Figure 9 shows the histogram of the particle size to contact force at t = 0.5 s. It is obvious that the larger the particle size is, the greater the contact force particles would be. The small particles (diameter less than 2.4 mm) had a fairly weak but even contact force compared to the large particles. Although the contact force of the large particles increased significantly due to the greater surface area and mass, the force was rather uneven, especially for those particles whose diameters ranged from 2.4 to 4.0 mm.

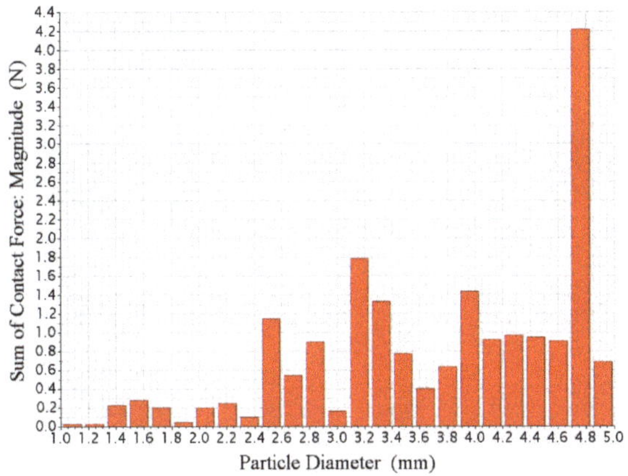

**Figure 9.** Histogram of the particle size to contact force.

### 3.3. Different-Shaped Particles

### 3.3.1. Particle Trajectory and Distribution

Figure 10 shows the trajectories and distribution of three distinct-shaped particles at the same particle flow rate. In the comparison of these three figures, it can be seen that the trajectory of cylindrical particles in the impeller was close to the pressure side (see Figure 10a). After acquiring a significant speed from the impeller, cylindrical particles tended to scatter in the volute passage. With regard to the trajectory of the pie-shaped particles, it was similar compared to that of the cylindrical particles but closer to the pressure side of the impeller passage (see Figure 10b). Meanwhile, the pie-shaped particles were distributed more uniformly than the cylindrical particles in the volute passage. In contrast, the distribution of spherical particles became most uniform in the impeller and volute compared to the other two cases (see Figure 10c). This may indicate that the motion of spherical particles would be mainly under the influence of the liquid phase so that they can disperse uniformly in the passage.

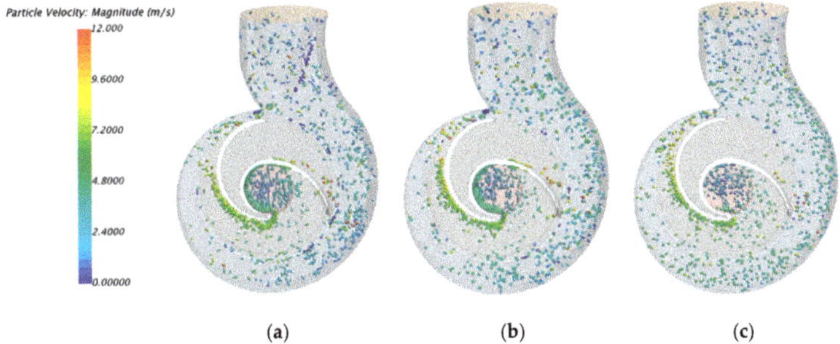


**Figure 10.** Trajectory and distribution of three distinct-shaped particles (**a**) cylinder practice; (**b**) pie-shaped particle; (**c**) spherical particle.

### 3.3.2. Velocity Field of Liquid Phase

Figure 11 shows the relative velocity distribution of the liquid phase at the mid-span of the impeller for the three shape cases. It can be seen that there is little difference in the relative velocity fields of the three cases, which may reveal the subtle effect on liquid exerted by particles with different shapes under this particle flow rate. Additionally, it can be clearly seen that a wide low-velocity region is generated in the vicinity of the pump inlet. This can be explained by the fact that the inlet angle of the blade is pretty small, causing stagnation points both on the pressure side and suction side near the inlet edge. Thus, a certain range of flow separation occurs along the downstream direction of the inlet edge. Simultaneously, there is a vortex field in the hollow area of the blade, causing a backflow at the interface of the impeller volute. These second flows would significantly impact the pump performance.

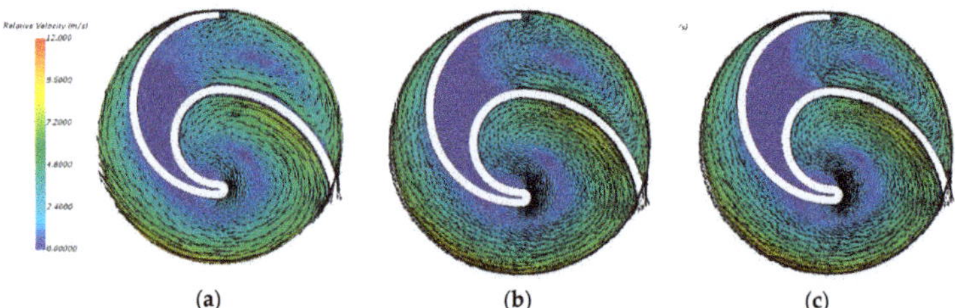

**Figure 11.** Relative velocity field of the liquid phase at the mid-span of the impeller for the three shape cases: (**a**) cylinder particle, (**b**) pie-shaped particle, (**c**) spherical particle.

On the other hand, compared with the results of the particle distribution, it can be reasoned that the liquid phase might possess the dominant influence on particles. Especially for those particles in the hollow area of the blade, the backflow would carry the particles back to the hollow area.

### 3.3.3. Collision Between Particle and Wall

The following figure illustrates the particle–wall contact force for three shape cases (refer to Figure 12). It is clear that the pie-shaped particles have the most serious collision, and spherical particles have the least in total. When it comes to the collision of particles with hub and shroud, it can be obtained that both the hub and shroud suffer a minor contact force, whose value was under around 1.0 N, by all the three types of particles. On the contrary, the blade and volute wall both sustained considerable contact force from particles, especially from the cylindrical and pie-shaped particles. This indicates that these solid particles may cause severe wear problems on the blade and volute areas. Hence, in the further optimization of the single-channel pump, the blades and volute can be designed to have more wear-resistance than the other parts. Moreover, before entering the pump, the particles or foreign bodies could be made as round as possible to reduce wear.

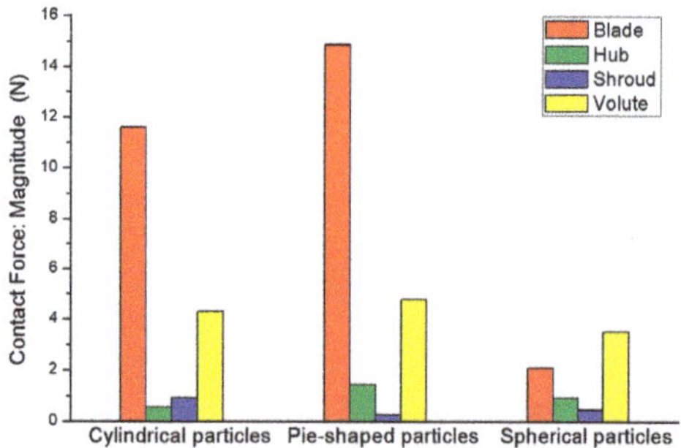

**Figure 12.** Particle–wall contact force for the three shape cases.

## 4. Conclusions

In this paper, the CFD-DEM coupling method was applied to study the distribution and motion features of various-sized and three different-shaped particles in a single-channel pump. The results are compared with the experimental ones, which prove that the CFD-DEM method used in this paper is reliable. The conclusions obtained from this research are as follows:

1.  Particles tended to maintain a steady trajectory towards the volute that corresponds with the shape of the impeller blade. The smaller particles had a much more uniform distribution in the passages of the impeller and volute than the larger ones.
2.  The smaller-sized particles possessed a greater velocity distribution range and velocity peak but a smaller contact force compared with the larger particles. Besides this, there were apparent slip velocities between the liquid and solid-phase flows inside the pump.
3.  The trajectories of the cylindrical and pie-shaped particles in the impeller were close to the pressure side. However, the spherical particles were dispersed more uniformly in the impeller and volute than the other two cases.
4.  The pie-shaped particles had the most severe collisions, and the spherical particles had the least in total. The hub and shroud wall suffered a minor contact force, but the blade and volute wall both sustained a considerable contact force.
5.  In order to reduce wear, the blades and volute can be designed to have more wear resistance than other parts. Moreover, before entering the pump, the particles or foreign bodies can be made as round as possible to reduce wear.
6.  According to the limited particle model and flow parameters in this research, the other shapes and rigidity of particles with different fluid viscosities and other fluid properties could be factored into the CFD-DEM coupling method for future research.

**Author Contributions:** Conceptualization, C.T. and Y.-J.K.; methodology, C.T. and Y.-J.K.; software, C.T.; validation, C.T. and Y.-J.K.; formal analysis, C.T.; investigation, C.T.; resources, C.T. and Y.-J.K.; data curation, C.T.; writing—original draft preparation, C.T. and Y.-J.K.; writing—review and editing, C.T. and Y.-J.K.; supervision, Y.-J.K. All authors have read and agreed to the published version of the manuscript.

**Funding:** This work was supported by the Korea Agency for Infrastructure Technology Advancement (KAIA) grant funded by the Ministry of Land, Infrastructure and Transport (Grant 20CTAP-C157760-01).

**Conflicts of Interest:** The authors declare no conflict of interest.

## References

1.  AOKI, M. Instantaneous interblade pressure distributions and fluctuating radial thrust in a single-blade centrifugal pump. *Bull. JSME* **1984**, *27*, 2413–2420. [CrossRef]
2.  Benra, F.; Dohmen, H. Numerical and experimental evaluation of the time-variant flow field in a single-blade centrifugal pump. In Proceedings of the 5th International Conference on Heat Transfer, Fluid Mechanics and Thermodynamics, Sun City, South Africa, 1–4 July 2007.
3.  Benra, F.-K.; Dohmen, H.J. Investigation on the time-variant flow in a single-blade centrifugal pump. In Proceedings of the 5th WSEAS International Conference on Fluid Mechanics (FLUIDS'08), Acapulco, Mexico, 25–27 January 2008.
4.  Nishi, Y.; Matsuo, N.; Fukutomi, J. A study on internal flow in a new type of sewage pump. *J. Fluid Sci. Technol.* **2009**, *4*, 648–660. [CrossRef]
5.  Auvinen, M.; Ala-Juusela, J.; Pedersen, N.; Siikonen, T. Time-accurate turbomachinery simulations with Open-Source®CFD: Flow analysis of a single-channel pump with OpenFOAM®. In Proceedings of the ECCOMAS CFD, Lisbon, Portugal, 14–17 June 2010.
6.  Tsuji, Y.; Kawaguchi, T.; Tanaka, T. Discrete particle simulation of two-dimensional fluidized bed. *Powder Technol.* **1993**, *77*, 79–87. [CrossRef]
7.  Kafui, K.; Thornton, C.; Adams, M. Discrete particle-continuum fluid modelling of gas–solid fluidised beds. *Chem. Eng. Sci.* **2002**, *57*, 2395–2410. [CrossRef]

8.	Cleary, P.W.; Sawley, M.L. DEM modelling of industrial granular flows: 3D case studies and the effect of particle shape on hopper discharge. *Appl. Math. Model.* **2002**, *26*, 89–111. [CrossRef]

9.	Santamarina, J.; Cho, G.-C. Soil behaviour: The role of particle shape. In *Advances in Geotechnical Engineering: The Skempton Conference: Proceedings of a Three Day Conference on Advances in Geotechnical Engineering, Organised by the Institution of Civil Engineers and Held at the Royal Geographical Society, London, UK, on 29–31 March 2004*; Thomas Telford: London, UK, 2015; pp. 604–617.

10.	Pena, A.; Garcia-Rojo, R.; Herrmann, H.J. Influence of particle shape on sheared dense granular media. *Granul. Matter* **2007**, *9*, 279–291. [CrossRef]

11.	Huang, S.; Yang, F.; Su, X. Unsteady Numerical Simulation for Solid-Liquid Two-Phase Flow in Centrifugal Pump by CFD-DEM Coupling. *Sci. Technol. Rev.* **2014**, *27*, 15.

12.	Liu, D.; Tang, C.; Ding, S.; Fu, B. CFD-DEM Simulation for Distribution and Motion Feature of Crystal Particles in Centrifugal Pump. *Int. J. Fluid Mach. Syst.* **2017**, *10*, 378–384. [CrossRef]

13.	Yuanwen, L.; Shaojun, L.; Xiaozhou, H. Research on reflux in deep-sea mining pump based on DEM-CFD. *Mar. Georesour. Geotechnol.* **2020**, *38*, 744–752. [CrossRef]

14.	Lee, J.-G.; Kim, Y.-J. Effect of The Impeller Discharge Angle on the Performance of a Spurt Vacuum Pump. *Appl. Sci. Converg. Technol.* **2017**, *26*, 1–5. [CrossRef]

15.	Chu, K.; Yu, A. Numerical simulation of complex particle–fluid flows. *Powder Technol.* **2008**, *179*, 104–114. [CrossRef]

16.	Huang, S.; Su, X.; Qiu, G. Transient numerical simulation for solid-liquid flow in a centrifugal pump by DEM-CFD coupling. *Eng. Appl. Comput. Fluid Mech.* **2015**, *9*, 411–418. [CrossRef]

17.	Mindlin, R.D. Compliance of elastic bodies in contact. *J. Appl. Mech. ASME* **1949**, *16*, 259–268.

18.	Hertz, H. Ueber die Berührung fester elastischer Körper. *J. Für Die Reine Und Angew. Math.* **1882**, *1882*, 156–171.

© 2020 by the authors. Licensee MDPI, Basel, Switzerland. This article is an open access article distributed under the terms and conditions of the Creative Commons Attribution (CC BY) license (http://creativecommons.org/licenses/by/4.0/).

*Article*

# Reduction of Entrained Vortices in Submersible Pump Suction Lines Using Numerical Simulations

**Virgel M. Arocena, Binoe E. Abuan, Joseph Gerard T. Reyes, Paul L. Rodgers and Louis Angelo M. Danao ***

Department of Mechanical Engineering, University of the Philippines, Diliman, Quezon City 1101, Philippines; vmarocena@up.edu.ph (V.M.A.); beabuan@up.edu.ph (B.E.A.); jtreyes2@up.edu.ph (J.G.T.R.); paul112464@yahoo.com (P.L.R.)
* Correspondence: louisdanao@up.edu.ph

Received: 19 August 2020; Accepted: 19 November 2020; Published: 23 November 2020

**Abstract:** Pump intake structure design is one area where physical models still remain as the only acceptable method that can provide reliable engineering results. Ensuring the amount of turbulence, entrained air vortices, and swirl are kept within acceptable limits requires site-specific, expensive, and time-consuming physical model studies. This study aims to investigate the viability of Computational Fluid Dynamics (CFD) as an alternative tool for pump intake design thus reducing the need for extensive physical experiments. In this study, a transient multiphase simulation of a 530 mm wide rectangular intake sump housing a 116 m$^3$/h pump is presented. The flow conditions, vortex formation and inlet swirl are compared to an existing 1:10 reduced scaled physical model test. For the baseline test, the predicted surface and submerged vortices agreed well with those observed in the physical model. Both the physical model test and the numerical model showed that the initial geometry of the pump sump is unacceptable as per ANSI/HI 9.8 criteria. Strong type 2 to type 3 submerged vortices were observed at the floor of the pump and behind the pump. Consequently, numerical simulations of proposed sump design modification are further investigated. Two CFD models with different fillet-splitter designs are evaluated and compared based on the vortex formation and swirl. In the study, it was seen that a trident-shaped splitter design was able to prevent flow separation and vortex suppression as compared to a cross-baffle design based on ANSI/HI 9.8. CFD results for the cross-baffle design showed that backwall and floor vortices were still present and additional turbulence was observed due to the cross-flow caused by the geometry. Conversely, CFD results for the trident-shaped fillet-splitter design showed stable flow and minimized the floor and wall vortices previously observed in the first two models.

**Keywords:** intake structures; physical hydraulic model; free surface flow; free surface vortices; vertical pump; CFD

## 1. Introduction

Large-scale axial-flow and mixed flow pumps are typically used for a variety of purposes such as irrigation, drainage, water treatment, thermal and nuclear power plants, steelworks, petrochemical plants and even in the shipbuilding industry. Developed specifically for large-capacity low-head applications, these pumps' operating conditions are highly influenced by flow conditions in the intake structures. Unfortunately, the proper design of these intake structures is also the most overlooked aspect when designing a pumping station. Poorly designed intake structures are those that fail to control any possible harmful formation of free-surface and submerged vortices. These vortices tend to result in energy loss, reduced flow rate, vibration, surging, structural damage, cavitation and safety hazards. A 3% to 4% air entrainment due to these vortices may produce a small but continuous decrease in pump efficiency. Fundamentally, for a 1% drop in efficiency, only a small amount of entrained air is necessary [1]. A loss in efficiency by this relatively small rate may lead to losses in profit which, in a few years, can exceed the initial capital cost of the pump [2].

Specifically, for pump bays, vortices are caused by the swirl that is formed at the hydraulic intakes due to a non-uniform approach flow. This swirl can be defined as the tendency of the fluid to move with a twisting or rotating motion. By itself, this swirling motion is oftentimes unavoidable and is not considered an engineering problem. Rather it is the degree of this swirling motion that determines the detrimental effect and possibility of vortex formation.

For pump installations experiencing these problems, the most commonly suggested solution is to increase the submergence of the vertical pump's inlet bellmouth. In most cases, this solution often results in largely oversized and expensive structures. Since the cost of a typical pump structure grows directly with its size, site excavation issues and economic constraints requires pump intake size to be kept as small as possible. This naturally limits the application of this solution. Reducing the pump speed is still another common remedy. Although this implies sacrificing pump efficiency by operating the structure below its rated flow capacity. This in turn increases the long-term operating costs.

In general, no amount of engineering can produce an ideal design that ensures that the intake will be free from any swirl or vortices. As a solution, the American National Standard Institute Hydraulic Institute Standard for Intake Design (ANSI/HI) [3] established strict conditions on when pump stations designs should undergo physical model testing prior to construction or rehabilitation. Among the conditions that necessitate a physical model test are when:

- an individual pump or total station flow exceeds 9085 m$^3$/h (40,000 gpm) and 22,710 m$^3$/h (100,000 gpm) respectively;
- intake or pump bay designs that deviate from standards, pump compartments with non-symmetrical approach flows;
- pump stations whose operation is critical and prolonged outages due to maintenance are unacceptable.

These physical models allow visual observation of the flow as well as collection of data such as velocity distribution, pressure gradients, depth of flow, and prerotation. Such a test presents a reliable method to identify unacceptable flow patterns. Unfortunately, these physical models are also site-specific, time-consuming and costly to perform and often add very low economic value to the project. Therefore, development of alternative tools or methods for evaluating sump performance is highly demanded in the pump design industry. One such tool that deserves attention is the numerical simulation of computerized models representing the system that needs to be studied. Such simulation, termed Computational Fluid Dynamics (CFD), uses the general fluid flow equations to predict the flow field, turbulence, mass transfer, and other related hydraulic phenomena. In contrast to the cost of conducting a scaled physical model experiment, the lower operating cost together with the current advances in numerical simulations position CFD as an ideal alternative tool for pump designers.

For the past few decades, the ever lower cost coupled with the advancement in computing technology had constantly driven the pump industry to look into Computational Fluid Dynamics as an alternative means of developing better, least expensive, and more reliable pumps [4–6]. CFD coupled with stress analysis had been efficiently used in the design of various pump components like shafts, seals, impellers, diffusers and casings among others. But for intake structure design, physical model experiments had still remained as a primary mandatory requirement as per existing codes and standards. The capability of CFD to consistently provide information about the vortex strength and temporal variation in these structures had remained a debatable topic. Also, additional difficulties associated with modelling free surfaces and predicting vortex phenomena oftentimes forces designers and CFD analysts to avoid the use of multiphase flow models and instead enforce a free-slip wall on the free-surface. For this reason, various investigations and researches had been conducted aiming to validate the accuracy and suitability of numerical models as compared to physical model studies.

Among the early studies in using CFD for the investigation of flow problems at pump intakes were those conducted by Constantinescu and Patel [7]. A numerical model of a simple water-intake bay was developed to simulate the three-dimensional flow field and to study the formation of free-surface and submerged vortices. The analysis solves the Reynolds-averaged Navier-Stokes equation with a two-layer $k - \varepsilon$ turbulence model. Symmetric vortex formations were observed in the numerical solution. It was highlighted that this symmetry is rarely observed in reality and was only present in the numerical results due to the idealized flow and boundary conditions. It was reported that the CFD model was able to predict in detail the location, size and strength of the vortices.

Later, as part of an extensive experimental study of pump-bay flow phenomena, Rajendran and Patel [8], conducted a simplified laboratory experiment specifically to validate the numerical model presented by Constantinescu and Patel. A model of a 0.003 m$^3$/s rectangular pump sump was constructed and velocity fields were measured using particle-image velocimetry (PIV). Comparing the CFD results, it was confirmed that the results for the position, number and overall structure for both the free-surface and subsurface vortices were in good agreement with the physical model. However, with the exception of the strongest vortex, the calculated vortices were more diffused and less intense than the vortices observed in the experiment.

Since both the physical experiments [8] and the numerical model [7] was conducted using simplified laboratory model, several limitations were noted. Among these are:

- no inlet suction bell was used in the experimental sump, instead a straight vertical column was used;
- the flow condition was limited to a very low Reynolds number ($Re = 60{,}000$);
- the numerical model was not able to handle flows with high Reynolds numbers;
- the intake column was modelled using zero-thickness walls since the numerical model was not able to handle complicated geometries (suction bellmouths);
- the numerical results did not report neither the velocity distribution nor the swirl angle at the pipe column which are both vital for inlet structure design.

To address these limitations, Li et al. [9] conducted a CFD model study based on an actual water pump intake structure. Their study applied higher Reynolds number to mimic a more practical pump-station. The simulation involved a more complex intake bellmouth geometry based on the 1:10 undistorted model of Union Electric's Labadie Power Plant on the Missouri River near St. Louis, MO, USA. The model was based on the works of Lai et al. [10] utilizing finite-volume-based unstructured grid technology that allowed the use of flexible mesh cell shapes. Similar to the previous studies, the simulation solves the RANS equation with the $k - \varepsilon$ turbulence model with wall functions. Two incoming flow conditions, designated as "cross-flow" and "no-cross-flow" were simulated to eliminate the limitations present in the first study [8]. It was reported that the pertinent flow patterns in the forebay for both "no-cross-flow" and "cross-flow" conditions observed in the scaled model experiment were well captured by the numerical model. For "no-cross-flow" conditions, the calculated axial velocity at the throat of the suction bell showed good agreement with experimental data except for points near the pipe wall. Inversely, for "cross-flow" conditions, the steady state solution gave relatively low agreement with experiment data. As such, an unsteady-state solution is recommended for such scenario. Taking these issues into considerations, the study concluded that CFD may be used as a cost-effective tool for preliminary engineering designs.

Recognizing the impact of conducting physical model studies on the development cost of pumps, Okamura et al. [11], carried out a study on the accuracy and reliability of various CFD codes in predicting vortices in sumps. The assessment was carried out by validating the results obtained from current commercially available CFD codes like STAR-CD, ANSYS CFX, Virtual Fluid Systems 3D and SCRYU/TETRA against results from a physical sump model. The benchmarks were conducted under three different discharge conditions and submergence level. Due to the difference in software capability, the CFD models varied in grid structure and mesh density. The turbulence model also varied across all numerical model with STAR-CD using $k - \varepsilon$ RNG, $k - \varepsilon$ for CFX, and $k - \omega$ SST being used for SCRYU/TETRA. Point velocities from numerical results were compared from experiment results acquired through PIV and LLS. Stream lines and vortex core lines taken using video and still cameras were also compared to those obtained from the numerical models. It was concluded that some CFD codes are able to predict the vortex formation with enough accuracy for industrial applications [11]. The results for both the physical model and the CFD code agrees qualitatively in terms of velocity distributions in the intake bellmouth. However, the agreement is poor in terms of magnitude and distribution patterns for the vorticity.

Wicklein et al. [12] are among those who have successfully utilized steady state RANS models to optimize the design of a wastewater treatment plant influent pump station. The original proposed pump station design was developed using extensive scale physical model test to verify hydraulic performance. Unfortunately, subsequent changes to the pump station's operating condition required a revised influent sewer design. With the goal of evaluating the effect of the proposed upstream sewer changes, Wicklean et al. utilized CFD to verify and refine the hydraulic design of the proposed pump station. Numerical results showed that surface vortex formation was very dependent on geometry. For this reason, proposed modifications were simulated using CFD. The aim of which is improving sump performance by reducing the potential for vortices to develop, improve velocity distribution and reduced pre-swirl. For this pump station, CFD models were used for design optimization and later for additional changes at the time of construction. Satisfactory results were reported in using CFD highlighting its advantage over physical model studies. One major advantage being that results produced are digital and can be kept to investigate changes at time of construction or any point in the future.

Similarly, the use of CFD in evaluating pump performance is being continuously developed and investigated in line with advancements in numerical methods. One such study was made by Shukla and Kshirsagar [13] on a vertical axis, single stage centrifugal pump. The study compared the numerical results with those obtained from a physical model test of a pump rated at 0.508 m$^3$/s at 60 m head running at 1450 rpm with an impeller eye of 330 m. The multiphase flow was modelled using Eulerian approach while the mass transfer through cavitation used Rayleigh-Plesset equation. Standard $k - \varepsilon$ turbulence model with scalable wall functions was selected for the numerical analysis. NPSH results obtained from Ansys CFX showed a good matching trend with those obtained from the physical experiment. Furthermore, the numerical model was able to predict the formation and growth of vapor bubbles on the impeller making CFD a viable tool in predicting pump performance deterioration caused by cavitation

Nagahara et al. [14] investigated a detailed velocity distribution around the submerged vortex cavitation in a pump intake by means of PIV utilizing a pressurized tank to control the main inlet velocity. They believed that there have been no quantitative data concerning submerged vortex cavitation in particular. Thus, it is necessary to investigate its effects to establish reliable guidelines for the design of trouble-free pumps and intakes.

Most previous studies in predicting vortex formation deals with either treating the calculation domain as a single phase model applying symmetry boundary condition on the free surface [15,16] or through high-fidelity multiphase simulations requiring highly intensive calculations [17,18] which becomes unacceptable for industrial application. The goal of this paper is to provide a practical CFD method to augment existing pump intake design procedures in terms for predicting and minimizing vortex formation and swirl. The method should be optimized in terms of computational efforts but with sufficient accuracy as compared with physical model test results.

In this study, an implicit volume of fluid (VOF) multiphase numerical model of a 530 mm wide rectangular intake sump housing a 116 m$^3$/h pump with a 260 mm diameter inlet bellmouth is analyzed using ANSYS Fluent. The flow conditions, vortex formation and inlet swirl are compared to the results obtained from a physical model test. The aim of this study is to validate the use of CFD as an alternative method of evaluating flow behavior in pump sumps concentrating mainly on vortex prediction, anti-vortex devices and related prerotation.

The novelty of this work is the investigation of various sump floor configurations showing their effects on reducing the entrained vortices developed within the suction lines.

## 2. Experimental Setup

In conducting any CFD simulations, it is vital that numerical accuracy be demonstrated by either comparing results to a well-established analytical model or to the results acquired from conducting a physical model experiment. In this paper, photographs and plot data taken from the baseline test of a 1:10 undistorted scale hydraulic model is presented as reference in evaluating the numerical results. These data are lifted from a recent pump project supplying cooling water to a thermal power plant and are presented here with implicit permission from Hitachi Plant Technologies, Ltd. PBO. (Makati City, Philippines). The prototype model consisted of two 5.3 m wide by 14 m long pump bay and one 2.0 m wide auxiliary channel. The 5.3 m wide pump bay feeds two vertical axis mixed flow pumps each rated at 36,700 m$^3$/h with a total dynamic head (TDH) of 15 m. The auxiliary channel feeds a smaller auxiliary pump rated at 4400 m$^3$/h against a TDH of 12.5 m. For this study, the focus will be on the main pumps since these are the crucial components for this pump station. For open channel flows such as these sump, gravity and inertial forces play a more dominant role than viscous or turbulent shear forces. As such, dynamic similarity during the test was maintained by keeping the Froude number ($Fr = 0.38$) between the model and the prototype constant. Furthermore, ANSI/HI recommends a minimum value for both the Reynolds number ($Re$) and Weber number ($We$) to avoid any scale effects and surface tension effects in the model. A minimum $Re$ is necessary in order to ensure that the flow condition in the model is as turbulent as that of the prototype. While a minimum $We$ is recommended in order to avoid surface tension effects particularly in fully developed stage where the vortices start to draw in air from the surface. Table 1 shows the $Re$ and $We$ numbers calculated at the 260 mm diameter suction bell. These values justify and prove that the selected scale (1:10) is sufficient for the physical model test. Using this scale, the model capacity based on the 36,700 m$^3$/h maximum flow capacity of the prototype is taken as 116.07 m$^3$/h.

**Table 1.** Calculated Reynolds number and Weber number at the suction bell.

| Criteria | Value | Minimum | Acceptable |
|----------|-------|---------|------------|
| $Re$ | $1.44 \times 10^5$ | $6 \times 10^4$ | ok |
| $We$ | $1.37 \times 10^3$ | 240 | ok |

In order to protect proprietary data, a 3D representation of the hydraulic test model is presented in Figure 1 in lieu of a picture of the actual setup. A centrifugal pump was used to recirculate water through a diffuser that spreads the flow over the entire width of the forebay. Orifice plates and control valves were installed to control the individual pump flows as well as the total model flow. Straightening devices were installed in the model head-box representing the trash racks and travelling screens in the prototype. This is to ensure that flow entering from the forebay is as uniform and as steady as possible. Typically for hydraulic model studies, impeller induced flows are not considered. This is mainly due to the fact that the main focus of the test is to verify the flow conditions and vortex formations in the sump as the fluid enters the pump and not the performance of the pump. Hence in this case, the 116 m$^3$/h vertical axis semi-axial pump is represented using a 130 mm diameter vertical pipe with a 260 mm diameter suction bellmouth. The bellmouth was fabricated from transparent polyvinyl chloride to facilitate visual observation.

All other aspects of the hydraulic model test comply with the latest ANSI/HI 9.8 test standards in terms of acceptance criteria, scale selection, data collection and instrumentation. Specifically, the pump model flow rates were determined using an ASME standard orifice meter with an accuracy of ±2%. The water level in the pump sump were recorded with a staff gauge referenced from the sump floor with a minimum accuracy of 3-mm. Velocity probe with a repeatability of ±2% was installed to measure point velocities at specific points along the throat of the suction bell. Typically, measurements of swirl in sump model test are done through visual inspection. The number of revolutions made by the swirl meter are counted and related to the flow rate. For the physical model test discussed in this paper, a swirl meter consisting of four straight vanes mounted on a shaft with low friction bearings was installed at a height of four suction pipe diameters downstream from the bell mouth to measure the level of pre-swirl as flow enters the pump.

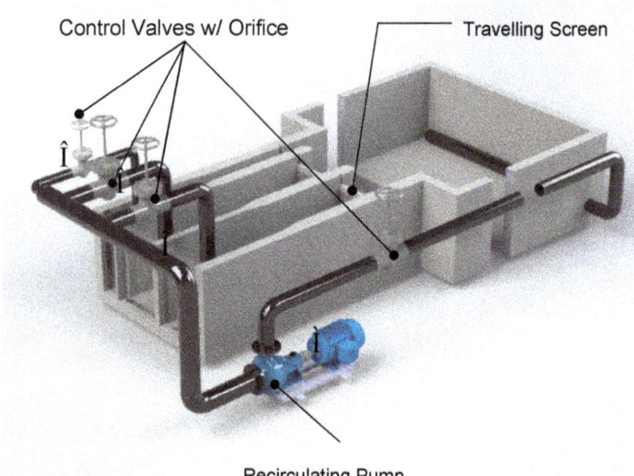


**Figure 1.** 3D rendering of the 1:10 undistorted scale hydraulic model test.

The swirl angle is calculated by:

$$\theta = \tan^{-1}\left(\frac{\pi d n}{u}\right) \tag{1}$$

where $u$ = average axial velocity, $d$ = diameter of the pipe in which the swirl meter is installed and $n$ = revolutions per second of the swirl meter

The design specifications for the physical model is outlined in Table 2.

**Table 2.** Flow parameters for the physical model.

| Parameter | Value |
|---|---|
| Flow rate (m$^3$/h) | 116 |
| Suction Bell Diameter (mm) | 260 |
| Bellmouth Throat Diameter (mm) | 125.7 |
| Bellmouth Submergence (mm) | 418.4 |
| Floor Clearance (mm) | 130 |

## 3. Numerical Model

In order to validate the suitability of CFD in predicting flow patterns and vortex formation in pump sumps, the conditions used for the hydraulic model must be exactly replicated in the numerical model. As such, the numerical models used in this paper are also a 1:10 undistorted reduce scale model of the prototype. Parameters such as flow capacity and water level were also based on the variables used for the hydraulic model test. The only difference is that for the CFD simulation, only one pump bay was modelled due to the symmetrical layout of the sump. The auxiliary pump is operated separately during both the hydraulic model test and during normal operating conditions. This means that the auxiliary pump will not cause any cross-flow during the test or during normal operating conditions. Also, it is expected that both sumps will perform similarly during operation justifying the use of only one pump compartment for the CFD analysis. The total water volume was dimensioned to mirror the low water level (*LWL*) used in the hydraulic model test while the air volume above the free surface of the water was set at a height of 200 mm. The sump dimension and pump location used for the numerical model are shown in Figure 2. For all cases, the numerical solutions for the Reynolds-Averaged Navier-Stokes (RANS) equations were performed using ANSYS Fluent 2019. Turbulence flow properties were described using $k - \omega$ Shear Stress Transport model. The free surface is tracked by means of the Volume of Fluid (VOF) multiphase model.

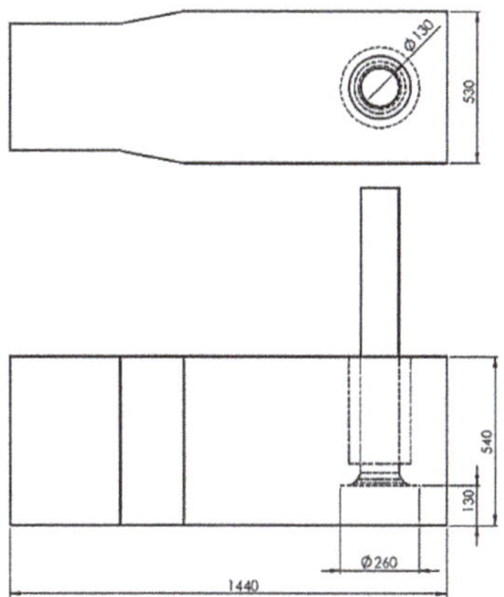

**Figure 2.** Pump sump model for numerical analysis.

### 3.1. Boundary Condition and Solver Parameters

Boundary conditions were calculated based in a total flow of 116 m$^3$/h and with a water level of 540 mm. A velocity flow inlet with negative velocity magnitude (outflow) was prescribed as outlet boundary condition at the end of the discharge pipe. The rectangular section serving as the entry point for the sump was prescribed as a pressure outlet boundary condition. Multiphase open channel condition is also prescribed on this surface with the pressure specification method set as free surface level. The free-surface level is set at 0.542592 m and the bottom level set to 0 m. Backflow pressure is specified as total pressure. The boundary condition for the air surface 200 mm over the water surface was also specified as pressure outlet boundary with zero backflow volume fraction indicating that only air can pass through this boundary. Figure 3 shows an overview of the boundary conditions as used throughout the analysis. No slip velocity conditions were used at the walls. Boundary roughness was not taken into consideration since walls were assumed to be smooth. For the purpose of this study, a constant value for density was specified for the entire model. The free-surface level was set as the reference pressure location (0.542952 m) and the specified operating density fixed as 1.225 kg/m$^3$. Calculations were carried out using Eulerian multiphase volume fraction method (VOF) transient conditions with water at 25 °C as the secondary phase and air as the primary phase. The effect of surface tension along the interface between each phase is added in the VOF model by specifying a constant surface tension coefficient (71.2 mN·m$^{-1}$). Flow is incompressible and isothermal with constant fluid properties. Turbulence was modelled using the $k - \omega$ shear-stress transport (SST). SST $k - \omega$ had been found to be suitable for numerical modelling of free-surface vortices [19,20] and exhibits better performance in predicting flows at walls and adverse pressure gradients as compared to other eddy-viscosity models [21]. The pressure-based coupled solver was applied. Second order discretization scheme were used for pressure, momentum, and turbulence equations. Converged solution from a steady state simulation was used for the initial conditions. The non-iterative time-advancement (NITA) scheme was used for temporal discretization. This scheme speeds up transient simulations by performing only a single outer iteration per time-step. Overall time-accuracy is preserved not by reducing the splitting error to zero but instead by maintaining it in the same order as the truncation error [22]. The initial time step has been as chosen as 0.001 s, small enough to ensure the correct vortex generation and convergence.

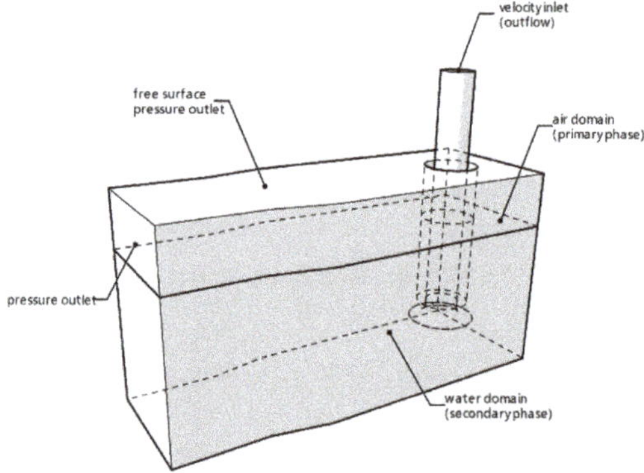


**Figure 3.** Overview of the boundary conditions and fluid domains for the numerical analysis.

### 3.2. CFD Mesh Generation

Of primary importance in this study is the flow pattern around areas with high velocity and high velocity gradient like the suction bellmouth and the free surface just above the suction pipe. This area is where vortices are expected to occur during operation thus focus is given to the mesh size and quality within these areas in order to ensure accurate results. The initial attempt to model the whole structure with a homogenous mesh size yielded a very large model size that easily exceeded the capacity of the current equipment used in this study. As such it was necessary for the model to be divided into four regions as shown in Figure 4. The first region extends from the backwall of the sump up to 220 mm upstream of the pump and covers the full height of the water and air domains. This region effectively covers the bellmouth, the throat and any region where the velocity gradient is expected to be high. The second and third regions contains the discharge pipe and the last region contains the area just upstream of the intake (i.e., forebay).

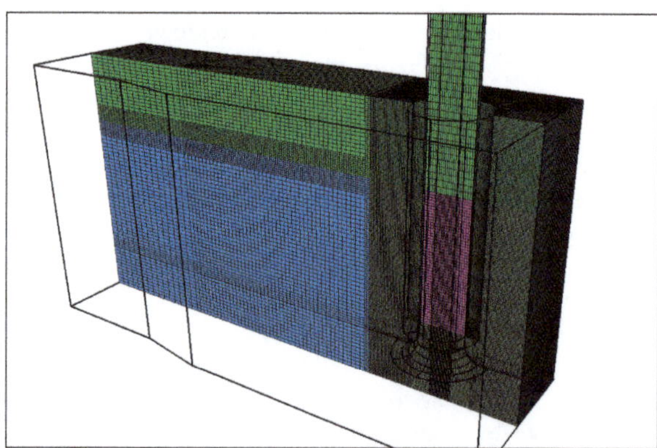

**Figure 4.** Cutaway view of the meshed model showing 4 regions with varying mesh densities to reduce calculation time.

In order to optimize computational time and study the effect of grid resolutions, a preliminary grid independence study had been conducted comparing results for three CFD models. Additionally, the grid independence study ensures that the results are due to the boundary conditions and physics used and not by the mesh resolution. Three hexahedral mapped meshed models labeled as fine, medium, and course mesh with 1.9, 2.4 and 3.2 million cells, respectively, were generated for this purpose. For the models used in this paper, the mesh refinement didn't follow the usual half/double element size since refining the mesh by a factor of 2 will result to an 8-fold increase in problem size which is unacceptable for engineering design purposes. The three CFD models differ in the first region which was modelled with three different mesh densities since this is the area where the analysis will be focused on. Since the final goal of this paper is to predict vortex formation, the 2.4 million mesh model was prepared first, keeping the element size between 0.8 to 2 mm near the suction bell and free surface. For the other regions with lower velocity gradients and where knowledge of the flow pattern was not so important in the analysis, a courser and similar mesh was applied so as to reduce overall computational time. Non-overlapping mesh interfaces were adapted in ANSYS Fluent to combine the three regions and independence between areas with different mesh densities were improved by using the automatic mesh refinement tool feature of the software. In this case, two to three mesh layers were automatically redefined providing an improved 4:1 cell face ratio between the master and the slave faces. A higher and more uniform element count could had been generated but it should be noted that this paper focuses on an optimized method that would result in reduced computational efforts in order for CFD to merit its use in industrial applications. A cross sectional view of the course, medium and fine meshed models are shown in Figure 5.

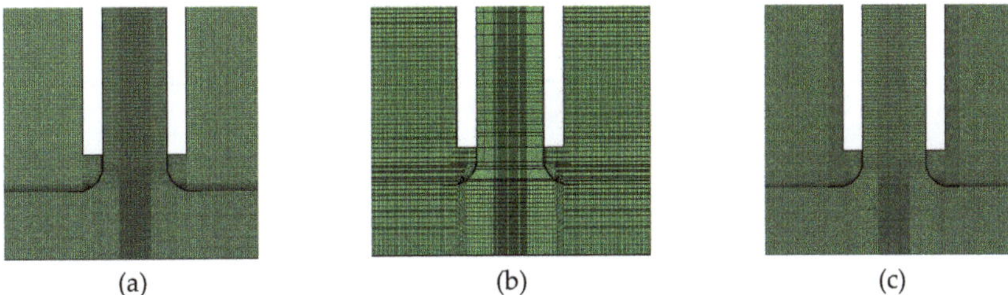

(a)           (b)           (c)

**Figure 5.** CFD Mesh for Grid Independence Study. (**a**) Coarse Mesh, (**b**) Medium Mesh, (**c**) Fine Mesh.

For the mesh independence study, it is crucial to have a prior knowledge of the flow phenomenon in order to apply the correct solution methodology. In this case, during the hydraulic model test, it was observed that the forebay was able to provide steady and uniform approach flow. Also, no significant free surface vortices with air-core were observed under the specified flow conditions. These observations coupled with the simple geometry of the sump makes it safe to consider a steady state solution for the mesh independence study. For all three numerical models, convergence was defined where the mean velocity and the amplitude of the fluctuating field does not vary for more than 1% for each iteration and the target residual errors kept below $10^{-4}$. Axial velocity ($z$) at several points along a line just below the entrance of the suction bell (el. = 0.12 m) perpendicular to the flow direction is plotted as shown in Figure 6. Additionally, in Figure 7, the velocity distribution along a central line that runs parallel to the flow direction and at 0.12 m elevation from the sump floor is plotted for all 3 numerical models. Based on these plots, the simulation results qualitatively show the same structure with the point velocities showing a maximum variation of 0.5% from the average velocity value occurring at the center of the suction bell as expected. The resemblance of these plots for all three cases shows that these simulations can be considered relatively grid-independent based on the presented mesh densities. Generally, for such cases, the coarse mesh is the best alternative for succeeding simulations in terms of minimizing computational time but in this paper, considering how close the results are between the fine mesh and the medium size mesh, and to more fully resolve both the free-surface vortices and submerged vortices originating from the bottom and sidewalls of the sump, the medium sized mesh model was used.

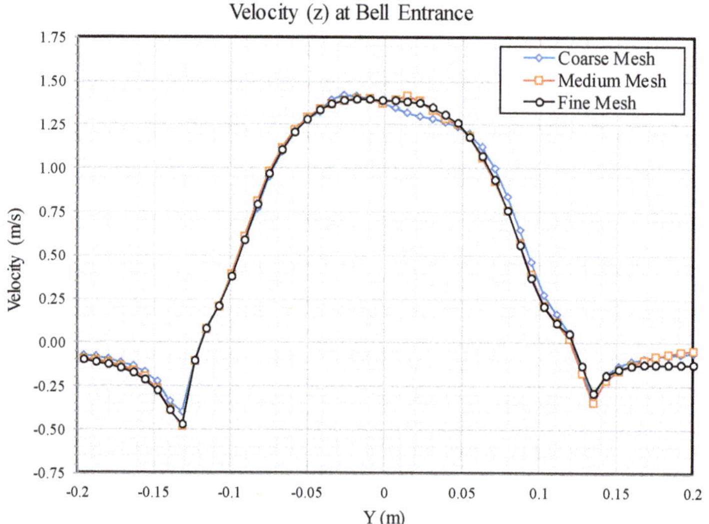

**Figure 6.** Axial Velocity at Bell Inlet.

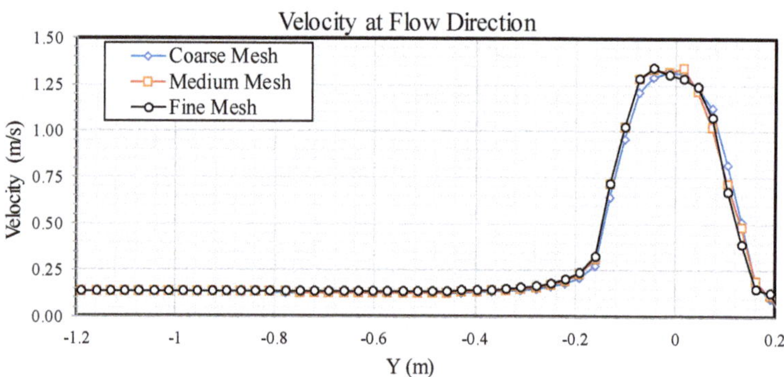

**Figure 7.** Velocity at line parallel to flow.

## 4. Results and Discussions

### 4.1. Approach Flow Pattern

For the physical model test, the hydraulic behavior in the sump were assessed by measuring the swirl-angle and by observation of the flow patterns and vortex phenomena. The approach flow condition from the forebay entering the pump compartment were uniform and stable as indicated by the dye patterns in Figure 8. Even though pre-swirl is slightly high and unsteady, the flow from the far end is stable and approaches the pump without much turbulence. Dead water zones were not observed.

Similarly, for the numerical model, the approach flow pattern for the sump is steady and uniform as shown in Figure 9. No unwanted swirl or circulation as water flows from the inlet of the domain and towards the suction bellmouth. This flow pattern is identical to the pattern observed during the physical model test previously. This verifies that the selected forebay length for the numerical model is sufficient in providing a stable velocity gradient from the inlet of the domain without any unnecessary turbulence which could influence any free surface or subsurface vortex formation downstream of the forebay.

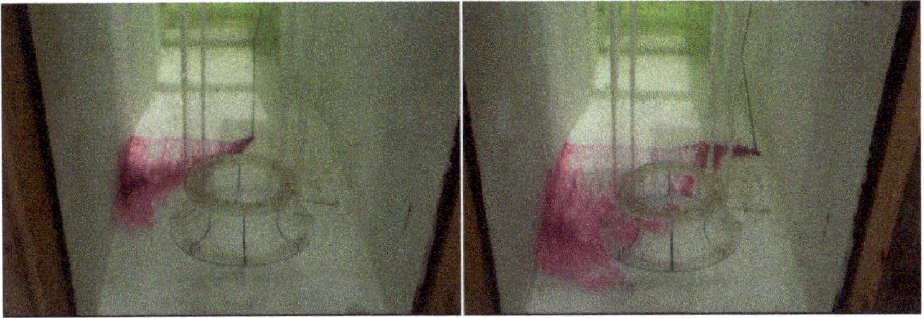

**Figure 8.** Injecting dye across the width of the pump bay shows a uniform and stable approach flow.

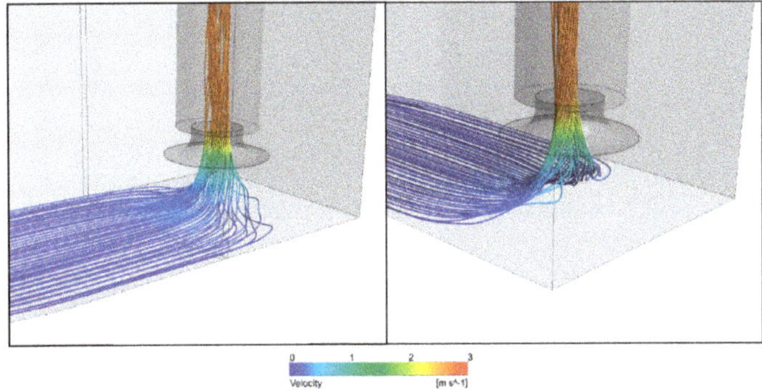

**Figure 9.** Flow streamline near the bottom of the sump indicating steady and uniform approach flow towards the pumps.

*4.2. Free Surface and Sub-Surface Vortex Prediction*

Figure 10 shows the presence of a strong floor vortex activity during the physical model test. In general, this phenomenon is expected given the absence of any floor cone. For this model, the vortex is a stationary type 2 vortex. In other installations, stronger and more dangerous floor vortices may be observed. This is when the pressure near the suction bellmouth falls below vapor pressure producing an air-core to develop at the center of the vortex which makes its way towards the impeller. Submerged vortices are very sensitive to the floor clearance or the distance of the suction bell mouth from the sump floor. One solution is to increase this floor clearance, but oftentimes, this solution inadvertently produces dead zones below the pump, hence an alternative solution is required to suppress this phenomenon.

Similarly, intermittent type 2 sidewall vortex activities were also observed as shown in (Figure 11) while stronger vortices which occasionally develop to type 3 were observed on the backwall (Figure 12). The backwall vortices were stronger because of some instability behind the pump. The flow drifts from side to side and then switches directions as shown Figure 13. This may be attributed to slightly larger backwall clearance (2.5 m) in the pump as compared to the recommended value of ANSI/HI 9.8 (1.95 m). Also, the sidewall vortex combined with the floor vortex significantly increases the swirl in the pump suction pipe.

**Figure 10.** Strong type 2 floor vortex activity even with a maximum floor clearance of 0.5D (130 mm).

**Figure 11.** Intermittent side wall vortex. Type 1 to occasional type 3. Combined with the floor vortex increases the swirl along the pump column.

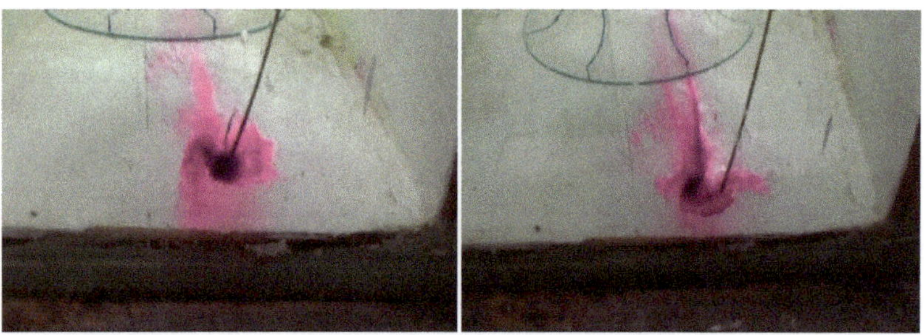

**Figure 12.** Backwall submerged vortex type 2 to type 3 as evidenced by the solid dye core.

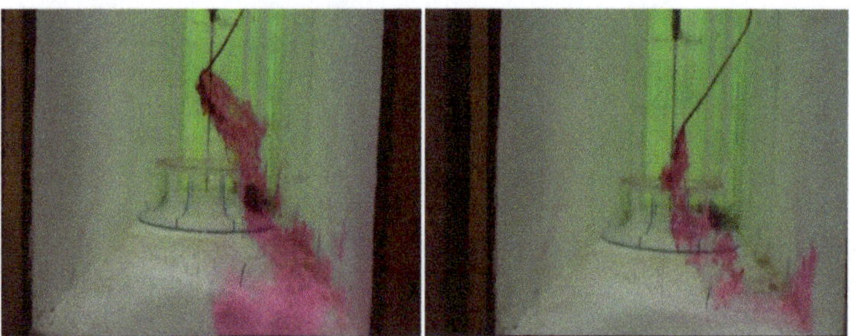

**Figure 13.** Flow instability and drifting behind the pump.

On the surface of the sump, type 1 to type 3 free surface vortices were also observed during the test. Figure 14 shows a surface vortex developing from a surface dimple (type 1) to a full dye-core (type 2) and then dissipating. Due to their transient nature, the vortices form and dissipate without any predictable pattern. For the given flow condition and submergence, the vortices were too weak to draw in some air and/or debris (type 4). As per ANSI/HI 9.8 criteria, type 3 vortices are allowed only when they occur for less than 10% of the 15-min test duration. For this particular case, these surface vortices are acceptable and the need for additional modification such as curtain walls or false ceilings are unnecessary.

**Figure 14.** Image shows a free-surface vortex developing from a type 1 surface dimple (**left**) to a type 3 full dye core (**right**) vortex.

Using the same flow conditions for the analysis, similar phenomena were observed in the CFD simulation. In Figure 15a, a strong backwall vortex is represented by the swirling streamline. It can be seen that at the backwall, there is downward flow towards the center of the sump from two separate directions. This causes the flow to form a strong swirling pattern (Figure 15c) just behind the pump which consequently leads to an organized submerged vortex.

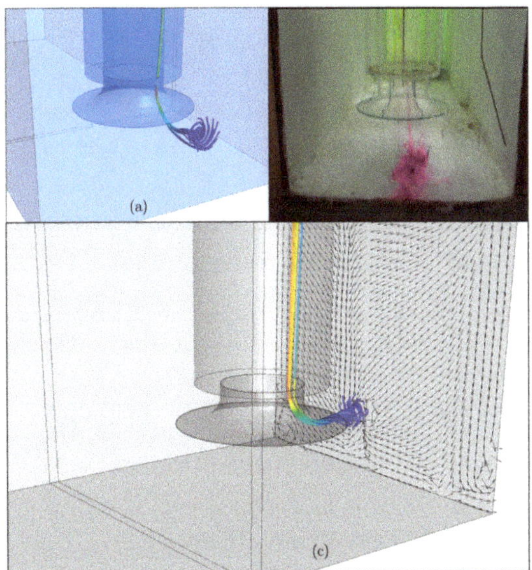


**Figure 15.** (**a**) strong sub-surface vortex attached to the backwall of the sump. (**b**) floor vortex as photographed during the physical model test. (**c**) vector plot showing flow direction near the backwall.

Likewise, similar conditions can be seen in Figure 16 for the sidewall and floor vortices. It is observed that as flow is forced to rotate behind the pump, circular eddies are created in the dead spots allowing vortices to form as the flow separates from the walls of the sumps. This flow separation is what contributes to the swirling motion as the fluid enters the bell mouth.

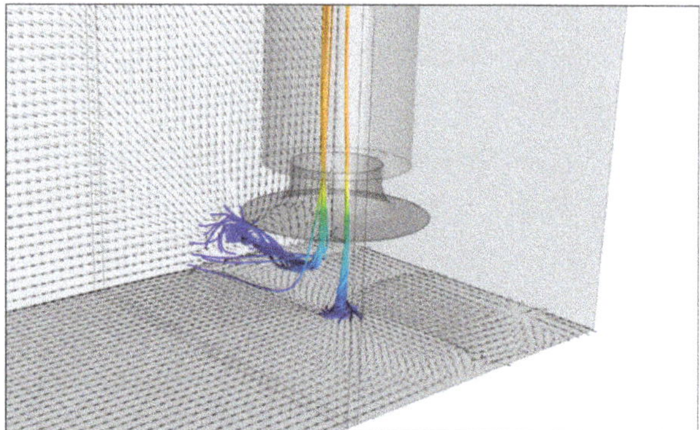

**Figure 16.** Floor vortex and sidewall vortex.

Lastly, Figure 17 below shows the CFD results for the sump wherein the presence of a surface vortex is indicated by the tight curling streamlines. It can be seen the dominant free-surface vortex matches perfectly to that shown in the hydraulic model. It is clearly seen via CFD that the vortex was at most type 3 and was not able to draw in any air as can be seen from the air-volume fraction (3%) isosurface plot in the same figure.

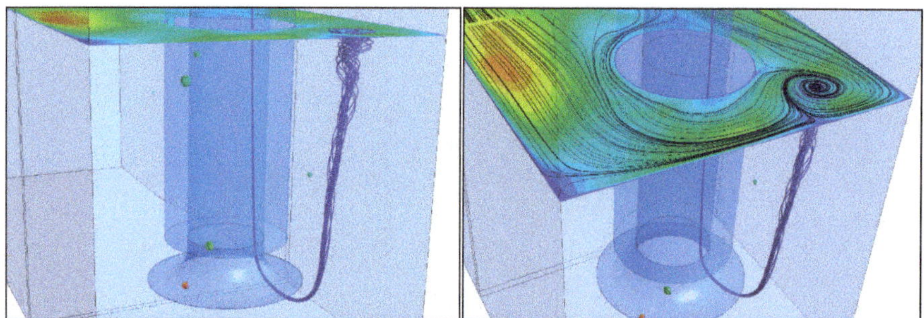

**Figure 17.** CFD results showing streamline indicating the presence of a surface vortex.

Although the strength of the vortex cannot be accurately determined through the CFD results, a qualitative validation can be made by comparing the location and number of vortices predicted by the simulation with those observed during the physical model test. For this study, it can be seen that numerical results tend to agree with the results of the physical model test.

Based on these results, it is clear that the sump fails to meet the acceptance criteria set forth by ANSI/HI and would need to be optimized based on the following observed hydraulic phenomena:

- strong type 2 submerged vortex at the floor of the sump;
- strong type 2 to type 3 submerged vortex behind the pump.

As such, two sets of numerical models are further investigated employing the use of published remedial measures specifically focusing on fillet-splitter designs in order to evaluate performance by improving flow patterns and vortex suppression. Since both the physical model and the numerical model showed allowable intermittent free-surface vortices, the succeeding analysis will be directed towards reducing the formation of sub-surface vortices within the sump.

ANSI/HI 9.8 provides several fillet-splitter designs as a guide for designers in improving sump performance. The most common of which is as shown in Figure 18a employing a trapezoidal cross-baffle along the floor of the sump and a vertical baffle along the backwall. It is important to note however that these recommendation from ANSI/HI are not part of any standards and are not mandatory but instead are presented only to assist engineers in considering factors beyond the standard sump design. Additionally, Figure 18b shows an alternative fillet-splitter design showing a trident shaped floor splitter with triangular profiles along the floor of the sump and 45° chamfers on all corners of the sump near the pump column (sidewall-backwall, backwall-floor, and sidewall-floor corners).

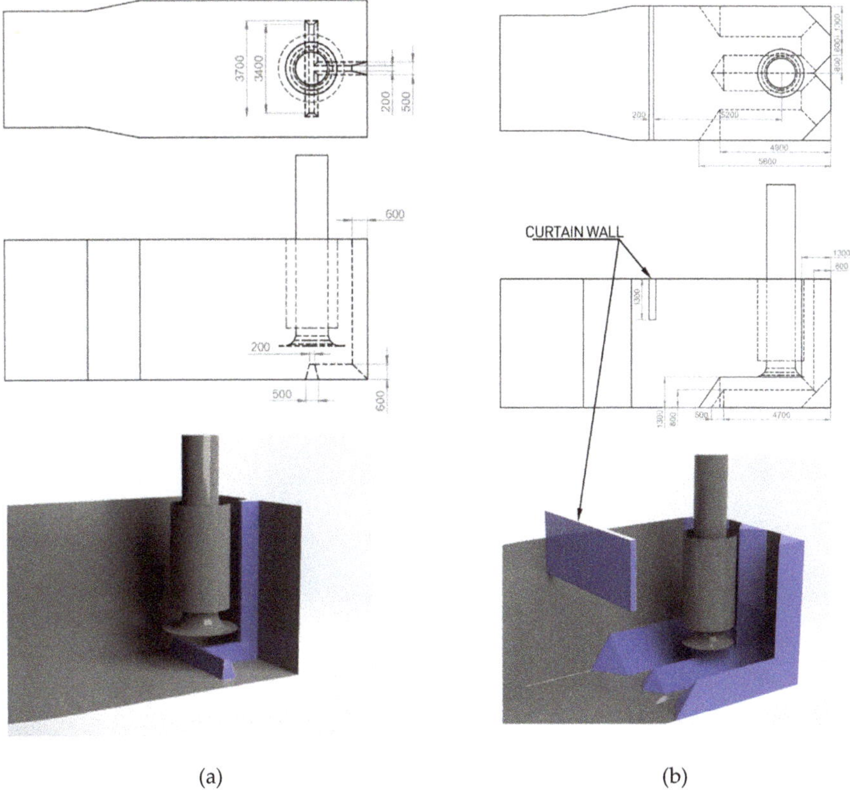


(a)           (b)

**Figure 18.** Proposed fillet-splitter design to reduce vortex formation: (**a**) trapezoidal-shaped cross floor baffle with vertical backwall splitter; (**b**) trident-shaped triangular floor baffle and 45° corner fillets.

The CFD results (Figure 19) for the cross-baffle splitter design which is based on ANSI/HI 9.8 recommendation shows that the proposed design was not able to suppress the subsurface vortices observed in the unmodified pump sump. Backwall and sidewall vortices were still observed as a result of the flow separating from the walls as soon as it reaches the shortest route to the suction bell.

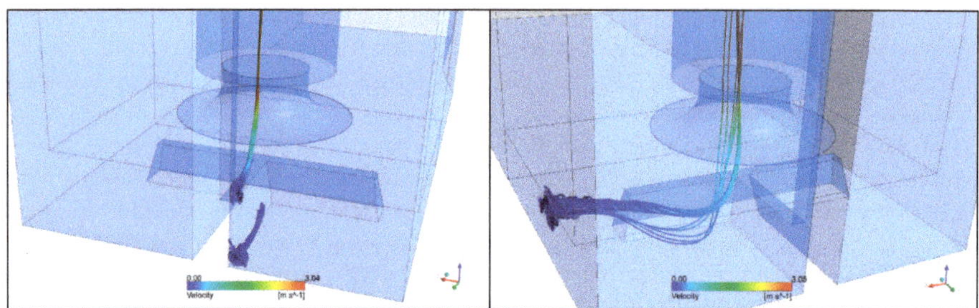

**Figure 19.** Sidewall and backwall vortices observed on numerical results.

Conversely, for the trident-shaped fillet-splitter design, it was observed that flow is very stable. The diverging flows on the sidewall as seen in the baseline and initial splitter models were not present in the optimized design as shown in Figure 20. No circular eddies can be observed on the velocity vector plot along the sidewall and flow remains attached to the wall and fillet as it approaches the pump.

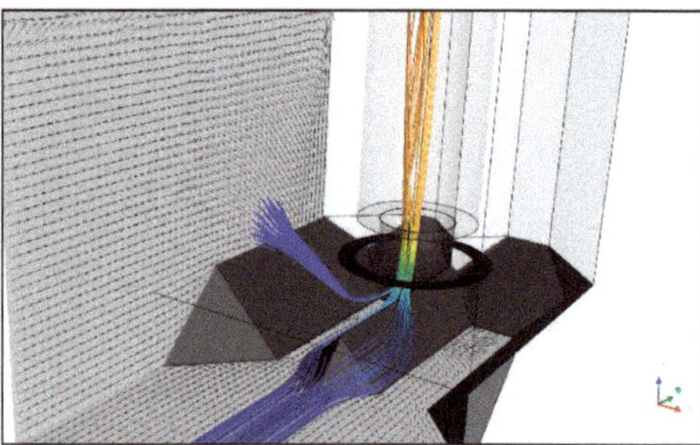

**Figure 20.** Vector plot showing uniform flow pattern on the sidewalls and sump floor.

Attributed to the increased fillet angle and removal of the horizontal floor splitter, streamlines on all sides of the sumps remains attached to the walls (Figure 21). No flow separation and unwanted swirls. Sub-surface vortices were very minimal and only occurred near the entrance of the suction bellmouth.

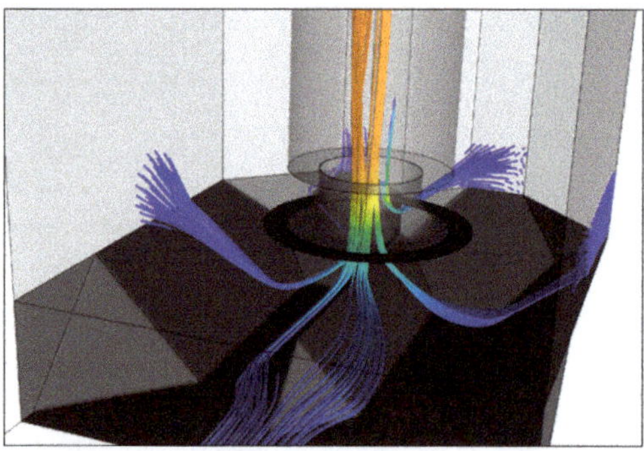

**Figure 21.** Flow remains attached to the wall and the surface of the splitters. No flow separation observed.

### 4.3. Swirl Angle Prediction

ANSI/HI 9.8 provides two criteria for assessing swirl angle. One is the long-term average swirl angle measured/observed for a 10 min duration. The other is the short-term maximum swirl angle observed for 30 s duration. The sump should not exhibit a swirl angle greater than 5° for both criteria in order to be considered acceptable. Among the two, the short-term maximum swirl angle is the more stringent and problematic criteria since this measurement gives an indication of the instantaneous swirl. For this same reason, the paper focused on measuring the pumps sump's max short-term swirl angle.

The CFD swirl measurements were taken at the similar location as that of the rotameters in the physical model test. Specifically, at a distance of 503 mm (approximately four times the pipe diameter) from the bellmouth. Figure 22 shows the CFD prediction for the swirl angle for all three sump geometries over a period of 30 s.

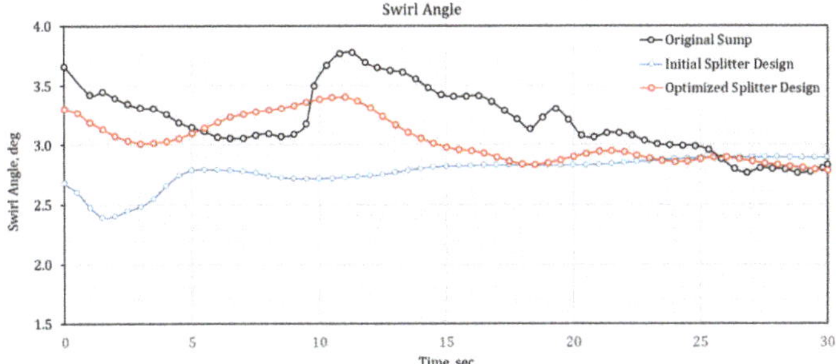


**Figure 22.** CFD results of swirl angle measurements.

For the original unmodified sump geometry, numerical results show a maximum 30-s swirl of 3.75° which is lower than the maximum short-term swirl obtained from the physical model test. Table 3 shows the results obtained from the physical model test. As shown in the table, a maximum number of 26 rotations made by the swirl meter for a 30 s duration was observed. This data, yields a maximum short-term swirl of 7.5°. Similarly, the table shows that about 86% of the time, the short-term swirl angle was observed to be greater than 5°. Similarly, an average of 13 rotations per minute was observed over a 900-s duration yielding an average long-term swirl angle of 1.88°. With regards to the maximum short-term swirl angle, the discrepancy between the numerical and the experimental results may be attributed to the difference in the measurement method used between in the physical model test and the CFD analysis. The physical model test measures swirl angle using a rotameter which can change rotation direction depending on the flow. In general, any change in direction of rotation introduces errors or uncertainties in determining the average causing a lower observed value. On the contrary, assuming correct tangential and axial velocities, CFD results provides exact average conditions.

**Table 3.** Maximum short-term and average long-term swirl angles for the physical model test.

| Rotation Direction | Clockwise | Counterclockwise |
| --- | --- | --- |
| Max. short-term pre-rotation (rounds per 30 s) | 20 | 6 |
| Short term swirl angle | 7.5° | |
| •   % time swirl is above 5° | 86.7% | |
| •   % time swirl is above 7° | 68.4% | |
| Ave long-term pre-rotation (per 900 s) | 9 | 4 |
| Long term swirl angle | 1.88° | |

Model Flow = 116.07 m³/h, $D_i$ = 0.1257 m.

For the initial and optimized splitter designs, the CFD predicted maximum swirl is 3.4° and 2.9°, respectively. If these values are considered, since ANSI/HI 9.8 mandates that both the short-term maximum and the long-term average swirl angle should not exceed 5°, it may be inferred that both fillet-splitter design meets the criteria for acceptable swirl angle.

### 4.4. Point Velocities

For the physical model, the velocity profiles are measured by means of velocity probes installed at the throat of the suction bellmouth. Figure 23 shows the velocity profile as measured across the plane of the impeller eye.

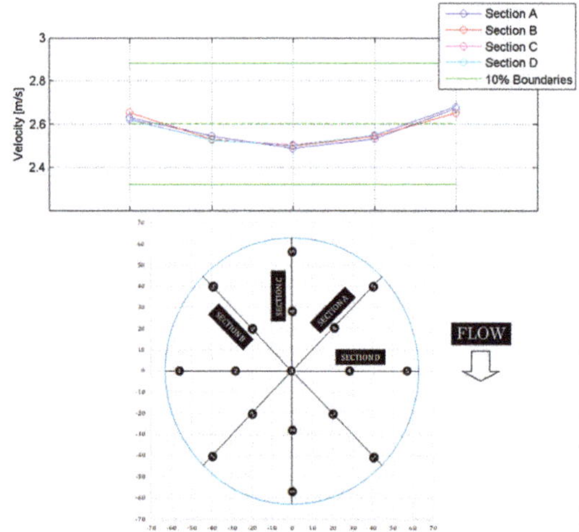

**Figure 23.** Point velocity profile for the physical model test.

The spatial variation varies between −4.0% and 3.6% of the mean velocity indicating that the flow velocity for the unmodified sump falls within the 10% acceptance criteria.

Similarly, for the numerical models, temporal velocity profile taken for a period of 30 s for the same points monitored during the physical model test are shown in Figure 24. The spatial variation for all three cases varies between 0.4% and 4.8% of the mean velocity indicating that the flow velocity for all the sump geometries is within the acceptance criteria. In comparison, it can be seen that the CFD results were able to match the trends and the magnitude of the axial velocities as obtained from the physical model test.

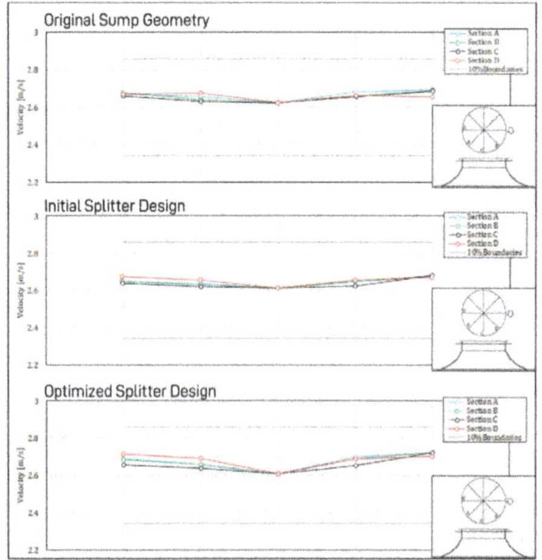

**Figure 24.** Point velocity taken on points on a plane across the impeller eye (CFD).

## 5. Conclusions

This paper presented results of numerical simulation as compared with data from reduced-scale physical model test of pump sump focusing on vortex prediction and swirl angle at intakes.

The numerical model was able to accurately predict the formation, size and location of the type 3 surface vortex that was observed in the physical experiment. Similar to the experiment, the predicted surface vortex was intermittent and was not able to draw in any air to the suction bell mouth.

The strong backwall and floor vortices observed in the physical model test were also replicated in the numerical results. Numerical results showed that the flow separation along the side walls of the sump as well as high turbulence at the back of the pump to be the primary cause of this submerged vortices.

Point velocity data from the numerical analysis showed a good agreement with those measured at several points along the bellmouth throat during the physical model test. Both numerical data and experimental data displayed acceptable similarity in terms of magnitude and trend.

On the other hand, numerical results showed a 30-s maximum short-term swirl angle of 3.75° at a location where the pump impeller was supposed to be. This value is lower than the maximum short-term swirl of 7.5° obtained from the physical model test. The discrepancy between the numerical and the experimental results may be attributed to the difference in the measurement method used. The physical model test measures swirl angle using a rotameter which can change rotation direction depending on the flow. In general, any change in direction of rotation introduces errors or uncertainties. On the contrary, assuming correct tangential and axial velocities, CFD results provides exact conditions.

For the model used for the baseline test, the strength of the submerged side wall and floor vortices observed both in the physical model test and CFD simulation rendered the initial design unacceptable as per established performance criteria. However, additional CFD simulation showed that the strong vortices can be successfully suppressed by the installation of a trident-shaped triangular floor baffle and 45° corner fillets.

Based on the comparison of these results, it can be concluded that CFD simulation can serve as a viable means of evaluating sump performance. CFD could provide the necessary insight in the flow performance within pump sump thereby possibly reducing the need for extensive physical experiments.

Lastly, it was observed that CFD could provide results within a shorter period of time with a lower financial impact but the speed at which design revision can be made in CFD compared with the physical model is debatable. CFD design revision often requires new geometries and meshes and various pre-processing steps which are considered as the most time-consuming process. While the physical model test merely requires the installation of dummy geometries (e.g., fillet, splitters, AVDs) during each test iteration. Nevertheless, even at least from a financial perspective, CFD may still be a viable option in developing optimum intake designs.

**Author Contributions:** Conceptualization, V.M.A.; methodology, V.M.A. and L.A.M.D.; software, V.M.A., B.E.A. and L.A.M.D.; validation, V.M.A. and L.A.M.D.; formal analysis, V.M.A. and L.A.M.D.; investigation, V.M.A.; resources, V.M.A.; data curation, B.E.A., P.L.R., and J.G.T.R.; writing—original draft preparation, V.M.A. and L.A.M.D.; writing—review and editing, B.E.A., P.L.R., and J.G.T.R.; visualization, V.M.A. and L.A.M.D.; supervision, B.E.A., P.L.R., and J.G.T.R., L.A.M.D.; funding acquisition, L.A.M.D. All authors have read and agreed to the published version of the manuscript.

**Funding:** This research was funded by the Department of Science and Technology (DOST) through the Engineering Research and Development for Technology (ERDT) Program—Local Graduate Scholarships. The APC was funded by DOST-ERDT Faculty Research Dissemination Grant.

**Acknowledgments:** The authors would like to thank Hitachi Plant Technologies, Ltd. Philippine Branch Office for granting permission to use the pump physical model test data presented as well as the meshing software used in this paper.

**Conflicts of Interest:** The authors declare no conflict of interest.

## References

1.    Knauss, J. *Swirling Flow Problems at Intakes*; Balkema, A.A., Ed.; International Association for Hydraulic Research: Rotterdam, The Netherlands, 1987.

2.  Chang, E. *Experimental Data on the Hydraulic Design of Intakes and Pump Sumps*; Report RR1518; British Hydromechanics Reseach Association: Bedfordshire, UK, 1979.

3.  American National Standard. *Pump Intake Design*; ANSI/HI 9.8-1998; Hydraulic Institute: Parsippany, NJ, USA, 2012.

4.  Pascoa, J.; Mendes, A.; Gato, L. A fast iterative inverse method for turbomachinery blade design. *Mech. Res. Commun.* **2009**, *36*, 630–637. [CrossRef]

5.  Keck, H.; Weiss, T.; Michler, W.; Sick, M. Recent developments in the dynamic analysis of water turbines. In Proceedings of the 2nd IAHR International Meeting of the Workgroup on Cavitation and Dynamic Problems in Hydraulic Machinery and Systems, Timisoara, Romania, 24–26 October 2007; pp. 9–20.

6.  Keck, H.; Sick, M. Thirty years of numerical flow simulation in hydraulic turbomachines. *Acta Mech.* **2008**, *201*, 211. [CrossRef]

7.  Constantinescu, G.S.; Patel, V.C. Numerical Model for Simulation of Pump-Intake Flow and Vortices. *J. Hydraul. Eng.* **1998**, *124*, 123–134. [CrossRef]

8.  Rajendran, V.P.; Constantinescu, G.; Patel, V.C. Experimental Validation of Numerical Model of Flow in Pump-Intake Bays. *J. Hydraul. Eng.* **1999**, *125*, 1119–1125. [CrossRef]

9.  Li, S.; Silva, J.M.; Lai, Y.; Weber, L.J.; Patel, V.C. Three-dimensional simulation of flows in practical water-pump intakes. *J. Hydroinformatics* **2006**, *8*, 111–124. [CrossRef]

10. Lai, Y.G.; Weber, L.J.; Patel, V.C. A non-hydrostatic three-dimensional numerical model for hydraulic flow simulation—Part II: Validation and application. *J. Hydraul. Eng.* **2003**, *129*, 206–214. [CrossRef]

11. Okamura, T.; Kamemoto, K.; Matsui, J. CFD Prediction and model Experiment on Suction Vortices in pump sump. In Proceedings of the 9th Asian Conference on Fluid Machinery, Jeju, Korea, 16–19 October 2007.

12. Wicklein, E.; Sweeney, C.; Senon, C.; Hattersley, D.; Schultz, B.; Naef, R. Computation Fluid Dynamic Modeling of a Proposed Influent Pump Station. *Proc. Water Environ. Fed.* **2006**, *2006*, 7094–7114. [CrossRef]

13. Shukla, S.N.; Kshirsagar, J. Numerical prediction of cavitation in model pump. In Proceedings of the ASME 2008 International Mechanical Engineering Congress and Exposition, Boston, MA, USA, 31 October–6 November 2008.

14. Nagahara, T.; Sato, T.; Kawabata, S.; Okamura, T. Effect of submerged vortex cavitation in pump suction intakes on mixed flow pump impeller. *Turbomach. Soc. Jpn.* **2002**, *30*, 70–75.

15. Tang, X.; Wang, F.J.; Li, Y.; Cong, G.H.; Shi, X.Y.; Wu, Y.L.; Qi, L.Y. Numerical investigations of vortex flows and vortex suppression schemes in a large pumping-station sump. *Proc. Inst. Mech. Eng. Part C: J. Mech. Eng. Sci.* **2011**, *225*, 1459–1480. [CrossRef]

16. Kanemori, Y.; Pan, Y. Surface vortex prediction and preventing technique in pump intake sump. *Turbomachinery* **2015**, *43*, 90–98. (In Japanese)

17. Yamade, Y.; Kato, C.; Nagahara, T.; Matsui, J. Numerical investigations of submerged vortices in a model pump sump by using Large Eddy Simulation. *IOP Conf. Series: Earth Environ. Sci.* **2016**, *49*, 32003. [CrossRef]

18. Ohyama, S.; Tsukamoto, H.; Oshikawa, T.; Miyazaki, K.; Ohishi, M. Visualization and numerical analysis of unsteady flow in pump sump. *Turbomachinery* **2008**, *36*, 746–755. (In Japanese)

19. Qian, Z.; Wu, P.; Guo, Z.; Huai, W.-X. Numerical simulation of air entrainment and suppression in pump sump. *Sci. China Technol. Sci.* **2016**, *59*, 1847–1855. [CrossRef]

20. Ahn, S.-H.; Xiao, Y.; Wang, Z.; Zhou, X.; Luo, Y. Numerical prediction on the effect of free surface vortex on intake flow characteristics for tidal power station. *Renew. Energy* **2017**, *101*, 617–628. [CrossRef]

21. Menter, F. Review of the shear-stress transport turbulence model experience from an industrial perspective. *Int. J. Comput. Fluid Dyn.* **2009**, *23*, 305–316. [CrossRef]

22. ANSYS, Inc. *ANSYS Fluent Theory Guide*; ANSYS, Inc.: Canonsburg, PA, USA, 2018.

**Publisher's Note:** MDPI stays neutral with regard to jurisdictional claims in published maps and institutional affiliations.

© 2020 by the authors. Licensee MDPI, Basel, Switzerland. This article is an open access article distributed under the terms and conditions of the Creative Commons Attribution (CC BY) license (http://creativecommons.org/licenses/by/4.0/).

*Article*

# Effect of Fins on the Internal Flow Characteristics in the Draft Tube of a Francis Turbine Model

Seung-Jun Kim [1,2], Young-Seok Choi [1,2], Yong Cho [3], Jong-Woong Choi [3], Jung-Jae Hyun [3,4], Won-Gu Joo [4] and Jin-Hyuk Kim [1,2,*]

1   Industrial Technology (Green Process and Energy System Engineering), Korea University of Science and Technology, Daejeon 34113, Korea; kimsj617@kitech.re.kr (S.-J.K.); yschoi@kitech.re.kr (Y.-S.C.)
2   Clean Energy R&D Department, Korea Institute of Industrial Technology, Cheonan-si 31056, Korea
3   K-water Convergence Institute, Korea Water Resources Corporation, Daejeon 34045, Korea; ycho@kwater.or.kr (Y.C.); jwchoi@kwater.or.kr (J.-W.C.); coolguy04@kwater.or.kr (J.-J.H.)
4   Department of Mechanical Engineering, Yonsei University, Seoul 03722, Korea; joo_wg@yonsei.ac.kr
*   Correspondence: jinhyuk@kitech.re.kr; Tel.: +82-41-589-8447

Received: 9 May 2020; Accepted: 29 May 2020; Published: 1 June 2020

**Abstract:** Undesirable flow phenomena in Francis turbines are caused by pressure fluctuations induced under conditions of low flow rate; the resulting vortex ropes with precession in the draft tube (DT) can degrade performance and increase the instability of turbine operations. To suppress these DT flow instabilities, flow deflectors, grooves, or other structures are often added to the DT into which air or water is injected. This preliminary study investigates the effects of anti-cavity fins on the suppression of vortex ropes in DTs without air injection. Unsteady-state Reynolds-averaged Navier–Stokes analyses were conducted using a scale-adaptive simulation shear stress transport turbulence model to observe the unsteady internal flow and pressure characteristics by applying anti-cavity fins in the DT of a Francis turbine model. A vortex rope with precession was observed in the DT under conditions of low flow rate, and the anti-cavity fins were confirmed to affect the mitigation of the vortex rope. Moreover, at the low flow rate conditions under which the vortex rope developed, the application of anti-cavity fins was confirmed to reduce the maximum unsteady pressure.

**Keywords:** Francis turbine; anti-cavity fins; draft tube; vortex rope; low flow rates; internal flow characteristics; unsteady pressure

---

## 1. Introduction

Since hydropower technology has a low impact on the natural environment, there is significant potential for its efficient application in power generation. Traditionally, hydroelectric power generated by hydro turbines has been used to provide electrical energy during times of peak load. However, as solar and wind power technologies with intermittent power generation have recently been developed and added to the grid, the requirement for large and flexible energy adjustments for these instabilities has increased. Consequently, hydro turbines are used over range with off-design operating conditions to provide a base load to maintain a constant frequency and provide a stable power supply. However, because of variations in operating conditions, existing hydro turbines must operate with significant fluctuations in flow rates; this results in both the failure of the turbine systems and a reduction in service lifespan. Particularly, those operations under conditions of low flow rates generate a complicated flow in the draft tube (DT) as a vortex rope, resulting in severe noise and vibration [1–3].

Francis turbines exhibit the undesirable phenomenon of a vortex rope with precession, which induces pressure fluctuations in the DT under low flow rate conditions. A pressure fluctuation frequency close to the natural frequency of the system causes a resonance that seriously undermines the stability of the turbine system [4,5].

To examine these influences on turbine systems, many studies have attempted to investigate the flow mechanisms and characteristics of the vortex rope in the DT. Susan-Resiga et al. [6] used an axisymmetric flow model to numerically study a methodology for analyzing the swirling flows of a vortex rope. This axisymmetric flow solver provides a circumferential averaged flow field. Nicolet et al. [7] conducted an experimental investigation of pressure fluctuations with vortex ropes at the upper-part load; they used high-speed camera visualization with synchronized measurement of pressure fluctuations in a Francis turbine model with high specific speed. Zuo et al. [8] used a Francis turbine model to assess the stability of a vortex rope in a DT under different operating conditions; via unsteady simulations, they observed the relationships between the induced hydraulic instabilities and the characteristics of the vortex rope. Favrel et al. [5] explored the influence of flow rate on the structure and parameters of vortex ropes. The intensity of the excitation source was observed for the vortex rope dynamics using PIV measurements in a Francis turbine operating at part load.

To suppress the flow instabilities associated with the vortex rope in the DT, several previous studies have modified the tube's internal flow; modifications have included injecting air or water or adding internal flow deflectors or grooves to the DT's cone. Aeration and water injection are well-established control methods for suppressing pressure fluctuations within the DT. These injections work to decrease the turbulence intensity of the swirl flow by increasing the momentum of axial flow in the central region, thereby stabilizing the operation of the hydro turbine system [9].

Altimemy et al. [9] conducted a large eddy simulation at the design and partial loading stages of the Francis turbine to examine the influence of water injection on pressure fluctuations and flow behavior in the DT. Susan-Resiga et al. [10] performed an investigation to mitigate the severe flow fluctuations induced by the vortex rope in a Francis turbine operating at part load by numerically determining a method for injecting a water jet from the crown tip of the runner. WF et al. [11] applied numerical analyses to reduce pressure fluctuations by simulating the unsteady flow in the DT of a Francis turbine and incorporating air and water injections and flow deflectors within the DT. Feng et al. [4] conducted a numerical analysis to decrease the pressure fluctuations caused by the vortex rope in the DT of a Francis turbine by applying extended runner cones, damping gates, and flow deflectors in the DT. Anup et al. [12] numerically calculated the four different depths of the J-Grooves in the DT to minimize vortex rope characteristics and flow instabilities. Chen and Choi [13] used both experimental and numerical approaches to investigate the effects of the J-Groove on the DT wall in suppressing cavitation in the DT of a Francis turbine.

Anti-cavity fins with air injection are often applied to aid the suppression and minimization of the vortex rope by the DT wall. However, the effects of DT anti-cavity fins on the internal flow and the unsteady pressure characteristics without air injection have not been systematically elucidated under low flow rate conditions in a Francis turbine.

Therefore, this study preliminarily investigates the effects of anti-cavity fins for suppressing the vortex characteristics in the DT in a Francis turbine. Anti-cavity fins without air injection were applied in the DT of a Francis turbine model, and the internal flow and unsteady pressure characteristics at low flow rates were assessed. The investigation performs unsteady-state Reynolds-averaged Navier–Stokes (RANS) equations using a scale-adaptive simulation shear stress transport (SAS-SST) turbulence model to observe the unsteady internal flow and pressure fluctuation characteristics in the presence of vortex rope and anti-cavity fins under conditions of low flow rate. The research compares the magnitudes and locations of the vortex rope via the application of anti-cavity fins under low flow rate conditions, and the results are discussed concerning the effects of the fins. Furthermore, to investigate the unsteady pressure fluctuations induced by the vortex ropes with precession, pressure measurement points were applied on the wall of the DT to analyze the unsteady flow.

## 2. Francis Turbine Model and Anti-Cavity Fins

This study conducted three-dimensional (3-D) unsteady-state numerical analyses of a Francis turbine model with a specific speed of 270-class [m, kW, min$^{-1}$]. The model's specific speed was calculated using Equation (1). Figure 1 presents a 3-D model of the DT of the Francis turbine model, which shows the locations of the anti-cavity fins. The fins comprised two short fins and two long fins; the two short fins were the same length (0.7 $D_2$) and were arranged to face each other, whereas the two long fins also faced each other but were of different lengths (1.09 and 1.3 $D_2$, respectively). Here, $D_2$ is defined as the outlet diameter of the runner. The anti-cavity fins with the air injection outlet (the dotted red circles in Figure 1a) were located at 0.2 $D_2$ on the wall of the DT's cone, and the air injection outlet protruded from the anti-cavity fins. This study considered the shapes of the air injection outlets in the analysis domain to accurately reflect the shape of the DT without considering the effects of the air injection. The specifications of the Francis turbine model are listed in Table 1 and include the speed factor, discharge factor, and energy coefficient. These coefficients and factors are applied along with Equations (2)–(4) and are defined by IEC Standard 60193 [14].

$$N_s = \frac{N\sqrt{P}}{H^{\frac{5}{4}}} \tag{1}$$

$$n_{ED} = \frac{nD}{\sqrt{gH}} \tag{2}$$

$$Q_{ED} = \frac{Q}{D^2 E^{0.5}} \tag{3}$$

$$E_{nD} = \frac{gH}{n^2 D^2} \tag{4}$$

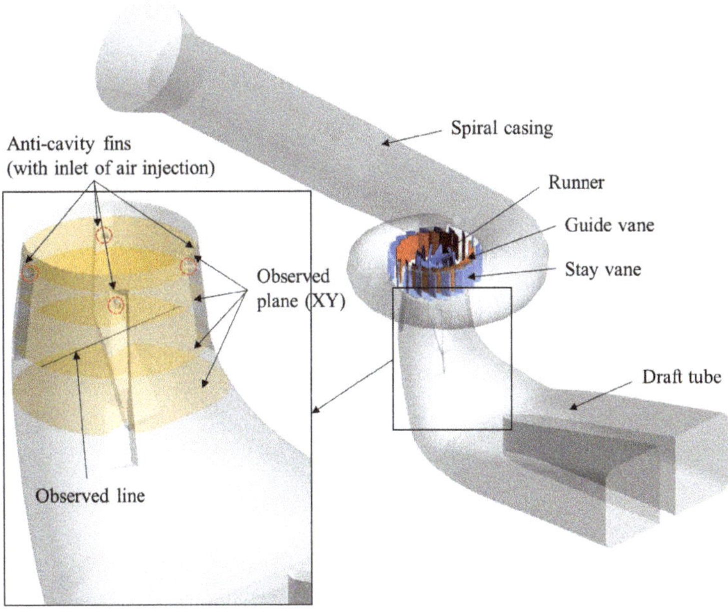


(a) 3-D model

**Figure 1.** *Cont.*

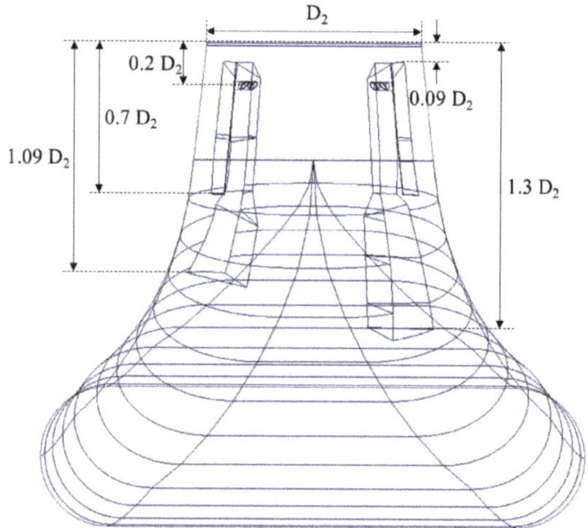


**(b)** Location of anti-cavity fins

**Figure 1.** (**a**) 3-D model of the DT with (**b**) location of anti-cavity fins in the Francis turbine model.

**Table 1.** Francis turbine model specifications.

| Specification | Value |
| --- | --- |
| Speed factor, $n_{ED}$ (–) | 0.48 |
| Discharge coefficient, $Q_{ED}$ (–) | 0.33 |
| Energy coefficient, $E_{nD}$ (–) | 4.35 |
| Diameter of runner outlet, $D_2$ (m) | 0.35 |
| Number of runner blades | 12 |
| Number of stay vanes | 20 |
| Number of guide vanes | 20 |

## 3. Numerical Analysis Methods

This study used ANSYS CFX-19.1 commercial software [15] to analyze the 3-D internal flow field of the Francis turbine model in terms of the unsteady-state calculations. The numerical grids for the runner and vane components were generated using the software's TurboGrid function, whereas the numerical grids of the spiral casing (SC) and DT were produced using the ANSYS Meshing and ICEM-CFD functions. The boundary conditions for the numerical analysis were set using CFD-Pre, and to solve the governing equations and to conduct the post-processing of the results, CFX-Solver and CFX-Post were used, respectively. The unsteady-state RANS equations for the incompressible flow behavior of the Francis turbine model were calculated using governing equations (Equations (5) and (6)), which were discretized via the finite volume method.

$$\frac{\partial \rho}{\partial t} = \frac{\partial(\rho u_i)}{\partial x_i} = 0 \tag{5}$$

$$\frac{\partial(\rho u_i)}{\partial t} + \frac{\partial(\rho u_i u_j)}{\partial x_j} = -\frac{\partial p}{\partial x_i} + \frac{\partial}{\partial x_j}\left(\mu \frac{\partial u_i}{\partial x_j}\right) + \frac{\partial \tau_{ij}}{\partial x_j} \tag{6}$$

where $\tau_{ij} = -\rho \overline{u_i u_j}$ is known as the Reynolds stress. To ensure the physical boundaries, the discretizations of the high-resolution and second-order backward Euler schemes were solved for the advection and transient schemes.

The numerical grids of the Francis turbine model are presented in Figure 2. The computational domains of the SC and DT were constructed using tetrahedral-type grids, whereas hexahedral-type grids were applied for the construction of the computational domains of the stay vanes (SVs), guide vanes (GVs), and runner blades. In the runner domain, O-type grids were used on the surface of the runner blade, and the y+ value for the first runner grid point was kept below five. The grid convergence index method was applied to select the optimum grids among the three observed grids to estimate the numerical uncertainty due to discretization error [16–18]. Figure 3 presents a comparison between the efficiency and flow rates for the three observed numerical grids. The $GCI_{fine}^{21}$ values of the efficiency and flow rates were calculated to be about 0.0022 and 0.0458, respectively, for the optimum grids of $14.74 \times 10^6$, as shown in Table 2. These values were subsequently chosen for the numerical analyses [19].

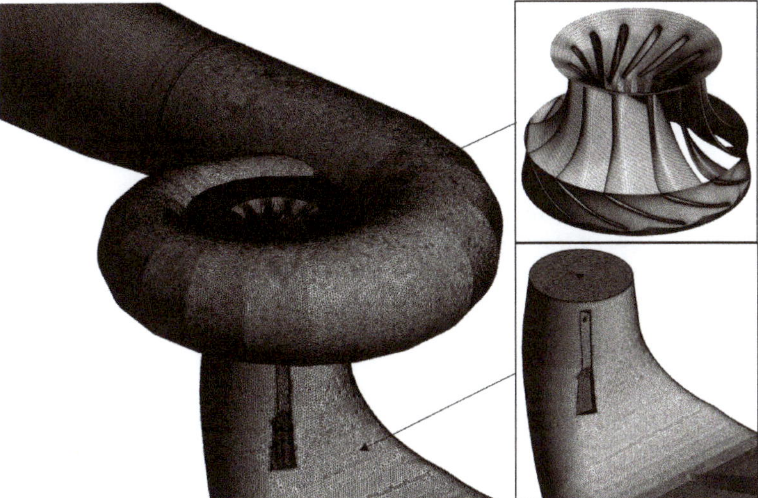

**Figure 2.** Numerical grids used in the Francis turbine model.

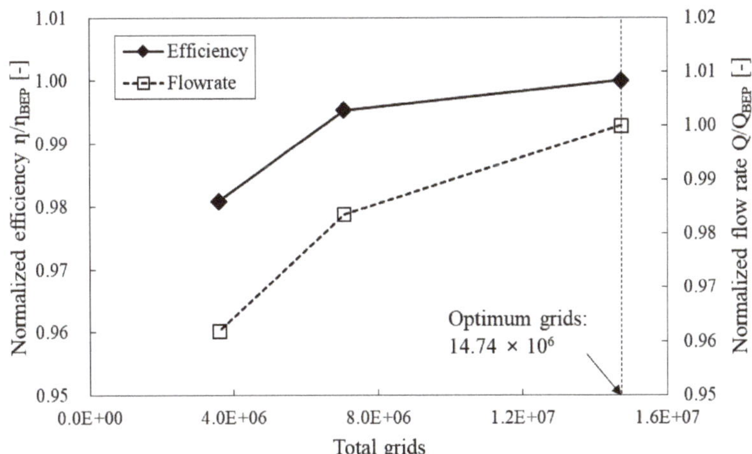


**Figure 3.** Comparisons of efficiency and flow rate with observed numerical grids in the Francis turbine model.

**Table 2.** Calculation of discretization error for the Francis turbine model.

|  | $\phi$ = **Efficiency** | $\phi$ = **Flow Rate** |
| --- | --- | --- |
| $N_1, N_2, N_3$ | $14.74 \times 10^6, 7.05 \times 10^6, 3.59 \times 10^6$ | |
| $r_{21}$ | 1.28 | |
| $r_{32}$ | 1.25 | |
| $\phi_1$ | 1 | 1 |
| $\phi_2$ | 0.9952 | 0.9841 |
| $\phi_3$ | 0.9808 | 0.9618 |
| $p$ | 5.20 | 1.51 |
| $GCI^{21}_{fine}$ | 0.0022 | 0.0458 |

The boundary conditions for numerical analyses are listed in Table 3. The working fluids (water and water vapor) were set at 25 °C to consider the cavitation phenomena using the Rayleigh–Plesset equation, which describes the growth and collapse of vapor bubbles in a liquid as a homogeneous model [20]. Here, the mean diameter of the cavitation bubble was established as $2.0 \times 10^{-6}$ m, and the water saturation pressure was set to 3169.9 Pa. The values for area-averaged total pressure and static pressure were established with consideration of the water levels of the upper and lower reservoirs at the inlet and outlet, respectively. To accurately calculate the influence of the flow separation phenomena, the SAS-SST turbulence model was employed, which provides the scale resolving simulation mode to the unsteady SST turbulence model. This is developed by including the von Karman length scale into the turbulence scale equation [21,22]. The transient rotor–stator condition was used to connect the interface between the rotating and stationary domains. Unsteady-state numerical analyses were performed at intervals of 1.5° over a total of eight revolutions of the runner at a time step of 0.0002272 s and a total time of 0.4364 s. To improve the convergence, the coefficient for the number of loops was set at 5. Finally, to investigate the time-averaged values and to avoid initial numerical noise, the values for the period of the runner's last three revolutions were averaged after a total of five revolutions.

**Table 3.** Boundary conditions of the Francis turbine model.

| Boundary Conditions | |
| --- | --- |
| Working fluid | Water and vapor at 25 °C |
| Cavitation model | Rayleigh-Plesset |
| Inlet | Total pressure (Level of upper reservoir) |
| Outlet | Static pressure (Level of lower reservoir) |
| Turbulence model | SAS-SST model |
| Interface condition | Transient rotor-stator |
| Unsteady calculation conditions | Total time: 8 revolutions of runner<br>Time step: 1.5° of runner<br>Loops coefficient: 5 |

Figure 4 presents the pressure measurement points in the Francis turbine model. To investigate the unsteady pressure characteristics caused by the vortex and internal flow phenomena, these measurement points were applied at regular intervals of 0.1 $D_2$ on the DT wall. The pressure measurement points (p1–p4 in Figure 4) were located at the same height on the DT wall.

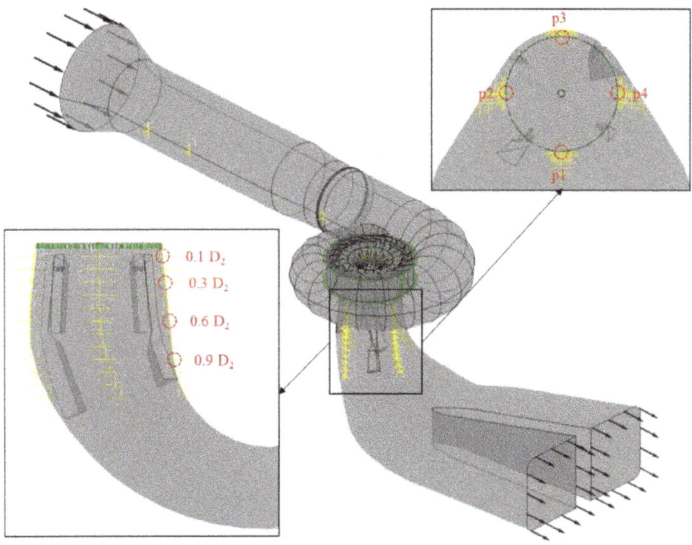

**Figure 4.** Pressure measurement points in the Francis turbine model.

## 4. Results and Discussion

### 4.1. Validation of the Numerical Analysis Results

To validate the unsteady-state numerical analysis results of the Francis turbine model, this study compared the results of the steady and unsteady-state RANS equations with the experimental results of the full-scale Francis turbine, as shown in Figure 5. The efficiencies were normalized by the maximum value of experimental efficiency. In the unsteady-state numerical analyses, two low flow rate conditions with GV angles at 16° and 12° were selected as the observed low flow rate conditions based on the best efficiency point (BEP) with a GV at 22°. The full-scale Francis turbine investigations were conducted using the pressure–time method with a measurement error of ±1.74% [23]. To compare the efficiencies between the model and the full-scale Francis turbine, the scale-up conversion of the hydraulic efficiency defined by IEC Standard 60193 was applied to the results of the Francis turbine model's analysis for both steady and unsteady-state RANS [14]. Equations (7)–(9) were applied as the formulae for scaling up the hydraulic efficiency, whereas Equation (7) was considered as the loss efficiency due to the model's geometrical scale. The equations of loss efficiency were calculated as a function of the Reynolds number along with Equations (8) and (9), as follows:

$$\eta_P = \eta_M + (\Delta\eta)_{M \to P} \tag{7}$$

$$(\Delta\eta)_{M \to P} = \delta_{ref}\left[\left(\frac{Re_{ref}}{Re_M}\right)^{0.16} - \left(\frac{Re_{ref}}{Re_P}\right)^{0.16}\right] \tag{8}$$

$$\delta_{ref} = \frac{1 - \eta_{opt.M}}{\left(\dfrac{Re_{ref}}{Re_{opt.M}}\right)^{0.16} - \dfrac{1-V_{ref}}{V_{ref}}} \tag{9}$$

where the subscripts of M and P indicate the model and the full-scale Francis turbine, respectively, and $Re_{ref} = 7 \times 10^6$, $V_{ref} = 0.7$ are the reference values of a radial turbine defined by IEC Standard 60193 [14]. The results of the unsteady-state numerical analyses were averaged during the last three revolutions of the runner to enable an efficiency comparison.

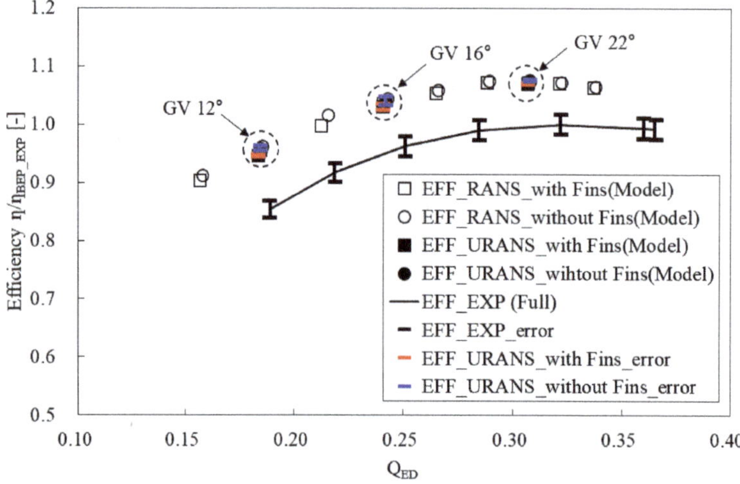


**Figure 5.** Comparison of the conversed numerical results and the experimental results of a full-scale Francis turbine.

The comparisons between the experimental and the conversed numerical results demonstrate a slight variation for each efficiency. However, the tendencies of the conversed efficiencies of the numerical analyses were similar to those exhibited by the experimental efficiencies; these variations in efficiency comparisons can be interpreted as the neglect of the mechanical loss and the surface roughness in the numerical analyses. Consequently, this study considered the numerical analysis results for the Francis turbine model to be valid. However, the addition of anti-cavity fins decreased the efficiencies across the entire range of observed flow rates. Particularly, in the unsteady-state analyses, the efficiencies of the GVs at 22°, 16°, and 12° decreased by 0.5%, 0.8%, and 1.0%, respectively, depending on the anti-cavity fins applied.

### 4.2. Internal Flow Characteristics Relative to the Anti-Cavity Fins in the Draft Tube

This study investigated the performance characteristics concerning the anti-cavity fins in the DT and the flow rate conditions by calculating the head losses of the main components of the Francis turbine model, as presented in Figure 6. The model's head losses were calculated using Equation (10) for the SC, SV, GV, and DT, whereas Equation (11) was applied to calculate the head loss of the runner [24].

$$H_{loss} = \frac{\Delta p_{total}}{\rho g} \tag{10}$$

$$H_{loss\ runner} = \frac{\Delta p_{total} - \frac{T\omega}{Q}}{\rho g} \tag{11}$$

Here, $H_{loss}$ represents the value of loss by a head, $\Delta p_{total}$ is the total pressure difference through each component, which was calculated with time-averaged total pressure in this study, $\rho$ is the water density, and $g$ is the acceleration due to gravity. In the $H_{loss\ runner}$, $T$ is the torque of the runner, which was measured by the force caused by a rotating axis, and $\omega$ is the angular velocity.

Similar head loss distributions were exhibited from the SC to the GV regions both with and without anti-cavity fins, whereas the inclusion of anti-cavity fins caused slight decreases in the runner's head losses. However, the application of anti-cavity fins in the DT increased the head losses; these losses were comparatively greater than the decreases in head loss in the runner region. In particular, the difference in head loss in the DT region due to the addition of anti-cavity fins was about 2.0% with a GV at 16°, which was relatively the highest under observed conditions of flow rate.

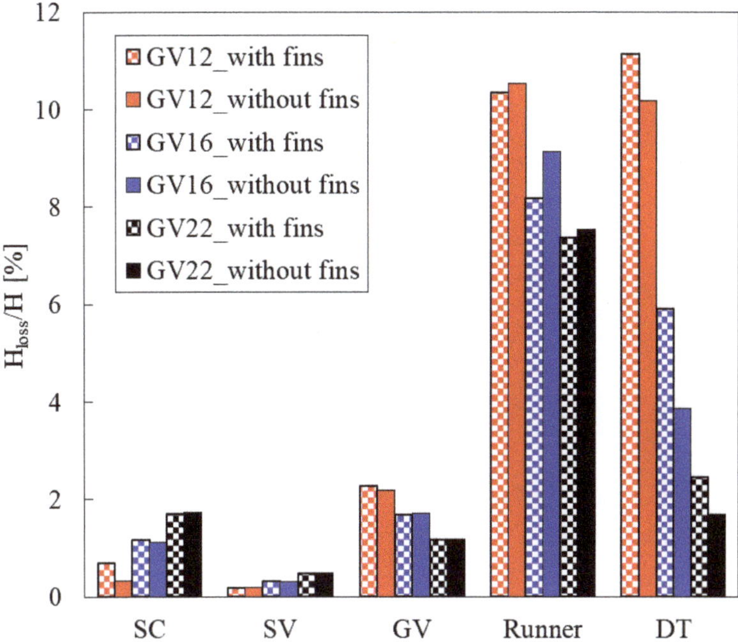


**Figure 6.** Head loss distribution in the main components of the Francis turbine model.

To investigate the qualitative effect of anti-cavity fins to the magnitude of vortex rope in DT according to the flow rate conditions, Figures 7–9 show the vortex rope in the DT by the iso-surface distributions of pressure during the last revolution of the runner for investigating the internal flow structures of GVs at 22°, 16°, and 12°. The iso-surface of pressure was determined as the relative water saturation pressure considering the water level of the lower reservoir. As Figure 7b reveals, with a GV at 22°, the vortex rope in the low-pressure region was not generated in the DT without anti-cavity fins. However, as can be seen in Figure 7a, the application of anti-cavity fins produced low-pressure regions in the DT. Due to the protrusion of the air injection outlets on the anti-cavity fins, the low-pressure regions occurred on the anti-cavity fins near the inlet of the DT (as detailed in Figure 1). Therefore, it is believed that inducing flow resistance at the sites of the air injection outlets generated the low-pressure regions. Figure 8 shows that the vortex rope was clearly developed in the DT with a GV at 16° (0.78 Q$_{BEP}$). The application of anti-cavity fins in the DT significantly decreased the vertical length of the vortex rope. Additionally, the low-pressure regions were found to occur near the anti-cavity fins. Figure 9 shows the condition of the GV at 12° (0.59 Q$_{BEP}$). Here, and similar to that shown in Figure 7a, without the addition of anti-cavity fins, no low-pressure regions in the DT were generated. However, Figure 9a reveals that the low-pressure regions were developed on the anti-cavity fins rather than those with a GV at 22° as the difference under both conditions of GV. In this way, the anti-cavity fins in the DT show the effect of reducing the vertical length of the vortex rope, but the shape itself, such as in the air injection outlets, acts as the factor that induces the low-pressure regions and impedes the flow in the DT. It can be regarded as the cause of the increased head loss by application of the anti-cavity fins in the DT, as shown in Figure 6.

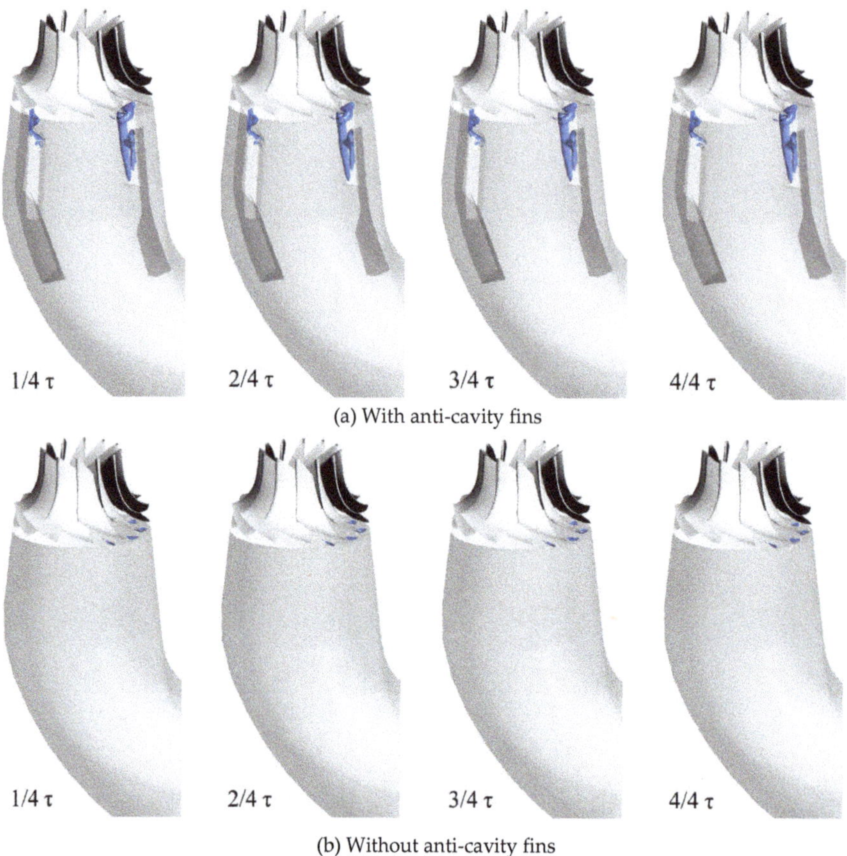



**Figure 7.** Iso-surface distributions of pressure in the DT during one runner revolution at a GV of 22°: (**a**) with anti-cavity fins; and (**b**) without anti-cavity fins.

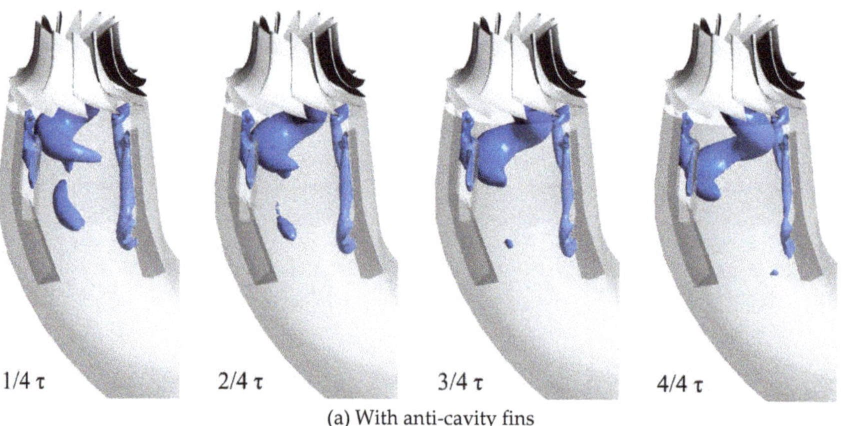



**Figure 8.** *Cont.*

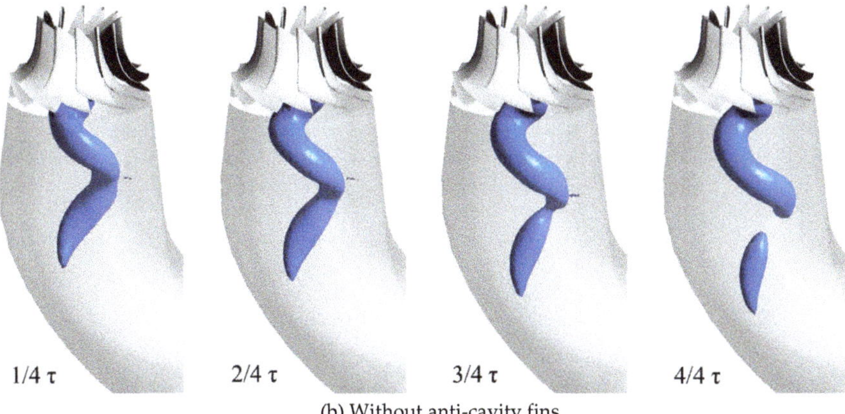



(b) Without anti-cavity fins

**Figure 8.** Iso-surface distributions of pressure in the DT during one runner revolution at a GV of 16°: (**a**) with anti-cavity fins; and (**b**) without anti-cavity fins.

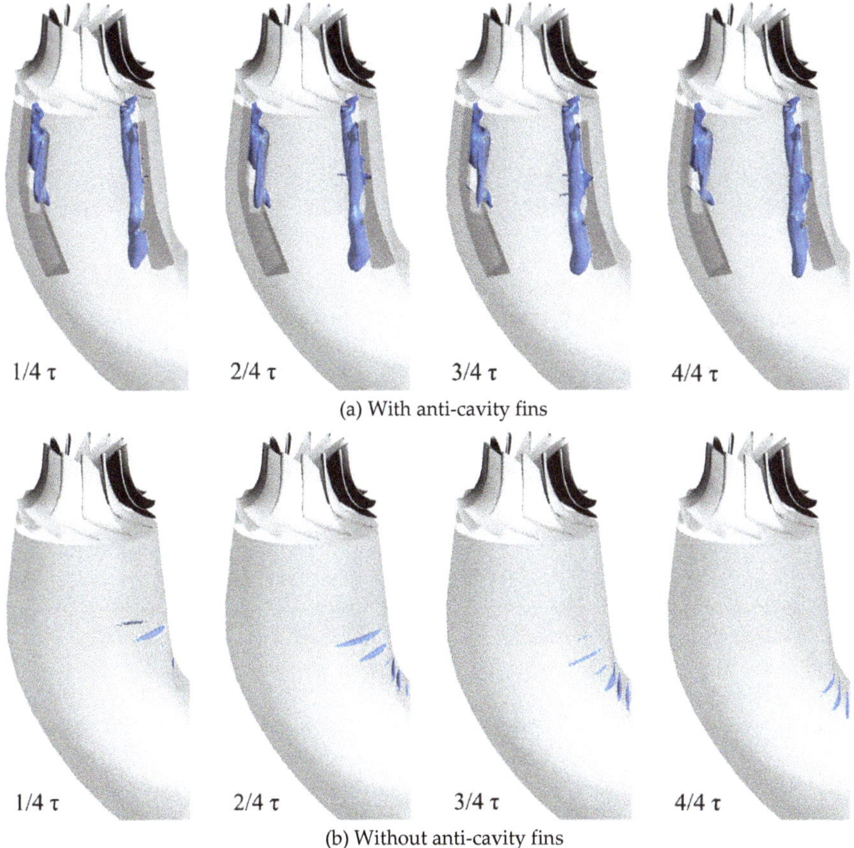



(a) With anti-cavity fins



(b) Without anti-cavity fins

**Figure 9.** Iso-surface distributions of pressure in the DT during one runner revolution at a GV of 12°: (**a**) with anti-cavity fins; and (**b**) without anti-cavity fins.

To observe the flow phenomena in the DT according to the flow rate, this study compared the velocity triangles at the runner outlet with GVs at 22°, 16°, and 12°, as illustrated in Figure 10. As the meridional velocity, $C_m$, decreased, the absolute flow angle, $\alpha_2$, gradually increased as the GV angle decreased. Thus, in the absolute velocity component, $C_2$, the radial velocity component increased as the $\alpha_2$ also increased. The increase in the swirl strength of the flow and the generation of both the complicated flow and the vortex rope in the DT can be due to the increase in the radial velocity component at the outlet of the runner under conditions of low flow rate.

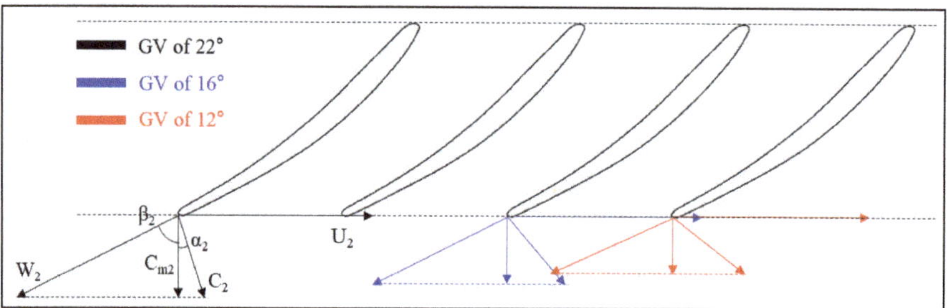


**Figure 10.** Velocity triangle on the mid-span at GVs of 22°, 16°, and 12°.

The streamline distributions of the time-averaged (Trnavg) velocity in the DT both with and without the anti-cavity fins are presented in Figure 11 for GVs at 22°, 16°, and 12°. The flow phenomena in the DT are shown with the associated complicated flow at low flow rates as confirmed by the flow characteristics at the runner outlet in the velocity triangles in Figure 10. Figure 11b presents the complicated flow with precession in the DT with a GV at 16°. As Figure 8 shows, the internal flow makes it possible to confirm the cause of the development of the vortex rope and the low-pressure region near the anti-cavity fins in the DT. Figure 11c presents a very complicated internal flow pattern without precession due to the increased radial velocity component with a GV at 12°. These flow characteristics can be confirmed as the reason for the development of the low-pressure regions on the anti-cavity fins, as shown in Figure 9a.

The angle distributions of the absolute and relative flows at the runner outlet along the spanwise direction from hub (0) to shroud (1) with GVs at 22°, 16°, and 12° are presented in Figure 12. The observed flow angles are normalized by the value of each maximum flow angle. The GVs at 22° and 16° demonstrated similar distributions of flow angles both with and without anti-cavity fins, whereas the GV at 12° produced a slight difference in flow angle distributions. This is due to the complicated internal flow characteristics induced by low flow rate conditions. The absolute flow angle distributions at the 22° GV are close to 0, and the absolute flow angles increase as the flow rates decrease. Additionally, compared with the other GV conditions, the contrasting trends are exhibited only at the GV at 12° near the spanwise range of 0–0.2. Similarly, the GV at 12° also demonstrates different tendency characteristics in relative flow angle distribution due to the complicated internal flow, as shown in Figure 12b, although the relative flow angle and the angle of the runner blade outlet generally appear similar.

The above analyses reveal that at the observed GV of 16°, the head loss difference in the DT was the greatest, and the effects of the anti-cavity fins were significantly observed. Figure 13 presents a more detailed analysis of the internal flow characteristics of the same GV; it investigates the influence of the anti-cavity fins by comparing the iso-surface distribution of the pressure in the DT during the last revolution of the runner with the top view of the runner. The pressure value was determined to be the same as that used in Figure 8. Depending on time and irrespective of the inclusion of anti-cavity fins, the vortex ropes were maintained at similar maximum rotating diameters. Therefore, the effects of the anti-cavity fins decreased the maximum vertical length of the vortex rope, whereas the maximum rotating diameters were not significantly affected by the inclusion of anti-cavity fins.

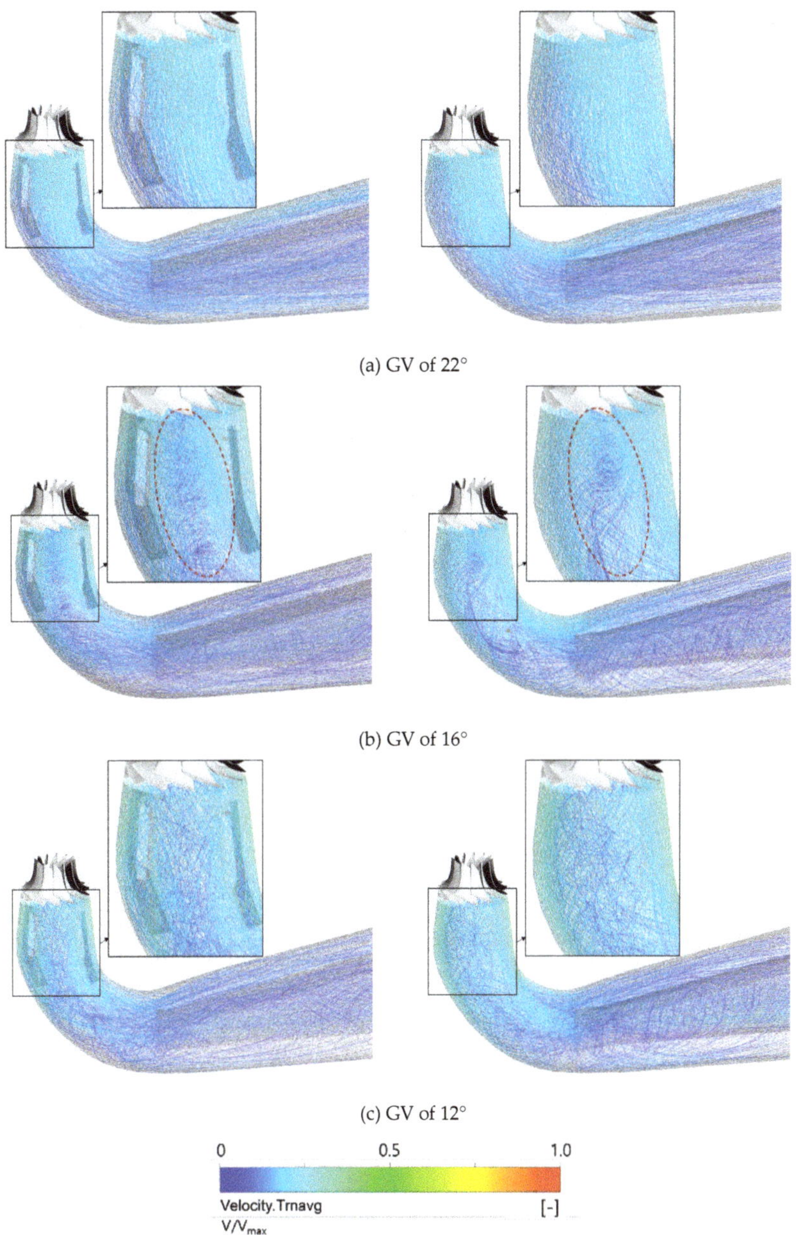

(a) GV of 22°

(b) GV of 16°

(c) GV of 12°



**Figure 11.** Streamline distributions in the DT (left column) with and (right column) without anti-cavity fins at GVs of (**a**) 22°, (**b**) 16°, and (**c**) 12°.

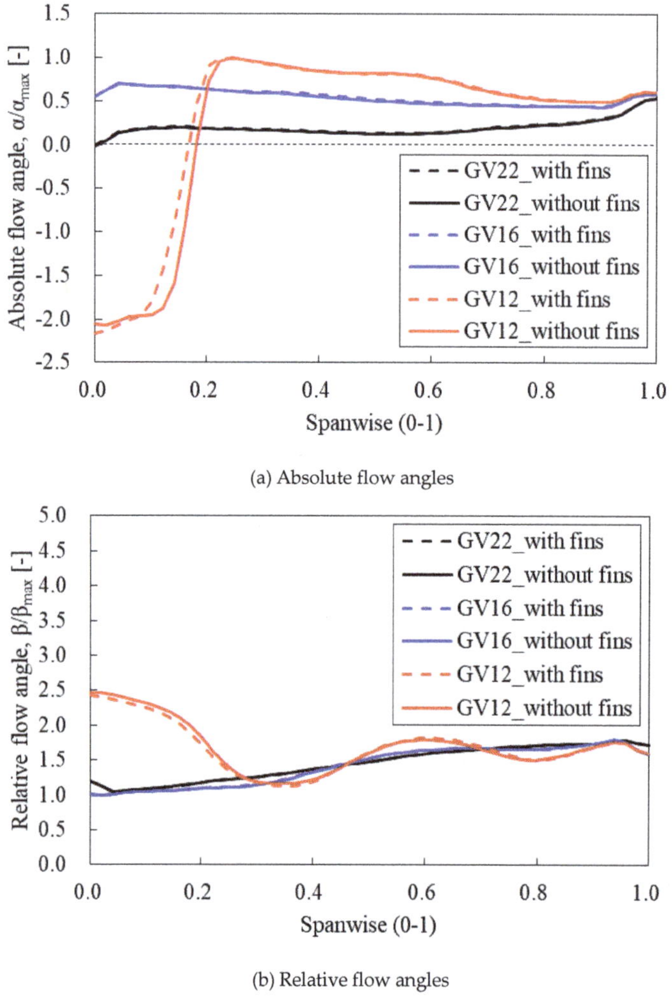

(a) Absolute flow angles

(b) Relative flow angles

**Figure 12.** (**a**) Absolute and (**b**) relative flow angle distributions along the spanwise direction at the outlet of the runner at GVs of 22°, 16°, and 12°.

To investigate the effect of the anti-cavity fins in detail, Figures 14 and 15 present the pressure distributions on the observed cross-sections (as shown in Figure 1) at the GV at 16° with and without anti-cavity fins, respectively. The cross-sections were examined at 0.1, 0.3, 0.6, and 0.9 $D_2$ in the direction of flow from the inlet of the DT, and the pressure values were normalized by the value of maximum pressure. The low-pressure regions in each cross-section of the pressure distributions were maintained with similar diameters, depending on time, and there was a tendency for the low-pressure regions to gradually decrease along the direction of flow. Furthermore, Figure 14 indicates that the low-pressure regions were clearly developed up to the cross-section of 0.3 $D_2$; however, because of the influence of the anti-cavity fins, these regions decreased considerably from 0.6 $D_2$. Meanwhile, the low-pressure regions were formed near the anti-cavity fins. It can be regarded that these regions were induced by the resistance of the anti-cavity fins to the tangential component of the absolute velocity, which increased as flow rate decreased as shown in Figure 10 (velocity triangle). Thus, via the low-pressure regions generated near the anti-cavity fins in Figure 14b, the formation sites of the low-pressure regions in

Figures 8 and 13 can be confirmed. Figure 15 clearly shows the low-pressure regions in the DT up to the cross-section of 0.9 $D_2$ without anti-cavity fins. Actually, the existing DT plays a role of recovering the static pressure in the flow; however in the low flow rate condition, the low-pressure regions indicated by the vortex rope with precession decreased with the anti-cavity fins rather than the decrease through the role of the DT itself. Therefore, it can be concluded that the anti-cavity fins effectively suppressed the generation of the vortex rope in the DT, particularly by decreasing the vertical length rather than by modifying the rotating diameters.

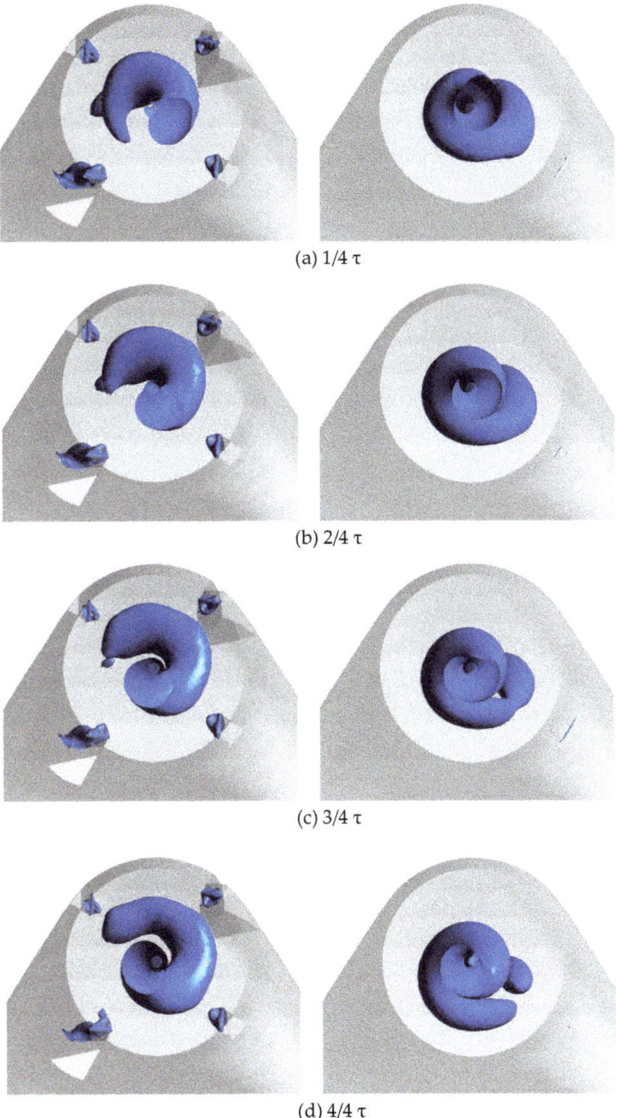

(a) 1/4 τ

(b) 2/4 τ

(c) 3/4 τ

(d) 4/4 τ

**Figure 13.** Iso-surface distributions of pressure in the DT by the top view from the runner at the GV at 16° (left column) with and (right column) without anti-cavity fins during one revolution of the runner. (**a**) 1/4, (**b**) 2/4, (**c**) 3/4, and (**d**) 4/4 τ.

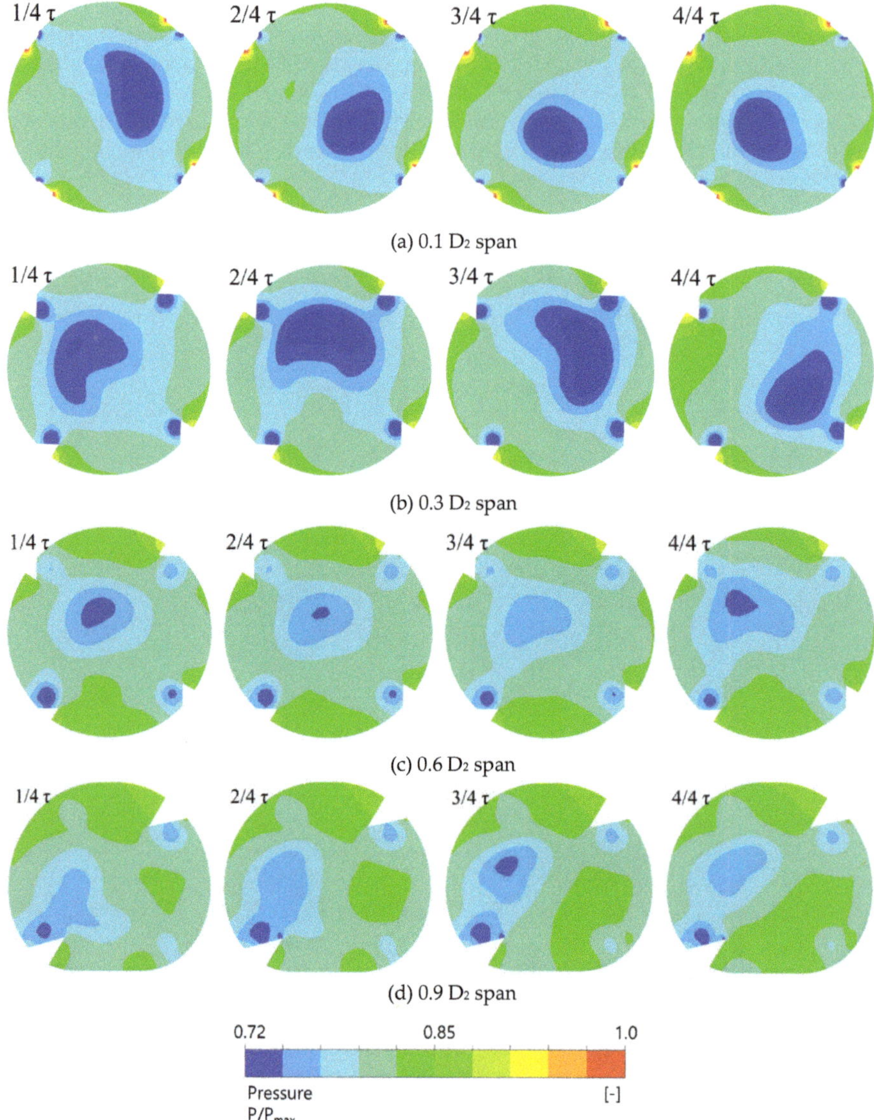



**Figure 14.** Pressure distributions on observed spans in the DT with anti-cavity fins at the GV at 16° during one revolution of the runner: span locations at (**a**) 0.1, (**b**) 0.3, (**c**) 0.6, and (**d**) 0.9 $D_2$.

To analyze the influence of the anti-cavity fins at GVs at 22°, 16°, and 12°, the time-averaged axial and circumferential velocity components were compared on the observed line of 0.6 $D_2$ (in Figure 1) in the DT, as shown in Figure 16. The abscissa indicates the measurement location relative to the diameter from the wall (0) to the wall (1) of the DT, and the velocity values were normalized by the value of maximum velocity. Without the addition of anti-cavity fins, the axial velocity of the GV at 22°, as shown in Figure 16a, decreased slightly near to the wall, whereas the relatively small velocity range altered according to the anti-cavity fins.

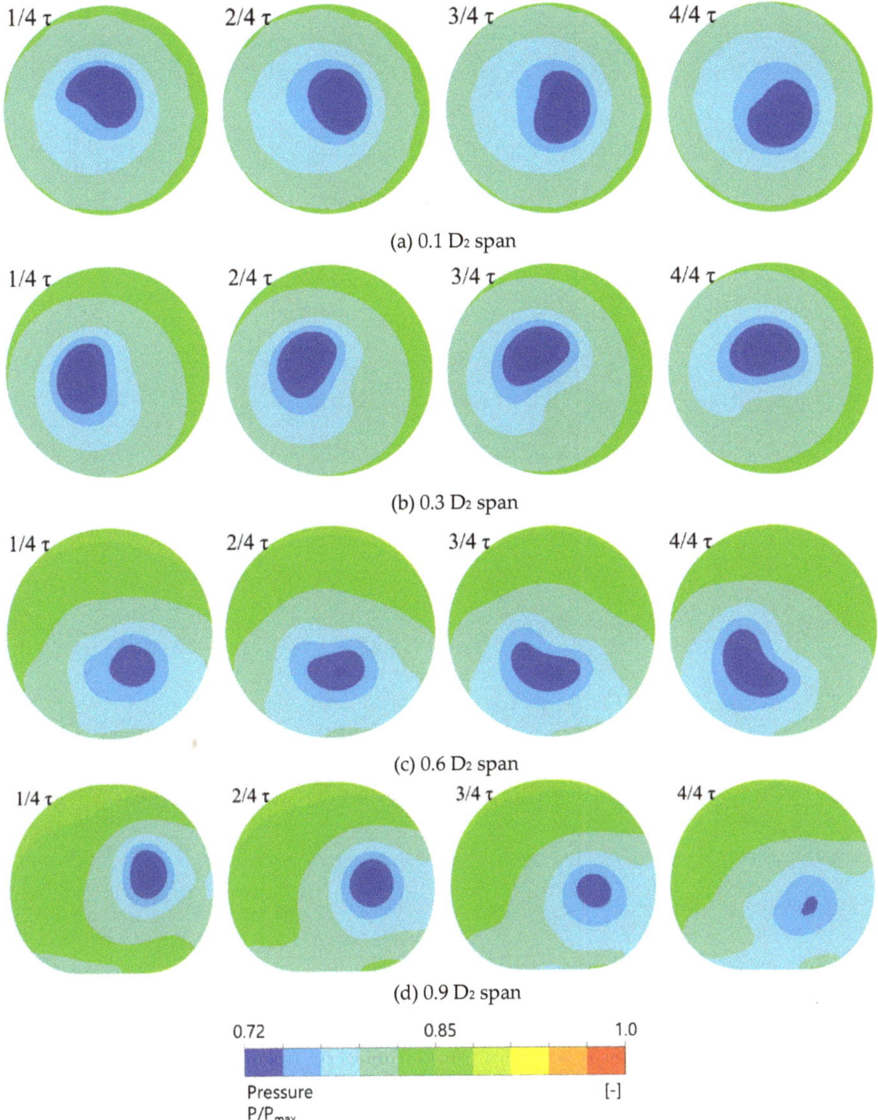



**Figure 15.** Pressure distributions on observed spans in the DT without anti-cavity fins at the GV at 16° during one revolution of the runner: span locations at (**a**) 0.1, (**b**) 0.3, (**c**) 0.6, and (**d**) 0.9 $D_2$.

However, the overall greatest difference in the axial velocity distribution was shown by the GV at 16° relative to the anti-cavity fins, and the backflow occurred near $d/D = 0.5$. With the GV at 12°, the backflow was generated in a relatively wide range of $d/D = 0.2$–0.8. A complicated internal flow without a vortex rope was demonstrated, and the difference in axial velocity was not shown to vary significantly according to the use of anti-cavity fins. Therefore, the anti-cavity fins had a relatively significant effect on the axial velocity at the GV at 16°; here, a vortex rope was formed, which was due to the shape characteristics of the anti-cavity fins installed in the axial direction concerning the flow direction.

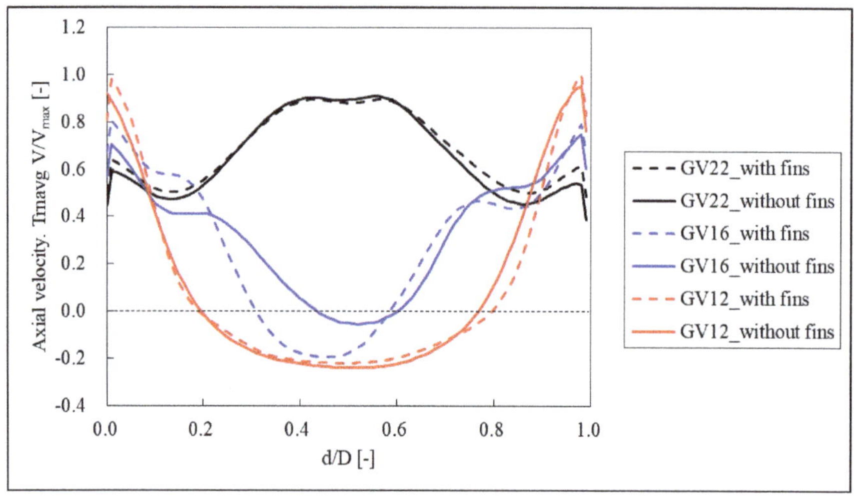

(a) Axial velocity distributions

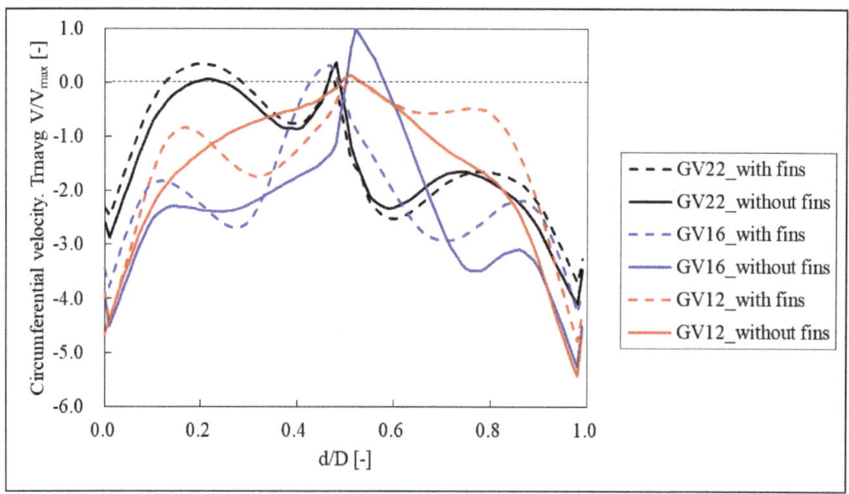

(b) Circumferential velocity distributions

**Figure 16.** (**a**) Axial and (**b**) circumferential velocity distributions on the observed line in the DT.

In terms of circumferential velocity distributions, the anti-cavity fins near the wall of the DT revealed a slight difference at the GV at 22°, as presented in Figure 16b. The difference between the maximum and the minimum circumferential velocity was relatively greater at the GV at 16° without the anti-cavity fins; the addition of anti-cavity fins effectively reduced this difference. Furthermore, because of the complicated internal flow, a difference in the circumferential velocity distribution was exhibited along with the anti-cavity fins at the GV at 12°. Therefore, this study considers that the anti-cavity fins influenced the velocity component in the circumferential direction rather than the axial direction in the DT; the internal flow was mainly influenced under the low flow rate conditions, during which the vortex rope was generated.

### 4.3. Unsteady Pressure Characteristics Relative to the Anti-Cavity Fins in the Draft Tube

To investigate the unsteady pressure characteristics according to the use of anti-cavity fins in the DT, this study compared the unsteady pressures obtained via fast Fourier transformation (FFT) analysis with GVs at 22°, 16°, and 12°, as shown in Figure 17. The pressure measuring points of 0.1, 0.3, 0.6, and 0.9 $D_2$ were used in the flow direction on the DT wall, and the value for the highest pressure was used for each height from the four measuring points (p1–p4), as indicated in Figure 4. The maximum magnitude normalized the pressure values, and the frequency was normalized by the rotational frequency, $f_n$, of the Francis turbine model. The highest first blade passing frequency (BPF) was shown on the 0.1 $D_2$ at the GV at 22° as the BEP condition. However, at the normalized frequency of 0.37 $f_n$ in the low-frequency region, relatively high-pressure characteristics were demonstrated before the first BPF at the GV at 16°, as the vortex rope with precession developed compared with other GV conditions. Previously, Kim et al. [25] numerically investigated similar unsteady pressure phenomena in the low-frequency region due to the vortex rope in the DT. Furthermore, the addition of anti-cavity fins at a measuring height of 0.1 $D_2$ slightly increased the unsteady pressure at the GV at 16°. This was due to the effect of the anti-cavity fins in reducing the length of the vortex rope of the GV at 16°, as can be seen in Figures 8, 14 and 15. However, the vortex rope remained near the inlet of the DT.

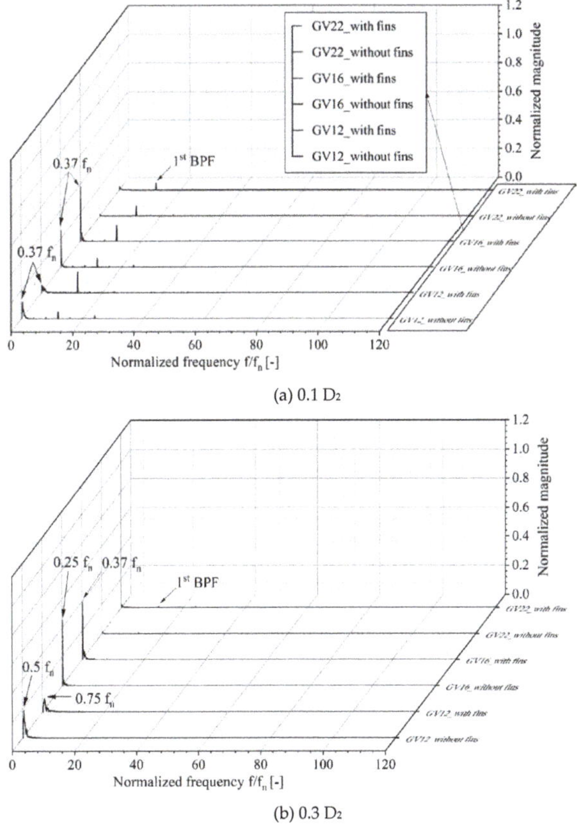


(a) 0.1 D₂



(b) 0.3 D₂

**Figure 17.** *Cont.*

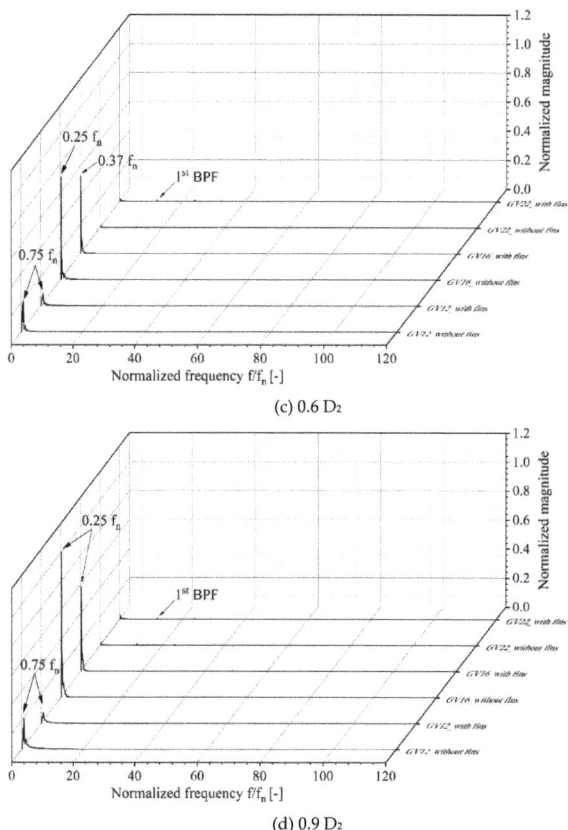

(c) 0.6 D₂

(d) 0.9 D₂

**Figure 17.** Normalized unsteady pressure characteristics along the flow direction in the DT: (**a**) 0.1, (**b**) 0.3, (**c**) 0.6, and (**d**) 0.9 $D_2$.

Unsteady pressure characteristics were exhibited at the normalized frequency of 0.37 $f_n$ and near the low-frequency regions before the first BPF at the GV at 12°, where a complicated internal flow without a vortex rope was evident. At all the observed GVs, the first BPF was gradually decreased from 0.3 $D_2$ (shown in Figure 17b) in the flow direction. Without the addition of anti-cavity fins, the GV at 16° at 0.3 $D_2$ demonstrated greater unsteady pressure characteristics at the normalized frequency of 0.25 $f_n$ than it did with them. Unsteady pressure at the 12° GV at 0.3 $D_2$ increased in the normalized frequency range of 0.5–0.75 $f_n$, which was greater than the magnitude at 0.1 $D_2$. In Figure 17c, both unsteady pressures at the GV at 16° increased more at the normalized frequency of 0.25–0.37 $f_n$ compared with those at 0.3 $D_2$; at the 12° GV, the normalized frequency of 0.75 $f_n$ was shown to be similar to the magnitude of 0.3 $D_2$. As is evident in Figure 17d, the GV at 12° exhibited similar unsteady pressure characteristics to those at 0.6 $D_2$.

The differences in the maximum magnitude of unsteady pressure due to anti-cavity fins were noticeable in the GV at 16°, and the magnitude of unsteady pressure with the anti-cavity fins was similar to the magnitude with anti-cavity fins at 0.6 $D_2$. Therefore, the unsteady pressure characteristics increased along the flow direction in the DT under conditions of low flow rate with a developed vortex rope; the use of anti-cavity fins effectively decreased the unsteady pressures. Particularly, the application of anti-cavity fins resulted in an approximate 41% reduction in the maximum magnitude of unsteady pressure at the 16° GV at a measured height of 0.9 $D_2$.

## 5. Conclusions

This study performed unsteady-state RANS analyses to investigate the effects of anti-cavity fins in the DT of a Francis turbine model on unsteady internal flow and pressure characteristics under low flow rate conditions. Reductions of around 0.5%–1.0% in hydraulic performance within the range of observed GV angles were observed, and the head losses of each component in the Francis turbine model were compared via the application of anti-cavity fins. Furthermore, the magnitudes and locations of the vortex ropes in the DT were verified via iso-surface distributions both with and without anti-cavity fins according to varying flow rate conditions. The existence of low-pressure regions due to flow resistance at the prominent air injection outlet on the anti-cavity fins was also confirmed.

The causes of the complicated flow phenomena in the DT respective to varying flow rates were confirmed via an analysis of the velocity triangles at the runner outlet. Furthermore, quantitative and qualitative investigations were conducted using the flow angle distributions at the runner outlet and the streamline distributions in the DT. A comparison of velocity distributions on the observed line confirmed the effects of the anti-cavity fins on the axial and circumferential velocity components in the DT. Furthermore, FFT analyses confirmed that the largest magnitudes of unsteady pressure were observed under low flow rate conditions with developed vortex ropes. There was a tendency for these magnitudes to gradually increase along the flow direction in the DT. Additionally, an approximate 41% reduction in the maximum unsteady pressure was confirmed with the application of anti-cavity fins.

Therefore, the use of anti-cavity fins was confirmed to affect the degradation of hydraulic performance under conditions of low flow rate, including BEP; however, the effects of reducing the unsteady pressure in the low-frequency regions were confirmed in the DT, which induced operational instability in the Francis turbine model. Thus, the anti-cavity fins can be presented as one alternative to suppress unsteady pressure fluctuations with vortex rope in the DT. Furthermore, by minimizing the loss induced by the anti-cavity fins through the optimization of the shape and length of the anti-cavity fins, it can be expected to improve the hydraulic performance with suppressing pressure fluctuation effectively. In the addition, based on the results of this research, the unsteady flow and pressure phenomena in the DT will be investigated in a future study by injecting air into the DT.

**Author Contributions:** Conceptualization, validation, investigation, data curation, S.-J.K.; Y.-S.C.; Y.C.; J.-W.C.; J.-J.H.; W.-G.J. and J.-H.K.; resources, Y.-S.C.; Y.C.; J.-W.C.; J.-J.H. and J.-H.K.; writing—original draft preparation, S.-J.K.; Y.-S.C. and J.-H.K.; writing—review and editing, supervision, project administration, J.-H.K.; funding acquisition, Y.C.; J.-W.C. and J.-J.H. All authors have read and agreed to the published version of the manuscript.

**Funding:** This research was funded by the Korea Agency for Infrastructure Technology Advancement under the Ministry of Land, Infrastructure, and Transport [grant number 20IFIP-B128593-04].

**Conflicts of Interest:** The authors declare no conflict of interest.

## Nomenclature

| | |
|---|---|
| $N_s$ | Specific speed |
| $N$ | Rotational speed (rpm) |
| $P$ | Power |
| $H$ | Head |
| $n_{ED}$ | Speed factor |
| $n$ | Rotational frequency (rps) |
| $D$ | Diameter of runner |
| $g$ | Acceleration due to gravity |
| $Q_{ED}$ | Discharge factor |
| $Q$ | Discharge |
| $E$ | Energy |
| $E_{nD}$ | Energy coefficient |
| $\eta$ | Efficiency |
| $\delta$ | Relative scalable losses |
| $\rho$ | Water density |
| $T$ | Torque of runner |
| $\omega$ | Angular velocity |

## References


1. Kim, S.J.; Choi, Y.S.; Cho, Y.; Choi, J.W.; Kim, J.H. Effect of Runner Blade Thickness on Flow Characteristics of a Francis Turbine Model at Low Flowrates. *J. Fluids Eng.* **2020**, *142*, 031104. [CrossRef]
2. Nishi, M.; Liu, S. An outlook on the draft-tube-surge study. *Int. J. Fluid Mach. Syst.* **2013**, *6*, 33–48. [CrossRef]
3. Eichhorn, M.; Taruffi, A.; Bauer, C. Expected load spectra of prototype Francis turbines in low-load operation using numerical simulations and site measurements. *J. Phys. Conf. Ser.* **2017**, *813*, 012052. [CrossRef]
4. Feng, J.J.; Li, W.F.; Wu, H.; Lu, J.L.; Liao, W.L.; Luo, X.Q. Investigation on pressure fluctuation in a Francis turbine with improvement measures. *IOP Conf. Ser. Earth Environ. Sci.* **2014**, *22*, 032006. [CrossRef]
5. Favrel, A.; Müller, A.; Landry, C.; Yamamoto, K.; Avellan, F. Study of the vortex-induced pressure excitation source in a Francis turbine draft tube by particle image velocimetry. *Exp. Fluids* **2015**, *56*, 215. [CrossRef]
6. Susan-Resiga, R.; Muntean, S.; Stein, P.; Avellan, F. Axisymmetric swirling flow simulation of the draft tube vortex in Francis turbines at partial discharge. *Int. J. Fluid Mach. Syst.* **2009**, *2*, 295–302. [CrossRef]
7. Nicolet, C.; Zobeiri, A.; Maruzewski, P.; Avellan, F. Experimental investigations on upper part load vortex rope pressure fluctuations in Francis turbine draft tube. *Int. J. Fluid Mach. Syst.* **2011**, *4*, 179–190. [CrossRef]
8. Zuo, Z.; Liu, S.; Liu, D.; Qin, D. Numerical predictions and stability analysis of cavitating draft tube vortices at high head in a model Francis turbine. *Sci. China Technol. Sci.* **2014**, *57*, 2106–2114. [CrossRef]
9. Altimemy, M.; Attiya, B.; Daskiran, C.; Liu, I.H.; Oztekin, A. Mitigation of flow-induced pressure fluctuations in a Francis turbine operating at the design and partial load regimes—LES simulations. *Int. J. Heat Fluid Flow* **2019**, *79*, 108444. [CrossRef]
10. Susan-Resiga, R.; Vu, T.C.; Muntean, S.; Ciocan, G.D.; Nennemann, B. Jet control of the draft tube vortex rope in Francis turbines at partial discharge. In Proceedings of the 23rd IAHR Symposium Conference, Yokohama, Japan, 17–21 October 2006; pp. 67–80.
11. Li, W.F.; Feng, J.J.; Wu, H.; Lu, J.L.; Liao, W.L. Numerical investigation of pressure fluctuation reducing in draft tube of Francis turbines. *Int. J. Fluid Mach. Syst.* **2015**, *8*, 202–208. [CrossRef]
12. Anup, K.C.; Lee, Y.H.; Thapa, B. CFD study on prediction of vortex shedding in draft tube of Francis turbine and vortex control techniques. *Renew. Energy* **2016**, *86*, 1406–1421.
13. Chen, Z.; Choi, Y.D. Suppression of cavitation in the draft tube of Francis turbine model by J-Groove. *Proc. Inst. Mech. Eng. Part C J. Mech. Eng. Sci.* **2019**, *233*, 3100–3110. [CrossRef]
14. International Electrotechnical Commission. *Hydraulic Turbines, Storage Pumps and Pump-Turbines—Model Acceptance Tests, Standard No. IEC 60193*; International Electrotechnical Commission: Geneva, Switzerland, 1999.
15. *ANSYS CFX-19.1, ANSYS CFX-Solver Theory Guide*; ANSYS Inc.: Canonsburg, PA, USA, 2018.
16. Richardson, L.F., IX. The approximate arithmetical solution by finite differences of physical problems involving differential equations, with an application to the stresses in a masonry dam. *Philos. Trans. R. Soc. Lond. Ser. A Contain. Pap. Math. Phys. Character* **1911**, *210*, 307–357.
17. Richardson, L.F.; Gaunt, J.A., VIII. The deferred approach to the limit. *Philos. Trans. R. Soc. Lond. Ser. A Contain. Pap. Math. Phys. Character* **1927**, *226*, 299–361.
18. Celik, I.B.; Ghia, U.; Roache, P.J.; Freitas, C.J.; Coleman, H.; Raad, P.E.; Celik, I.; Freitas, C.; Coleman, H.P. Procedure for estimation and reporting of uncertainty due to discretization in CFD applications. *J. Fluids Eng.* **2008**, *130*, 078001.
19. Kim, S.J.; Choi, Y.S.; Cho, Y.; Choi, J.W.; Hyun, J.J.; Joo, W.G.; Kim, J.H. Analysis of the Numerical Grids of a Francis Turbine Model through Grid Convergence Index Method. *KSFM J. Fluid Mach.* **2020**, *23*, 16–22. (In Korean) [CrossRef]
20. Zwart, P.J.; Gerber, A.G.; Belamri, T. A two-phase flow model for predicting cavitation dynamics. In Proceedings of the Fifth International Conference on Multiphase Flow, Yokohama, Japan, 30 May–4 June 2004.
21. Egorov, Y.; Menter, F. Development and application of SST-SAS turbulence model in the DESIDER project. *Adv. Hybrid RANS-LES Model* **2008**, 261–270. [CrossRef]
22. Menter, F.R.; Egorov, Y. The scale-adaptive simulation method for unsteady turbulent flow predictions. Part 1: Theory and model description. *Flow Turbul. Combust.* **2010**, *85*, 113–138. [CrossRef]
23. *Korea Agency for Infrastructure Technology Advancement, Report Development of Construction Technology for Medium Sized Hydropower Plant*; Report No. 17IFIP-B128593-01; Anyang-si, Korea, 2017; Available online: http://icities4greengrowth.in/people/agency/kaia-korea-agency-infrastructure-technology-advancement (accessed on 20 May 2020).

24. Chen, Z.; Singh, P.M.; Choi, Y.D. Francis turbine blade design on the basis of port area and loss analysis. *Energies* **2016**, *9*, 164. [CrossRef]
25. Kim, S.J.; Suh, J.W.; Choi, Y.S.; Park, J.; Park, N.H.; Kim, J.H. Inter-Blade Vortex and Vortex Rope Characteristics of a Pump-Turbine in Turbine Mode under Low Flow Rate Conditions. *Water* **2019**, *11*, 2554. [CrossRef]

© 2020 by the authors. Licensee MDPI, Basel, Switzerland. This article is an open access article distributed under the terms and conditions of the Creative Commons Attribution (CC BY) license (http://creativecommons.org/licenses/by/4.0/).

*Article*

# Two-Objective Optimization of a Kaplan Turbine Draft Tube Using a Response Surface Methodology

Riccardo Orso [1], Ernesto Benini [1,*], Moreno Minozzo [2], Riccardo Bergamin [2] and Andrea Magrini [1]

[1] Department of Industrial Engineering, University of Padua, Via Venezia 1, 35131 Padua, Italy; riccardo.orso@studenti.unipd.it (R.O.); andrea.magrini@phd.unipd.it (A.M.)
[2] ZECO S.r.l., via Astico 52/c, 36030 Fara Vicentino, Italy; moreno.minozzo@zeco.it (M.M.); riccardo.bergamin@zeco.it (R.B.)
* Correspondence: ernesto.benini@unipd.it

Received: 20 August 2020; Accepted: 16 September 2020; Published: 18 September 2020

**Abstract:** The overall cost of a hydropower plant is mainly due to the expenses of civil works, mechanical equipment (turbine and control units) and electrical components. The goal of a new draft tube design is to obtain a geometry that reduces investment costs, especially the excavation ones, but the primary driver is to increase overall machine efficiency, allowing for a reduced payback time. In the present study, an optimization study of the elbow-draft tube assembly of a Kaplan turbine was conducted. First, a CFD model for the complete turbine was developed and validated. Next, an optimization of the draft tube alone was performed using a design of experiments technique. Finally, several optimum solutions for the draft tube were obtained using a response surface technique aiming at maximizing pressure recovery and minimizing flow losses. A selection of optimized geometries was subsequently post-checked using the validated model of the entire turbine, and a detailed flow analysis on the obtained results made it possible to provide insights into the improved designs. It was observed that efficiency could be improved by 1% (in relative terms), and the mechanical power increased by 1.8% (in relative terms) with respect to the baseline turbine.

**Keywords:** Kaplan turbine; draft tube optimization; CFD analysis; DOE; response surface

---

## 1. Introduction

In recent decades, environmental policies have been oriented towards reducing energetic dependence on fossil fuels, leading to renewed and increased investments in the hydroelectric sector; such investments have been primarily directed to optimizing both new designs and existing installations, thus allowing for more efficient plants and reduced payback times. Kaplan hydro turbines have been adopted for a long time to deal with efficient energy production in the range of high specific speeds [1]. The use of movable vaned distributors and runner blades, in fact, allows to limit incidence flow losses and achieve high conversion efficiency over a wide range of flow rates, from 40% to 120% of the nominal value. The main drawback of Kaplan turbines is that their size increases as the flow rate gets higher, and the cost attributed to civil works becomes more and more substantial.

In these machines, the draft tube is an essential component as it makes it possible to recover a significant fraction of the kinetic energy leaving the runner by conversion into static pressure. This, in turn, leads to creating an additional suction head downstream of the runner, which allows to increase the effective head that the runner is able to deliver [2]. In addition, the draft tube plays a fundamental role in determining turbine efficiency since the height at which the tube is installed contributes to a large percentage of the total net head that can be recovered [3].

Moreover, it is one of the most challenging parts to analyze from a fluid flow perspective due to the interaction of many complex flow features, such as unsteadiness, turbulence, separation, streamline curvature, secondary flow, swirl, and vortex breakdown [4].

Studies involving Kaplan turbine components made use of both steady and unsteady Reynolds-Averaged Navier-Stokes (RANS), the Scale-Adaptive Simulation (SAS) and hybrid RANS-LES (Large Eddy Simulations) models, such as the Zonal Large Eddy Simulation (ZLES) or the Detached Eddy Simulations (DES). Jošt reported an improvement in characteristic parameter predictions for a Kaplan turbine using unsteady models and ZLES [5]. Marjavaara [4] performed steady and unsteady CFD simulations of a draft tube using RANS and DES. The characteristic quantities and the local flow features obtained were highly dependent on the model and the inflow boundary conditions. These models can accurately predict turbulent phenomena in the draft tube but require very large computational resources. As such, their adoption in optimization studies seems, today, still infeasible, while validated RANS approaches offer a very good compromise between accuracy and computational effort [3,5]. Li [6] used steady RANS to study the hydraulic performance of a Kaplan turbine's components. Brijkishore [7] evaluated the influence of runner solidity using the k-$\Omega$ SST turbulence model in steady computations. Liu [8] carried out transient analyses on the runner lifting-up during load rejection using the same RANS model.

Past studies on draft tube optimizations were limited in terms of representativeness. In fact, while the draft tube can be acceptably optimized as an isolated component around its design point, therefore neglecting the interference effects with the turbine runner, the performances of optimal configurations have not been post-checked using the entire turbine model for a posteriori validation of the complete turbine installation. Lyutov [9] described a coupled optimization of runner and draft tubes, where their geometries were varied simultaneously, allowing to achieve a 0.3% higher efficiency gain compared to single-runner optimization. Ciocan [10] presented a draft tube optimization from part to full load, using a parameterized velocity field at the tube entry based on a swirl-free velocity profile for the runner. Eisinger [11] reported an automatic draft tube optimization, but no coupling of the inflow conditions with the attached runner was present. Puente [12] employed RANS simulations with the k-$\varepsilon$ turbulence model for the same purpose, using normalized velocity profiles at the duct inlet, derived from previous turbine simulations. Sosa [13] analyzed the influence of draft tube modification using a steady computational model with a single-channel guide vane and runner connected to the tube via a mixing plane. The geometric modifications based on a two-parameter description were meant to represent erosion or material addition in the wall. Schiffer [14] performed steady-state simulations of a complete model, including distributors, runners, and a draft tube, at several operating points. The tube was then manually adapted to optimize the turbine's efficiency while keeping construction costs low.

In the present paper, a fully 3D Kaplan turbine RANS CFD model was first built and validated against available experimental data in terms of net head, flow rate, efficiency, and mechanical power.

Following this, the pressure and velocity distribution at the draft tube inlet were extracted from previous computations. Finally, a CFD model that contained only the draft tube and its outlet extension was created and used in the optimization process. Unlike previous studies, the optimization was carried out based on two-objectives which are of prominent interest to draft tube designers, i.e., its pressure recovery and its total pressure loss coefficient. Also, a noteworthy improvement with respect to past studies relies on the verification of the obtained results: in fact, once the optimal draft tube geometries had been found, they were tested back using the full machine CFD model to find out their impact on the turbine's efficiency and to analyze the sources of draft tube losses in detail.

## 2. Materials and Methods

### 2.1. Draft Tube Performance

The datum, or baseline, draft tube was connected to a Kaplan turbine runner installed in a 576 kW hydraulic power plant located in the Adda river (Italy), as depicted in Figure 1a. The original geometry adopted for the elbow draft tube is schematically represented in Figure 1b.

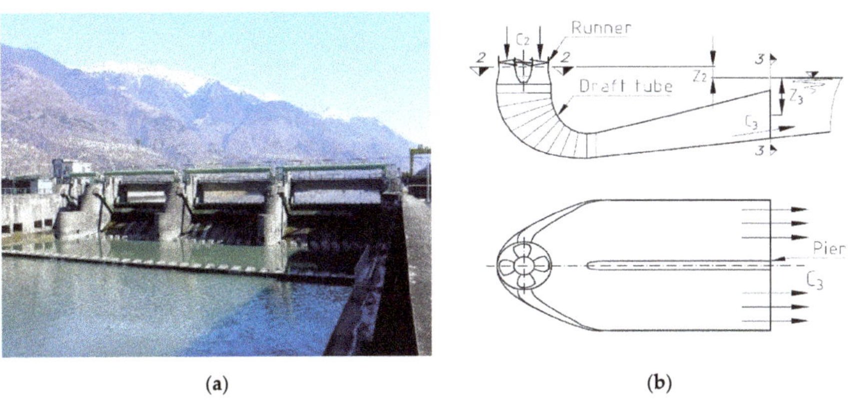


(a)                                            (b)

**Figure 1.** Original power plant: (**a**) overall external view; (**b**) baseline elbow-draft tube geometry.

As mentioned above, the draft tube behaves as a diffuser acting to recover part of the kinetic energy from Section 2 to 3, according to its prescribed area distribution along the equivalent centerline.

Two coefficients are used here to quantify the performance of a draft tube, namely a pressure recovery factor ($Cp$) and a drag coefficient ($Cd$), respectively defined as:

$$C_p = \frac{p_3 - p_2}{\frac{1}{2}\rho c_2^2}; \quad C_d = \frac{p_{tot-2} - p_{tot-3}}{\frac{1}{2}\rho c_2^2} \tag{1}$$

where $p$ is the static pressure and $p_{tot}$ the total pressure; subscripts 2 and 3 refer to the station downstream of the runner and on the draft tube outlet, respectively; $\rho$ is the water density; $c$ is the absolute velocity.

The values of these coefficients are strictly related to the area distribution and are notoriously conflicting in such a way that the more intense the diffusion is (high $Cp$), the bigger the head loss is (high $Cd$) that might occur as a result of flow separation, secondary losses, etc. Therefore, a good draft tube features the highest possible $Cp$ along with the minimum achievable $Cd$. These two coefficients were used as objective functions in the multi-objective optimization study described later.

### 2.2. CFD Model Setup and Validation

The steady-state RANS CFD model of the complete turbine adopted included complete spiral casing with stay vanes, guide vanes, and runner and draft tubes (Figure 2). The final grids were chosen based on a mesh sensitivity analysis and their details and statistics are provided in Table 1. The influence of mesh size on the draft tube performance parameters is shown in Figure 3, where a limited sensitivity to the grid size can be observed, starting from the third level, having 1.8 M cells. As for boundary conditions, a total pressure was defined at the inlet of the spiral casing and a static pressure equal to 0 Pa (relative to the atmospheric pressure value of 1 atm) was set at the outlet of the draft tube.

The total pressure at the inlet boundary corresponded to the total energy that the plant could process and was calculated from the net head and the flow rate that the turbine was subjected to. The flow solver was Ansys CFX 19.2. Since the runner was not modelled as a rotating mesh, a multiple-reference-frame approach was chosen, and a mixing-plane condition was set both at the distributor–runner and runner–draft tube interfaces. Among all the turbulence models (all the variants of both k-ε and k-ω available in Ansys CFX 19.2 were actually tested), it was observed that the k-ω SST [15] provided the most accurate predictions, but it was also the one that required the highest computational effort; for this reason, the k-ω SST was used in off-loop analyses on the complete machine model, while the standard k-ε was used in the optimization due to its cheaper requirement. A physical timescale corresponding to 1° of runner rotation and a high-resolution scheme were set.

In order to assure improved numerical stability, 50 iterations were run at first using first-order schemes. These results were adopted as initial values for the high-resolution scheme.

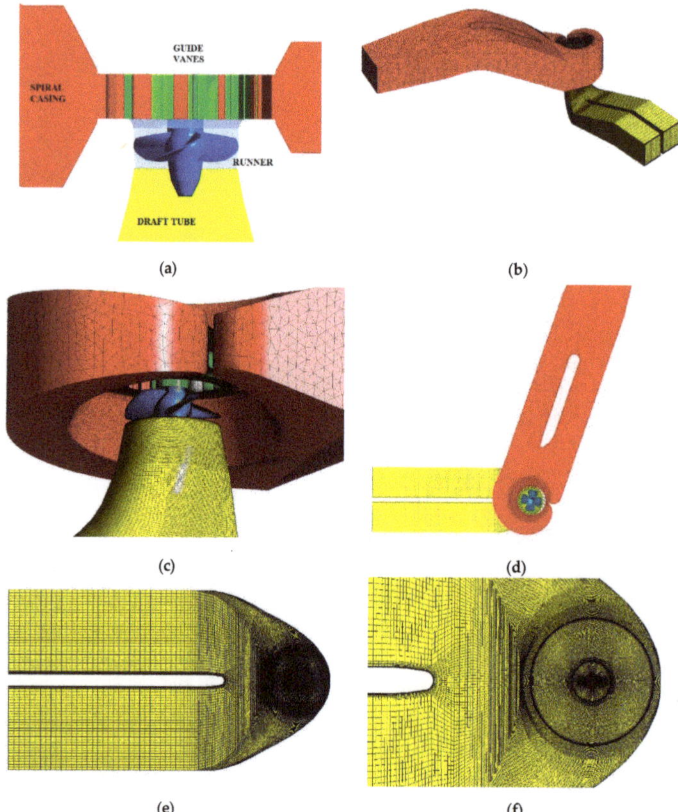


**Figure 2.** Turbine computational domains: (**a**) close up of the runner domain in meridional view; (**b**) 3D view of the overall domain; (**c**) close up of the computational mesh in the region close to the runner/draft tube interface; (**d**) top view of the overall domain; (**e**,**f**) details of the draft tube mesh.

**Table 1.** Details of the grids used in the subdomains.

| Subdomain | Meshing Tool | Mesh Type | Number of Nodes | Number of Elements | Average Y+ | Number of Boundary Layer Prism | Boundary Layer Growth Ratio |
|---|---|---|---|---|---|---|---|
| Spiral casing with stay vanes | ICEM-CFD | Tetra | 2,580,807 | 7,685,535 | 62 | 6 | 1.2 |
| Guide vanes | Turbogrid | Hexa | 1,939,968 | 1,781,112 | 95 | 40 | 1.1 |
| Distributor ring | ICEM-CFD | Hexa | 478,720 | 462,672 | 28 | 12 | 1.5 |
| Runner | Turbogrid | Hexa | 1,023,320 | 964,180 | 110 | 50 | 1.05 |
| Draft tube with extension | ICEM-CFD | Hexa | 1,882,944 | 1,842,398 | 65 | 25 | 1.2 |

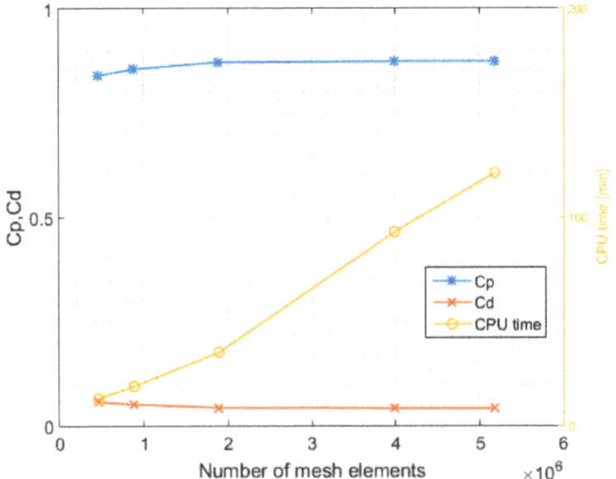

**Figure 3.** Mesh independency analysis for the draft tube.

During the convergence runs, several variables of interest were monitored, such as flow rate, efficiency, static pressure at draft tube inlet, and mechanical power output. Final results were available after 3000 iterations on average and almost 72 h of CPU calculations. All calculations were performed on a server equipped with Intel® Xeon® CPU x5650 processors using parallel solutions on 10 multiple cores. Convergence was assessed when all variables of interest showed a variation lower than 0.08%.

Results from both experiments and simulations are presented in Figure 4. All the experimental data (net head, flow rate, efficiency, and mechanical power) were taken in compliance with the European IEC EN 60041:1991 standard [16] and included the appropriate uncertainty bandwidth. Power and efficiency data from CFD were calculated as follows:

$$P = M \cdot \omega; \quad \eta = \frac{P}{\rho Q g H} \tag{2}$$

where $M$ is the total torque on the runner blades, $\omega$ is the rotational speed, $Q$ is the flow rate, and $H$ is the net head. As can be observed in Figure 4, the power and efficiency data were within the experimental measurement uncertainty range. The steady RANS CFD model set up appeared, therefore, to be able to reproduce the machine performance with sufficient accuracy and acceptable computational time. From this validated model, the runner–draft tube interface conditions, that is, the distributions of the $u$, $v$, $w$ (respectively tangential, absolute, and relative velocity components), were extracted to be used as inlet boundary conditions during the optimization process.

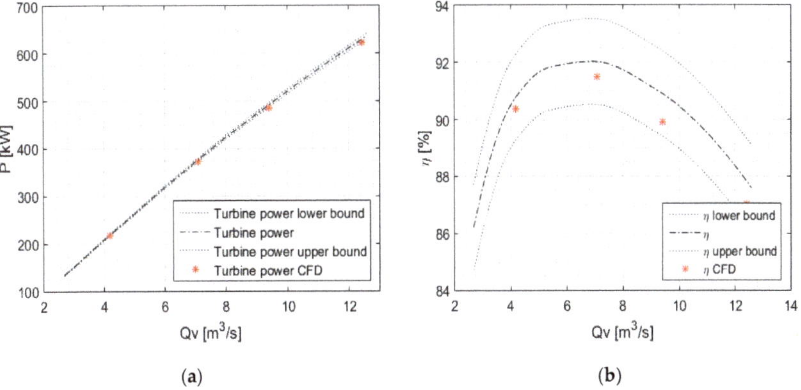

**Figure 4.** Validation results: (**a**) power data; (**b**) efficiency.

### 2.3. Draft Tube Geometry and Parameterization

The draft tube geometry of Kapan turbines is complicated by the different shapes of its inlet and outlet sections. The first is circular due to the interface with the runner, while the second is often rectangular. Moreover, the draft tube features a 90° elbow to minimize excavation costs and to improve powerhouse compactness. As a result, a large number of design parameters is necessary to provide successful geometry parameterization. In the present paper, the draft tube was parameterized using a mean line and a number of cross-sections stacked along it.

The mean line was designed to be composed by a first straight section, made up of two segments, one related to the divergent cone, and one to the cylindrical section; a second curvilinear line was related to the elbow shape and controlled by a three-point Bézier polygon; finally, a segment was connected to the last section of the draft tube.

The first straight part of the mean line was described by two parameters—the two segments' lengths. For the Bézier polygon, four additional variables were introduced—the coordinates of the polygon-points (the coordinates of the first Bézier polygon point were derived from the first straight part). The last parameter was the length of the last straight segment. In total, 7 variables were involved in the parameterization of the mean line. The area distribution was managed using a 4th-degree polynomial as a function of the mean line coordinate $x$:

$$A(x) = P1 \times \left(P2 \times ax^4 + bx^3 + cx^2 + dx\right) + e \tag{3}$$

In Figure 5, the area distribution of the original geometry is represented using dots along with the polynomial interpolation. From the original geometry, it was possible to obtain the polynomial coefficients ($a, b, c, d, e$) and use P1 and P2 as decision variables.

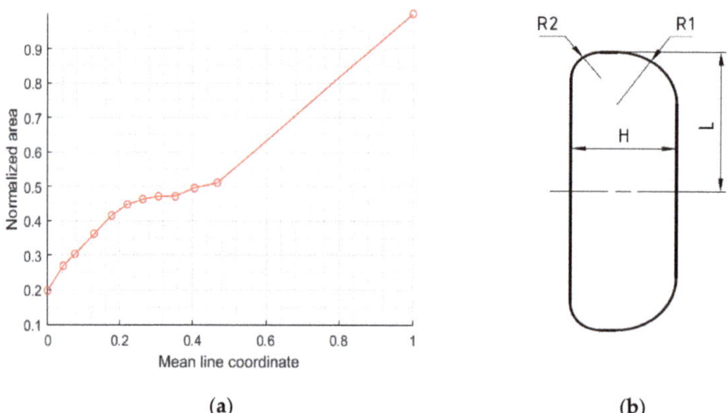


(a)      (b)

**Figure 5.** Draft tube geometry definition: (**a**) area distribution; (**b**) shape of a generic cross section.

Overall, the draft tube was parameterized using 9 decision variables. In order to avoid the need for further parameters to account for the shape change from inlet to exit sections, the geometric similarity with the baseline tube was preserved, keeping the geometric ratios R1/L, R2/L, and H/L for each section along the mean line equal to those of the datum.

### 2.4. Optimization: Problem Formulation and Tools

As previously mentioned, the optimization study was conducted on the draft tube alone by taking the inlet boundary conditions from the full machine model at the runner–draft tube interface. As a result, the flow domain included the runner discharge cone along with the runner tube interface (which form the inlet boundary), the first divergent part, the elbow, the last divergent part, and the draft tube prolongation leading to the outlet section.

The optimization was multi-objective, with the purpose of maximizing $Cp$ and minimizing $Cd$ at the same time using a Pareto approach. As for the constraints, the optimization had to lead to solutions that were interchangeable with the baseline case currently installed without any modifications regarding the runner. As a result, during the entire procedure, the baseline inlet and runner cone (both belonging to the runner) remained fixed. A further constraint was related to the global draft tube depth which affects the cost of civil works. The outlet depth could vary in all directions but with the limit of always remaining with a certain margin under the free water surface of the tailrace.

A response surface methodology for the optimization tool was used [17]. The first step was a design space exploration using design of experiments (DOE) in order to understand how the design parameters were related to each other and which were the most significant ones. A "central composite design" (CCD) DOE technique was adopted, which provided a screening set of 150 samples to determine the overall trends of the metamodel. The latter was built using the genetic aggregation method [18] and the Pareto optimal solutions were finally obtained (Figure 6a). A check-in-the-loop of optimal solutions was necessary to evaluate the response surface goodness using CFD (Figure 6b); the best-so-far optimum individuals from the response surface were taken and verified using a dedicated CFD simulation, then inserted back in the DOE, which, in turn, updated the response surface from which a new Pareto front was obtained. The loop was kept running until a stable and invariant Pareto optimal set was determined.

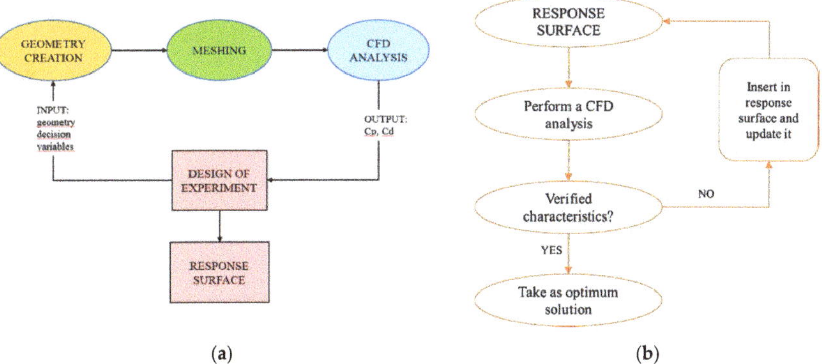


**Figure 6.** Optimization procedure: (**a**) general loop; (**b**) response surface update workflow.

It is worth mentioning that during the loop iterations, the response surface method allowed for an adaptive refinement in the search regions around maximum $C_P$ and minimum $C_D$ so as to make it possible to have a wider and more uniform Pareto front.

## 3. Results and Discussion

The final Pareto front is depicted in Figure 7, from which a subset of relevant configurations could be extracted: the baseline solution is marked with a black diamond, while the optimum $C_D$, the optimum $C_P$, and a good compromise between the $C_D$ and $C_P$ are marked using a red triangle, a yellow circle, and a green square, respectively. Dominated individuals are indicated as other points in the figure. The performance data of optimized individuals are given in Table 2. The solution corresponding to the minimum $C_D$ (Candidate 1) featured a reduction count $\Delta C_D = -0.0254$ (i.e., $-2.54$ percentage point reduction), while the one corresponding to the maximum $C_P$ (Candidate 3) showed $\Delta C_P = +0.0574$ (i.e., $+5.7$ percentage point improvement).

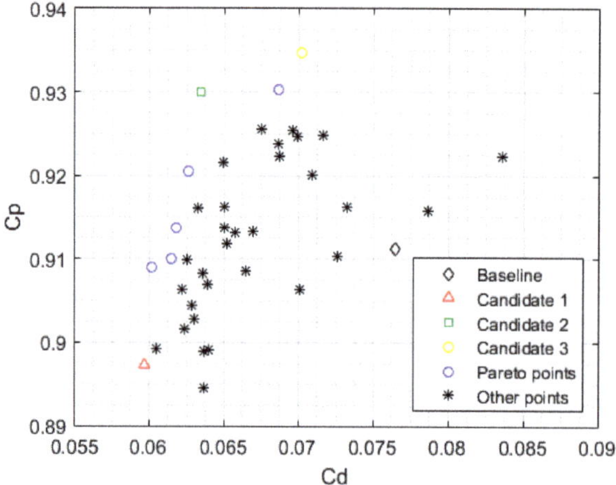

**Figure 7.** Optimization results: Pareto solutions.

**Table 2.** Optimization results: Pareto solutions.

| Performance Data | $C_d$ | $C_p$ |
|---|---|---|
| Baseline | 0.0851 | 0.9186 |
| Candidate 1 | 0.0597 | 0.8974 |
| Candidate 2 | 0.0634 | 0.9300 |
| Candidate 3 | 0.0702 | 0.9347 |

### 3.1. Optimal Draft Tube Configurations

In Figure 8, the area distribution along the mean line coordinate for the Pareto optimal subset of solutions compared to the baseline one is shown. It can be seen that optimal solutions exhibited a much lower increase in area distribution from the beginning of the draft the tube (mean line coordinate > 0.05, see Figure 8b); this is beneficial to limit adverse phenomena during flow diffusion in this zone as confirmed, also, by the distribution of Cd versus the mean line coordinate (Figure 10a), particularly visible in the case of Candidate 1.

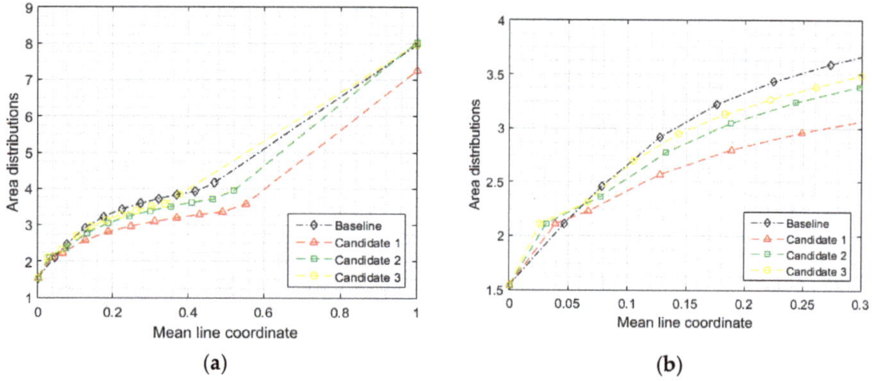

**Figure 8.** Optimization results: (**a**) area distribution (**b**) close up in the range $x \in [0\text{–}0.3]$.

In Figure 9, a detailed comparison between the optimal solutions and the baseline is presented. For the first draft tube part (until the 20% of the mean line), very similar characteristics can be observed, while the most remarkable differences are visible in the elbow and in the last part.

**Figure 9.** Comparison among geometries in meridional view: Baseline (grey); Candidate 1 (red); Candidate 2 (green); Candidate 3 (yellow).

Candidate 1 had a very similar geometry compared to the baseline, while Candidate 2 featured an elbow with a much higher radius of curvature. Candidate 3, on the other hand, had much larger dimensions than the baseline and this has a major impact on the excavation costs. As it will be seen later, Candidate 2 was the best in performance and this is due to the shape of the elbow, a fact that emphasizes the importance of producing a good design in the region where a big part of the diffusion is realized by the draft tube.

### 3.2. Post-Check Validation on Full Machine and Result Transposition

A number of Pareto optimal draft tube geometries were obtained from the optimization and were subsequently analyzed, a posteriori, by connecting them into the CFD model of the entire machine in order to assess their influence on critical plant characteristics, such as mechanical power and efficiency. In Figure 10, both the $C_d$ and $C_p$ of the optimal draft tube geometries computed using the full CFD model of the turbine are plotted as functions of the mean line coordinate.

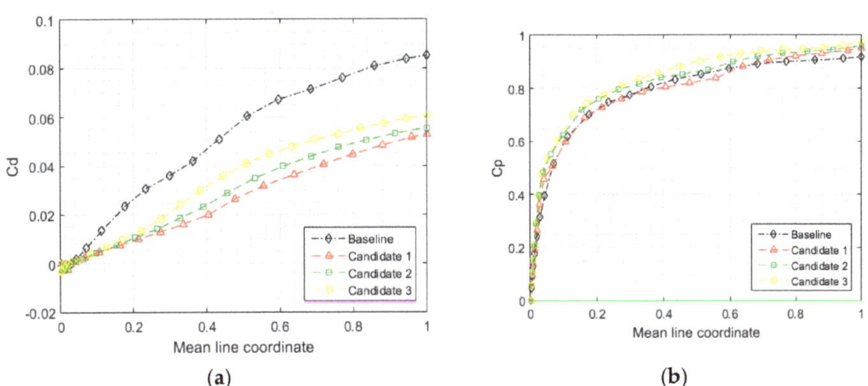


**Figure 10.** Post-check results on draft tube performance using full machine model: (**a**) $C_d$ distribution; (**b**) $C_p$ distribution.

The CFD analyses performed on the full machine returned different values for the turbine performance figures, especially for the net head and mass flow rate (Table 3). For a better comparison, a result transposition was carried out according to the European IEC EN 60041:1991 standard [16]. In particular, the mass flow rate and the power output data were transposed using the baseline net head. This was acceptable since the condition $0.99 < \sqrt{H}/\sqrt{H'} < 1.01$ on head $H'$ compared to baseline $H$ was met for each candidate. Optimal candidate solutions led to greater power and efficiency values compared to the baseline. In particular, Candidate 2 featured an increment of 1.8% on the produced power and 1% in efficiency ratio.

**Table 3.** Full machine CFD results (transposed based on baseline CFD results).

| CFD | $Q'_V$ [m$^3$/s] | H [m] | $\eta/\eta_{baseline}$ | $\eta$ Variation [%] | $P/P_{baseline}$ | P Variation [%] |
|---|---|---|---|---|---|---|
| Baseline | 9.251 | 5.88 | 100 | - | 100 | - |
| Candidate 1 | 9.286 | 5.88 | 100.57 | +0.57 | 100.957 | +0.96 |
| Candidate 2 | 9.323 | 5.88 | 101.00 | +1.00 | 101.790 | +1.79 |
| Candidate 3 | 9.281 | 5.88 | 100.53 | +0.53 | 100.856 | +0.86 |

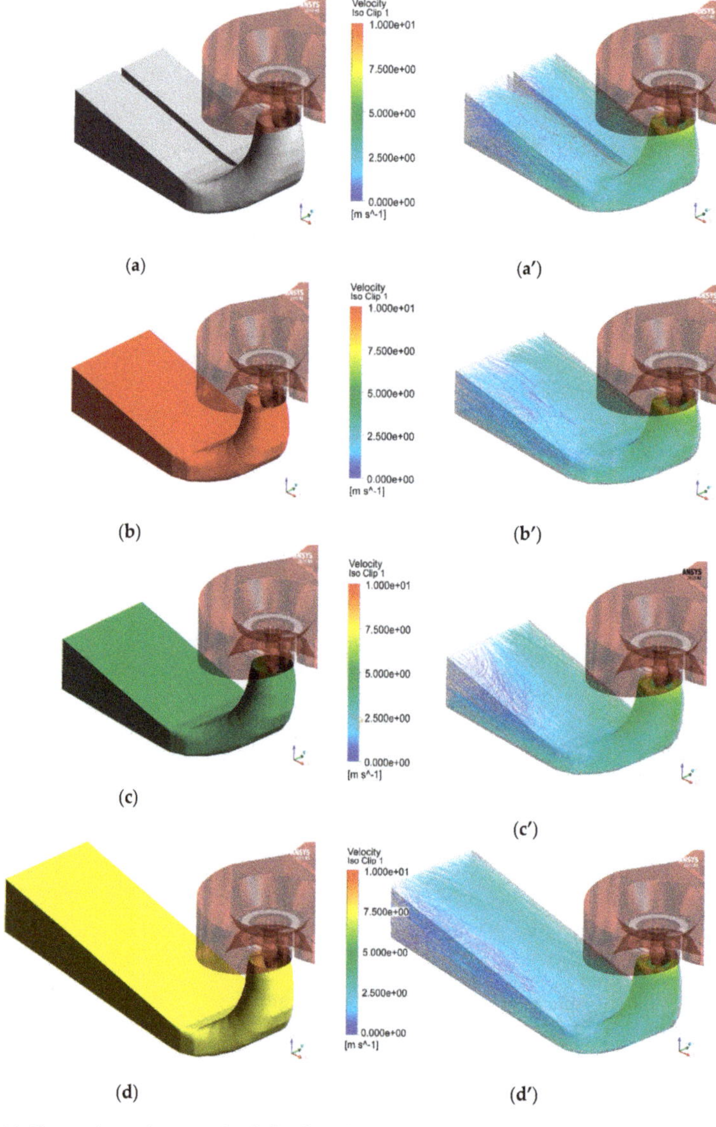

**Figure 11.** Comparison of geometries (left column), and streamlines colored by velocity (right column): (**a**,**a'**) Baseline; (**b**,**b'**) Candidate 1; (**c**,**c'**) Candidate 2; (**d**,**d'**) Candidate 3.

From a geometrical point of view (Figure 9), Candidates 1 and 2 were not very different from the baseline, although they clearly exhibited an elbow shape having a larger radius of curvature, which eventually caused the exit stations of both the elbows and the draft tubes to be placed further downstream with respect to the baseline. Additionally, Candidate 3 featured the same global tendencies but with a much longer draft tube, which, eventually, would lead to higher installation costs. In Figure 11 a 3D view of the several solutions is presented, with a comparison of the several streamlines. It can be seen that candidate 1 has the most uniform flow, a fact confirmed by his $C_d$ value (Figure 9).

From a fluid dynamic standpoint, it is apparent that the largest part of the pressure recovery took place in the first part of the diffuser (Figure 10b), which confirms the importance of having a very accurate design of the region upstream of the elbow.

As a matter of fact, approximately 70% of the pressure recovery is realized in the first 20% length of the diffuser. This also held true for the total pressure losses which were primarily generated in this region (see $C_d$ behavior in Figure 10, where the largest slope is evidenced in the $C_d$ distribution in the first 20% length of the baseline geometry); localized losses accumulated all along the draft tube mean line, finally leading to an overall large $C_d$ value in the baseline. Such an observation confirms what was previously pointed out in [5] and [13]. Furthermore, by looking at the total pressure distribution on consecutive station cuts along the draft tube (Figure 12, right column), it is evident that the rotational region at the lowest total pressure in the core of the discharge flow was of a much lower intensity in the optimized solutions compared to the baseline, despite the latter having a separation wall in the middle of the diffuser. In fact, the analyses carried out showed that the vortex rope greatly influences the draft tube's performance, creating instabilities and turbulent zones in the flow field.

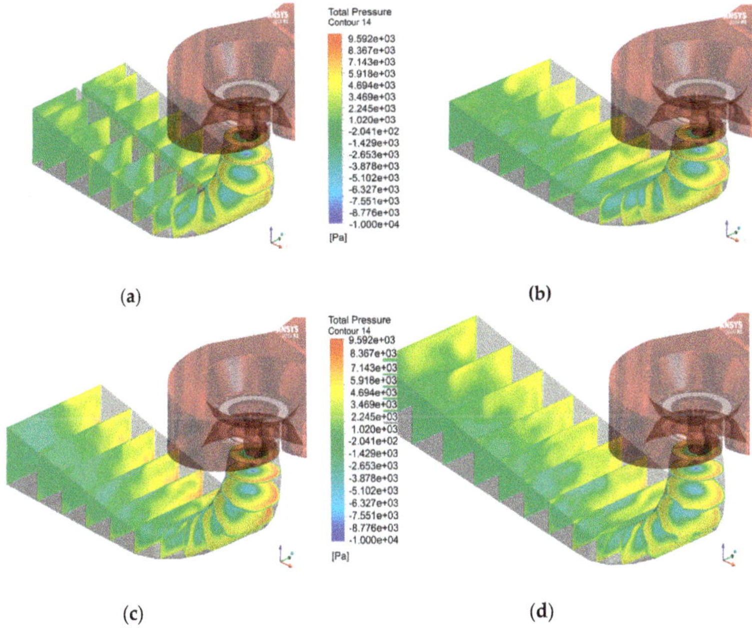

**Figure 12.** Comparison of relative total pressure distributions: (**a**) Baseline; (**b**) Candidate 1; (**c**) Candidate 2; (**d**) Candidate 3.

## 4. Conclusions

A validated CFD model of a draft tube of a Kaplan turbine was successfully implemented and used for a multi-objective optimization based on the construction of a response surface. The shape optimization involved only the draft tube domain, where a one-way coupling with the runner was included in the form of a prescribed velocity distribution extracted from the full-component baseline model.

Two Pareto optimal designs have been extracted from the final set which outperformed the baseline. Regarding the performance of the draft tube alone, the optimum $C_D$ featured a reduction count of −0.0254 (−2.54 percentage point reduction), while the optimum $C_P$ showed an incremental count of +0.0574 (i.e., +5.7 percentage point improvement) with respect to the baseline geometry.

A post-check carried out including the optimized draft tubes in the overall turbine confirmed the optimization trends, although the best improvements in terms of delivered hydraulic power—in the case where the overall turbine was considered—have been registered, including a third type of diffuser which exhibited an increment of 1.8% in the power produced.

The three best candidates overall were slightly deeper than the baseline in terms of excavation dimensions, so a cost/benefit assessment will be needed to determine which is the most appropriate draft tube for a given plant. In addition, future work will be related to extending the investigation to more operating points by performing a multi-objective multi-point optimization, where the Pareto solutions will account for the different conditions at the tube inflow along the turbine operating line.

**Author Contributions:** Conceptualization, E.B. and R.O.; methodology, R.O. and E.B.; software, R.O.; validation, R.O., M.M. and R.B.; formal analysis and writing—original draft preparation, R.O.; writing—review and editing, E.B.; supervision, E.B.; review and editing, A.M. All authors have read and agreed to the published version of the manuscript.

**Funding:** This research received no external funding.

**Conflicts of Interest:** The authors declare no conflict of interest.

## References

1. Kovalev, N.N. *Hydroturbines Design and Construction*; Israel Program for Scientific Translation: Jerusalem, Israel, 1965.
2. Krivchenko, G.I. *Hydraulic Machines Turbines and Pumps*, 2nd ed.; CRC-Press: Boca Raton, FL, USA, 1994.
3. McNabb, J.; Devals, C.; Kyriacou, S.A.; Murry, N.; Mullins, B.F. CFD based draft tube hydraulic design optimization. In Proceedings of the 27th IAHR Symposium on Hydraulic Machinery and Systems, Montreal, QC, Canada, 22–26 September 2014; Désy, N., Ed.; IOP Publishing: Bristol, UK, 2014.
4. Marjavaara, B.D. CFD Driven Optimization of Hydraulic Turbine Draft Tubes using Surrogate Models. Ph.D. Thesis, Lulea University of Technology, Lulea, Sweden, 2006.
5. Jošt, D.; Škerlavaj, A.; Lipej, A. Improvement of Efficiency prediction for a Kaplan Turbine with Advanced Turbulence Models. *Stroj. Vestn. J. Mech. Eng.* **2014**, *60*, 124–134. [CrossRef]
6. Li, Y.; Liu, Q. Analysis of hydraulic performance for Kaplan turbine components based on CFD simulation. *IOP Conf. Ser. Earth Environ. Sci.* **2020**, *510*, 022038. [CrossRef]
7. Brijkishore; Khare, R.; Prasad, V. Performance Evaluation of Kaplan Turbine with Different Runner Solidity Using CFD. In *Advanced Engineering Optimization through Intelligent Techniques*; Advances in Intelligent Systems and Computing; Springer: Singapore, 2020; Volume 949, pp. 757–767.
8. Liu, K.; Yang, F.; Yang, Z.; Zhu, Y.; Cheng, Y. Runner lifting-up during load rejection transients of a Kaplan turbine: Flow mechanism and solution. *Energies* **2019**, *12*, 4781. [CrossRef]
9. Lyutov, A.E.; Chirkov, D.V.; Skorospelov, V.A.; Turuk, P.A.; Cherny, S.G. Coupled Multipoint Shape Optimization of Runner and Draft Tube of Hydraulic Turbines. *J. Fluids Eng.* **2015**, *137*, 111302. [CrossRef]
10. Ciocan, T.; Susan-Resiga, R.; Muntean, S. Improving draft tube hydrodynamics over a wide operating range. *J. Hydraul. Res.* **2016**, *54*, 74–89. [CrossRef]
11. Eisinger, R.; Ruprecht, A. Automatic Shape Optimization of Hydro Turbine Components Based on CFD. 2002. Available online: http://kwk.ihs.uni-stuttgart.de/fileadmin/IHS-Startseite/veroeffentlichungen/v2001_05.pdf (accessed on 10 June 2020).
12. Puente, L.R.; Reggio, M.; Guibault, F. Automatic Shape Optimization of a Hydraulic Turbine Draft Tube. *IOP Conf. Ser. Earth Environ. Sci.* **2018**, *136*, 012019.
13. Sosa, J.B.; Urquiza, G.; Garcìa, J.C.; Castro, L.L. Computational fluid dynamics simulation and geometric design of hydraulic turbine draft tube. *Adv. Mech. Eng.* **2015**, *7*, 1–11. [CrossRef]

14. Schiffer, J.; Benigni, H.; Jaberg, H. An analysis of the impact of draft tube modifications on the performance of a Kaplan turbine by means of computational fluid dynamics. *Proc. Inst. Mech. Eng. Part C J. Mech. Eng. Sci.* **2017**, *232*, 1937–1952. [CrossRef]

15. ANSYS CFX. Theory Guide v 19.3. Available online: https://ansyshelp.ansys.com/account/secured?returnurl=/Views/Secured/corp/v193/cfx_thry/cfx_thry.html (accessed on 15 July 2020).

16. IEC 60041:1991. In *Field Acceptance Tests to Determine the Hydraulic Performance of Hydraulic Turbines, Storage Pumps and Pump-Turbines*, 3rd ed.; International Standard 1991-11; International Electrotechnical Commission: Geneva, Switzerland, 1991.

17. Kim, K.-Y.; Samad, A.; Benini, E. *Design Optimization of Fluid Machinery*; Wiley: Hoboken, NJ, USA, 2019.

18. Salem, M.B.; Roustant, O.; Gamboa, F.; Tomaso, L. Universal Prediction Distribution for Surrogate Models. 2015. Available online: https://arxiv.org/pdf/1512.07560.pdf (accessed on 10 June 2020).

© 2020 by the authors. Licensee MDPI, Basel, Switzerland. This article is an open access article distributed under the terms and conditions of the Creative Commons Attribution (CC BY) license (http://creativecommons.org/licenses/by/4.0/).

*Article*

# Fast Design Procedure for Turboexpanders in Pressure Energy Recovery Applications

**Gaetano Morgese [1], Francesco Fornarelli [1,2], Paolo Oresta [1], Tommaso Capurso [1],
Michele Stefanizzi [1], Sergio M. Camporeale [1] and Marco Torresi [1,*]**

[1]   Department of Mechanics, Mathematics and Management (DMMM), Polytechnic University of Bari,
      via Orabona 4, 70125 Bari, Italy; gaetano.morgese@poliba.it (G.M.); francesco.fornarelli@poliba.it (F.F.);
      paolo.oresta@poliba.it (P.O.); tommaso.capurso@poliba.it (T.C.); michele.stefanizzi@poliba.it (M.S.);
      sergio.camporeale@poliba.it (S.M.C.)
[2]   National Group of Mathematical Physics (GNFM) of the Italian National Institute of High
      Mathematics (INDAM), Piazzale Aldo Moro, 5, 00185 Rome, Italy.
*    Correspondence: marco.torresi@poliba.it; Tel.: +39-080-596-3577

Received: 13 May 2020; Accepted: 13 July 2020; Published: 16 July 2020

**Abstract:** Sustainable development can no longer neglect the growth of those technologies that look
at the recovery of any energy waste in industrial processes. For example, in almost every industrial
plant it happens that pressure energy is wasted in throttling devices for pressure and flow control
needs. Clearly, the recovery of this wasted energy can be considered as an opportunity to reach not
only a higher plant energy efficiency, but also the reduction of the plant Operating Expenditures
(OpEx). In recent years, it is getting common to replace throttling valves with turbine-based systems
(tuboexpander) thus getting both the pressure control and the energy recovery, for instance, producing
electricity. However, the wide range of possible operating conditions, technical requirements and
design constrains determine highly customized constructions of these turboexpanders. Furthermore,
manufacturers are interested in tools enabling them to rapidly get the design of their products.
For these reasons, in this work we propose an optimization design procedure, which is able to rapidly
come to the design of the turboexpander taking into account all the fluid dynamic and technical
requirements, considering the already obtained achievements of the scientific community in terms
of theory, experiments and numeric. In order to validate the proposed methodology, the case of
a single stage axial impulse turbine is considered. However, the methodology extension to other
turbomachines is straightforward. Specifically, the design requirements were expressed in terms
of maximum allowable expansion ratio and flow coefficient, while achieving at least a minimum
assigned value of the turbine loading factor. Actually, it is an iterative procedure, carried out up
to convergence, made of the following steps: (i) the different loss coefficients in the turbine are
set-up in order to estimate its main geometric parameters by means of a one dimensional (1D)
study; (ii) the 2D blade profiles are designed by means of an optimization algorithm based on a
"viscous/inviscid interaction" technique; (iii) 3D Computational Fluid Dynamic (CFD) simulations
are then carried out and the loss coefficients are computed and updated. Regarding the CFD
simulations, a preliminary model assessment has been performed against a reference case, chosen
in the literature. The above-mentioned procedure is implemented in such a way to speed up the
convergence, coupling analytical integral models of the 1D/2D approach with accurate local solutions
of the finite-volume 3D approach. The method is shown to be able to achieve consistent results,
allowing the determination of a turbine design respectful of the requirements more than doubling
the minimum required loading factor.

**Keywords:** energy recovery; turboexpander; throttling valves; CFD; modelling techniques

---

## 1. Introduction

The United Nations (UN) promotes the 17 Sustainable Development Goals (SDGs) which should be achieved by 2030. The UN SDGs lead the interplay between the social needs and the environmental impact through new strategies in energy production and management. Among others, companies and scientific communities are strongly involved in the decarbonization process with the improvement of new technologies for harvesting energy from renewable sources. The efficiency of the renewable energy production is crucial for dealing with the growing global energy demand, which rose by 2.2% in 2018. More details are described in the report edited by the International Energy Agency (IEA) [1].

How can the efficiency in energy production be increased? In order to try answering this key question, we focused our attention on energy recovery and, particularly, on the recovery of that wasted every time a throttling device is used to set the pressure of a working fluid in an industrial plant. In recent years, the throttling valves have been replaced by turbines getting both the pressure control and the energy recovery. The linked benefits to the former strategy are the reduction of both the $CO_2$ emissions and the operating expenditure (OpEx). For instance, natural gas is transported at high pressure (40–100 bar) in pipelines but its pressure needs to be reduced (2–20 bar) for delivery in the Gas Regulation Stations (GRS) by means of valves. However, Kuczyński et al. [2] shown that GRS are characterized by a wide range of parameters (e.g., pressure drop) and only some of them are suitable to replace the throttling valves with the turboexpander. Furthermore, boundary conditions such as irregular daily and seasonal temperatures make stringent conditions in using the turboexpander.

Kuczyński et al. [2] proposed a discounted payback period (DPP) equation which depends on the annual average natural gas flow rate through the analysed GRS, average annual level of gas expansion, average annual natural gas purchase price, average sale price of produced electrical energy and capital expenditure (CapEx). Moreover, the natural gas must be preheated in order to avoid methane-hydrate formation. For example, Borelli et al. [3] studied the possibility of integrating a pressure reduction station with low temperature heat sources, where the optimization of the heat exchangers can play a key role in the global efficiency of the system [4–6].

Lo Cascio et al. [7] proposed a configuration of the natural gas pressure reduction stations coupled with Concentrated Solar Plant (CSP) equipped with sun-tracking parabolic trough solar collectors and thermal energy storage [8,9] in order to allow the preheating of the natural gas. Lo Cascio et al. [10] proposed a set of key performance indicators for evaluating the effectiveness of the replacement in terms of energy recovery. Zabihi and Taghizadeh [11] analysed the case study of the Sari-Akand gate station (Iran), characterized by an annual pneumatic energy loss of about 7.1 GWh. In this case, the installation of a turboexpander could allow an energy production of 3.2 GWh. Another test case was presented by Naseli et al. [12], where the electricity to be recovered in one natural gas pressure reduction station in Izmir (Turkey) was calculated to be equal to 4.1 GWh. Howard et al. [13] examined the installation of a turboexpander at a small City Gate Station (CGS) in Canada. Poživi [14] simulated the installation of a turboexpander in parallel with the CGS regulator to recover energy and prevent its loss and concluded that the gas temperature drop was about of 15-20 °C every 1 MPa of pressure drop.

Golchoobian et al. [15] studied the feasibility of using a turboexpander instead of an expanding valve in a gas pressure reduction station of a gas turbine power plant. The recovered energy is supposed to be used for a cooling cycle in order to reduce the temperature of the air at the gas turbine compressor inlet.

For natural gas transportation over long distances, it is preferred to liquefy the gas because of economic-, technical-, and safety-related reasons. The natural gas liquefaction requires considerable costs up to 35% of CapEx and 50% of OpEx of the entire liquefaction plant [16].

In the Oil and Gas plants for processing Liquefied Natural Gas (LNG), expansion devices are mainly used for two purposes, that is, to reduce the LNG high stream pressure and for generating the cooling effect. Ancona et al. [17] proposed and investigated two different Liquefied Natural Gas (LNG) production layouts pointing out how the installation of a turboexpander allows the optimization of the LNG production process and the minimization of the process energy consumption. The same kind of work was performed by Taher et al. [18], who explained the benefits of utilizing a turboexpander in LNG applications.

Turboexpanders are also widely used in the reverse Brayton refrigerators in order to achieve lower refrigeration temperature, larger cooling capacity and more effective energy usage [19]. In this kind of applications, the efficiency of the turboexpanders is crucial since the power required to transport heat from the low temperature source to the high temperature sink increases as the temperature decreases. Zhang et al. [20] and Hou et al. [21] carried out analysis on turboexpanders in order to maximize their performance. Qyyum et al. [22] evaluated the effect of substituting the Joule-Thomson valve with a hydraulic turbine in the enhancement of a single mixed refrigerant process, that is, an energy saving of about 16.5 %.

Process engineering is another field where considerable pressure energy is wasted. Some examples are—washing plants, systems for ammonia synthesis, desalting plants based on reverse osmosis, refrigeration systems in mines, pipelines in the Oil and Gas. In these processes, the fluids are usually supplied at high pressure for making feasible and economic transportation. But then, these fluids need to be expanded to be used. Thus, the expansion energy could be converted in a PaT, that is, a conventional pump used in reverse mode as a turbine. In same cases, due to free-gases and dissolved gases in the liquid, the PaTs can work with two-phase flow. The gas expansion increases the energy extraction of the PaT with respect to the incompressible case. This additional power output is very interesting in terms of energy management improvement in process industries [23].

Several authors started investigating PaTs in the early 1980s. Inter alia, Laux [24] investigated reverse-running multistage pumps as energy recovery turbines in oil supply systems; Apfelbacher et al. [25] investigated the use of a PaT in a reduction pressure station (Aachen, Germany). Hamkins et al. [26] studied PaT under to two-phase flow conditions. Examples of PaT in reverse osmosis plants and in gas washing plants have been studied by by Gopalakrishnan [27], Bolliger [28] and Bolliger & Menin [29]. Slocum at al. [30] evaluated the integration of the Pumped Hydro with Reverse Osmosis systems. Recently, Capurso et al. [31] proposed a novel impeller geometry for double suction centrifugal pumps used as PaTs. Finally, Gutiérrez et al. [32] studied the potential of using Pressure Recovery Turbines (PRTs) in the geothermal field in order to exploit the pressure losses at the production orifice plate of geothermal wells which reduce the wellhead pressure to that of steam separation and transportation. The know-how of the companies involved in the manufacturing of the PRTs [33–35] coexists with the state of the art of the scientific literature whose key features are described below.

Fiaschi et al. [36], and Talluri and Lombardi [37] developed one-dimensional approaches. In the former the one-dimensional model includes correlations from the literature for the estimation of losses, whilst in the latter a Non-Isentropic Simple Radial Equilibrium model is included, in order to evaluate the thermodynamics and kinematics of the flow throughout the vanes.

A significant development in turboexpanders modelling for energy recovery has been reached with the growing success of Computational Fluid Dynamics (CFD). Indeed CFD can play an important role in understanding the behaviour of turboexpanders at various operating conditions. For example Ying et al. [38], with the help of CFD, studied several geometry modifications in order to learn their influence on turbine performance. Sam et al. [39] studied the flow field of a high-speed helium turboexpander by means of three-dimensional transient flow analysis using Ansys CFX. The aim of both works was the improvement of the efficiency of these machines. However, CFD is also used in the design process of turboexpanders, often coupled with one-dimensional formulations. In different works, like in Alshammari et al. [40] and in Dong et al. [41], the one-dimensional model is used in order to carry out a preliminary design of the turboexpander, whilst the three-dimensional CFD simulation is performed in order to optimize the turbine geometry. Other works, like Fiaschi et al. [42] and Sam et al. [43], use the 3D CFD simulations in order to correct the one-dimensional formulations. Particularly, Sam et al. [43] with this approach obtained a modified Balje's chart aimed at the design of efficient cryogenic microturbines for helium applications.

Turboexpanders are also subject of experimental investigations, as in [44–46]. Since turboexpanders for energy recovery are often machines to be design ad hoc, there is the need to define at least a general and effective procedure. A feature particularly appreciated by manufacturers is the rapid response of the design tool, which allows them to shorten the time to market. Under this point of view, our main objective has been the definition of a fast design procedure for turboexpanders. This is a a multi-step iterative procedure composed of: 1. a 1D analysis for the overall turbine design, based on the knowledge of the different loss coefficients; 2. a 2D viscous/inviscid interaction technique coupled with an optimization algorithm in order to define both stator and rotor blade profiles; 3. the use of a script file for the automatic generation of the mesh; 4. 3D CFD calculations for the estimation of the loss coefficients, to be used in the 1D analysis. Each step of the procedure is probably anything but innovative, however the entire procedure actually it is, as will be shown in the next chapter. Furthermore, the procedure proved to be very suited for the optimization algorithm used for the definition of the blade shape due to the low computational cost.

In order to validate the procedure, a turboexpander has been designed in order to replace a throttling device. The particular type of turboexpander to be designed, with its peculiar constraints, has been a challenging test. The device is a Pressure Reduction Valve (PRV) valve which determines a pressure ratio, $\beta_e$, equal to 1.144 with a non-dimensional mass flow rate, $\Gamma$, equal to 0.01312. The turbine should operate with—(i) a specific speed, $N_S$, equal to 32.3 rpm imposed by the coupling with the electric generator; (ii) a specific diameter, $D_S$, equal to 2.3 ft imposed by the dimension of the pipeline into which the machine should be mounted; (iii) a flow coefficient, $\phi$, equal to 0.31. With these non dimensional parameters, we checked on the Balje's diagram (Figure 1) which could be the best configuration for this application. The Balje's diagram suggests a full admission single stage axial turbine. This is considered a good choice, since the axial configuration allows a simpler insertion of this system in a pipeline in place of a Pressure Reduction Valve (PRV). Furthermore, the axial configuration allows a simpler integration of the magnets of the direct drive generator inside the rotor of the turbine since, actually, the turboexpander was supposed to embed the electric generator rotor [47–50], aiming at a robust design, with as few as possible components and avoiding any loss in the mechanical transmission of the torque. This configuration made unfeasible the use of conventional multistage axial turbomachines or centripetal turbines.

Furthermore, a significant advantage of the proposed procedure is that it can be used even when the operating conditions goes outside the conventional range, where the models available in the literature (e.g., References [51–54]) fail.

Finally, in order to make considerations about the impact of this technology, its application to a real gas pipeline, namely the Trans Austria Gas Pipeline (TAG) (Austria), has been taken into account (power absorbed by 5 compressor stations = 450 MW and annual flow rate of natural gas equal to 41 bmc/year) [55]. It is worth to highlight that, especially in mountain areas, compressor stations are often employed to overcome a hill and back pressure control valves are installed, in order to avoid slackline. This is another kind of valve that can be retrofitted with a turboexpander technology. Depending on the pressure ratio required along the line, for example, a pressure drop after a hill or before a branching line to connect the pipeline to a CGS, different sizes of turboexpander can be selected. Taking into account geometries similar to the one proposed in this work, one can see that for small pressure drops, as in the case of a valley site down to the hill, the power recovered can be about 16.9 MW (e.g., obtained with a turbomachine characterized by $\beta_e$ = 1.15, $D_s$ = 1.06, $N_s$ = 47.0); otherwise for larger pressure drops before a CGS, the generated power can easily reach 45 MW (e.g., obtained with a turbomachine characterized by $\beta_e$ = 1.5, $D_s$ = 1.39, $N_s$ = 20.9). The latter is more or less half the power required by one compressor station in the TAG pipeline system, a remarkable result which confirms the potentiality of this kind of technology. Finally, considering an emission factor of 400 $kg_{CO_2}$/MWh obtained by Italian national data for the last ten years [56], the $CO_2$ emission reduction can reach 59,300 and 157,000 $t_{CO_2}$/year for small and large pressure drop, respectively.

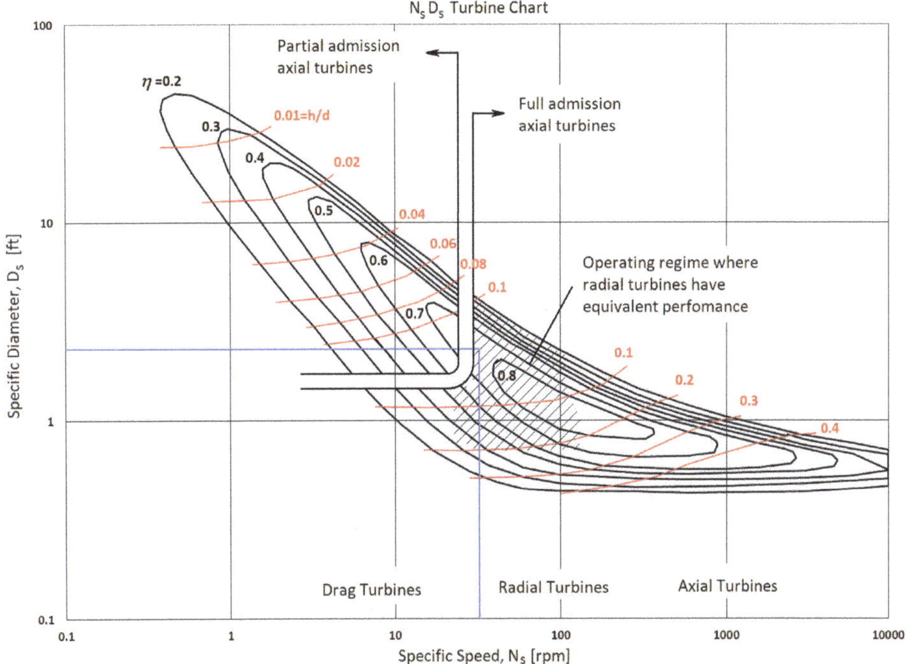



**Figure 1.** Working point on the Balje's diagram (diagram derived from Reference [50]).

## 2. Design Methodology Overview

The proposed methodology can be summarized by the flow chart in Figure 2. It begins with the definition of the turbine design point, from which the operating conditions are derived, that is, mass flow rate, $G$, total pressure, $p_0^0$, and total temperature, $T_0^0$, at the turbine inlet and the pressure drop, $p_0 - p_4$, across the turbine.

Loss coefficients are introduced, in order to take into account irreversibility. Actually, the very next step of the procedure involves the initialization of the loss coefficients with trial values, $\zeta_i^0$. As a first guess, an ideal calculation can be performed, considering the loss coefficients equal to either 1 or 0 (depending on their definitions). After that a 1D design procedure is applied by means of a script implemented in MATLAB®. The turbine is decomposed in its main parts (namely, the nozzle, the stator, the rotor, and the diffuser), and for each one a system of governing equations is solved. Each system of equations is composed by: the first law of thermodynamics; the continuity equation; the ideal gas law and the relations among velocity vectors according to the velocity diagrams. Thus, the one-dimensional calculation (as presented, for instance, in Reference [54] and in Reference [57]) provides the main operating and geometrical parameters of the machine such as the blade heights ($H_i, \forall i = 1, 2, 3$), the absolute and relative inlet and outlet angles ($\alpha_1, \alpha_2, \beta_2, \beta_3$), and the rotational speed, $N$, of the machine. It is important to emphasize that the use of an embedded permanent magnet direct drive generator allows the turbine to work always at its maximum efficiency.

Subsequently, a 2D calculation is performed to define the stator and rotor blade profiles. Regarding the stator blades, their camber lines are designed according to third order Bézier curves and their thicknesses are defined according to the ones of symmetric NACA four digit airfoils. This blade design criterium was defined in order to ensure the continuity of the surface curvatures, reducing the risk of flow separations. Regarding the rotor blades, an impulse turbine technology has been chosen, by considering a modified Curtis stage layout. After having assigned all the geometric and kinematic constraints, the set of independent variables is defined by means of the genetic algorithm NSGA-II, available in the multi-objective optimization software ModeFrontier®. The objective functions are derived by the minimization of the primary and secondary losses of the blades, and maximization of the rotor blade lift coefficient. For a fast evaluation of the force coefficients taking into account drag effects, a "one-way coupled viscous/inviscid interaction model" (see Reference [58]) is used. This is based on the vortex panel method shown in Reference [59] and the integral method for the solution of the boundary layer equations presented in Reference [60]. Moreover, the optimization algorithm allows to integrate not only fluid dynamic constraints but also operating and technological ones, for instance: maximum rotation speed of the electric machine and/or the minimum thickness of the trailing edge. At the end of this step, the proposed model provides all the geometric parameters for the turbine construction.

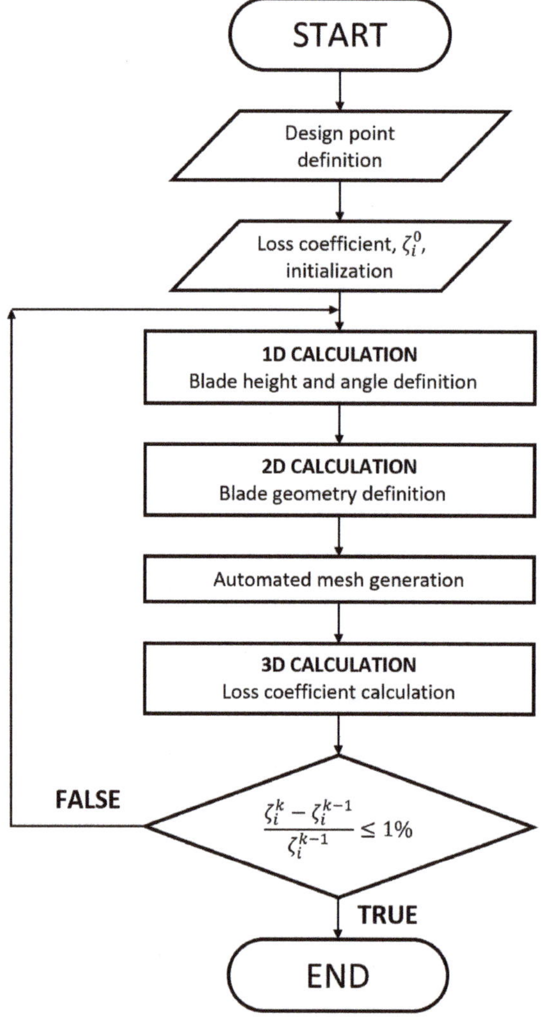


**Figure 2.** Flow chart of the proposed methodology.

Then, the computational domain and the mesh are automatically generated in order to minimize the time cost. This is carried out by means of a script file, (called *"replay"*) which is updated with the main parameters of the geometry and then run in the ANSYS ICEM CFD® grid generator. Due to the low aspect ratio of the blades, the inlet angle of the flow does not show significant variations from the hub to the tip, hence the blades are designed without twisting them along the radial direction.

The last step of the procedure is represented by a 3D RANS simulation of the flow through the turbine, which is performed in order to validate the design and evaluate the loss coefficients, to be updated and used in the next $(k+1)^{th}$ iteration ($\zeta_i^{k+1} = \zeta_i^k$).

Actually, the calculation of the loss coefficients from the 3D simulations is carried out considering the thermodynamic characteristics of the flow evaluated as mass-weighted averages and considering that the pressure drop across the machine is the same in both ideal and real cases.

The proposed methodology is necessarily iterative since the turbine geometry depends on the loss coefficients, but these depend on the turbine geometry. The procedure stops when the generic loss coefficient, $\zeta_i^k$, at the $k^{th}$ iteration, differ less than 1%, from the value at the previous one, $(k-1)^{th}$ (see Figure 2).

## 3. Three-Dimensional Computational Fluid Dynamic Modelling

3D viscous CFD calculations are carried out by means of the commercial software ANSYS FLUENT® in order to evaluate the performance of the turboexpander. In this way, the trial values used for the loss coefficients are checked against the numerical results.

The computational domain includes the whole machine since the different numbers of stator and rotor blades do not allow the application of circumferential periodic boundary conditions.

The domain consists of three regions connected by non-conformal interfaces. Fixed and moving components are placed in different sub-domains. The numerical solution is based on both the moving and the stationary reference frames.

The manner in which the equations are treated at the interface leads to three possible approaches, all supported in ANSYS FLUENT:

1.  Multiple Reference Frame model (MRF);
2.  Sliding Mesh Model (SMM);
3.  Mixing Plane Model (MPM).

Both MRF and MPM approaches are steady-state approximations and differ primarily in the manner in which data are treated and exchanged at the interfaces. On the other hand, the SMM approach is intrinsically unsteady due to the actual motion of the meshes.

In detail, the MRF model [61] is a steady-state approximation in which different rotational and/or translational speeds can be assigned to individual cell zones. The flow in each moving cell zone is solved using the moving reference frame equations. It is worth noting that the mesh relative motion of adjacent zones is not considered by the MRF approach. Hence, the moving parts are frozen and the flow field is calculated by neglecting the relative motion of the moving zones. The MRF model is suitable in nearly uniform flow at the interface between adjacent moving/stationary zones ("mixed out").

The SMM method is used when a time-accurate solution for rotor-stator interaction is desired instead of a time-averaged solution. Even though the SMM is the most computationally demanding, it is a good practice for unsteady flow simulations in multiple moving reference frames, especially to account for rotor-stator interaction in turbomachineries. In the SMM technique two or more cell zones are used and each of them actually slides along the grid interface in discrete steps. When the rotation or translation takes place, node alignment along the grid interface is not required.

The MPM provides an alternative to the MRF and SMM in order to simulate the flow through moving zones. In the MPM approach, a steady-state problem for each zone is solved considering the mixing plane interface between adjacent zones as a boundary where the flow-field boundary conditions are set with spatially averaged or mixed condition. This mixing keeps smooth the flow variations by means of a circumferential averaging at the interface (i.e., wakes, shock waves, separated flows), therefore yielding a steady-state result. Despite the approximation of the MPM model, the time-averaged flow field is acceptable in a wide range of flow field conditions.

### 3.1. CFD Model Validation and Grid Sensitivity Analysis

In order to assess the 3D CFD and, in particular, to verify the best approach for the analysis of the rotor/stator interaction, a reference case [62] has been considered for validation. The turbine under investigation is the air turbine stage of the Institute of Thermal Engineering (ITC) of Lodz, model TK9-TW3 [63] which is a typical HP turbine stage. It operates with short-height cylindrical blading and aft-loaded stator profiles with the following characteristics: span/chord 0.73 (stator) and 2.20 (rotor), pitch/chord 0.86 (stator) and 0.80 (rotor), span/diameter 0.08 (stator and rotor). In Figure 3 the turbine stages are reported on the blade-to-blade (a) and the throughflow planes (b).

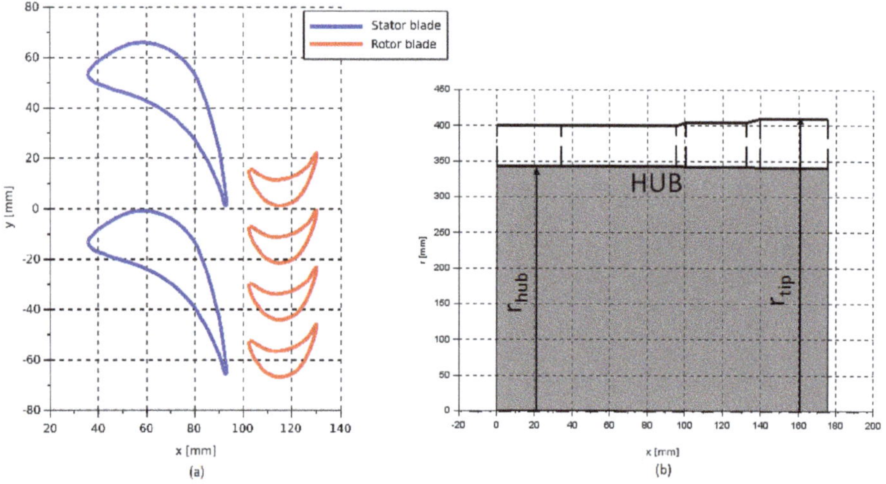


**Figure 3.** Blade geometry of the test case TK9-TW3 turbine: (a) blade-to-blade plane; (b) throughflow plane.

For each region of the domain, a customized structured multi block mesh has been generated, taking into account both the local geometry and the flow conditions.

In order to capture directly the wall boundary layer effects ($y^+ < 1$), the first cell height has been kept as small as possible however avoiding problems of negative volumes. Furthermore a settling chamber is added downstream of the computational domain in order to prevent any convergence problem related to "reversed flow".

The standard $k - \omega$ turbulence model [64] has been considered according to several previous numerical investigations [65–68].

The turbine under investigation operates with a mass flow rate $G = 4$ kg/s at an inlet static pressure $p_0 = 1$ bar and an inlet temperature $T_0 = 320$ K. The outlet pressure is $p_4 = 0.9$ bar. The rotational speed is $N = 1450$ rpm.

In order to simulate these operating conditions a mass flow rate has been imposed at the inlet section of the computational domain. At the outlet section, a pressure-outlet boundary condition has been imposed. At the inlet section, the turbulent intensity is set equal to 5% and the turbulent length scale equal to 1 mm. A grid independence study has been carried out using the moving reference frame model implementing the MRF approach. The study has been performed doubling each time the number of nodes on each edge. The total-to-total efficiency has been considered as the control parameter. In Table 1 the study is resumed, whilst in Figures 4–6 the meshes used in each simulation are shown.

**Table 1.** Results of grid refinement study on the TK9-TW3 geometry—MRF method.

| Number of Cells | 125 K | 1 M | 8 M |
|---|---|---|---|
| $\eta_{tt}$ | 0.8925 | 0.9297 | 0.9329 |
| Variation | — | 4.16% | 0.34% |

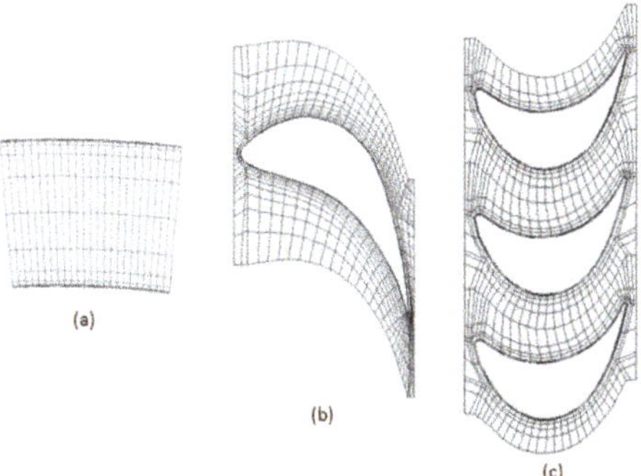

**Figure 4.** Mesh distribution for 125 k cells. (**a**) radial distribution (**b**) stator vane distribution (**c**) rotor vane distribution.

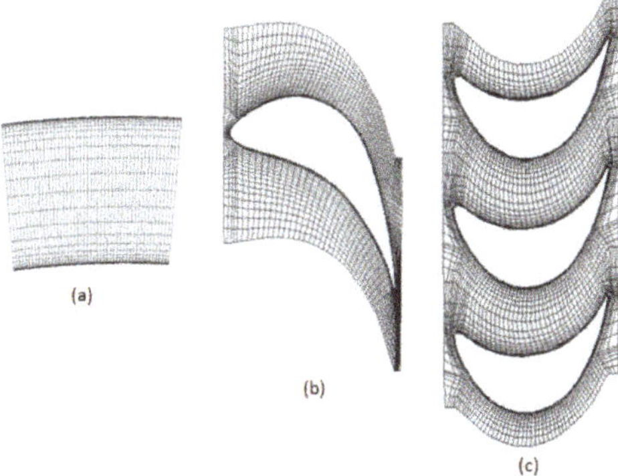

**Figure 5.** Mesh distribution for 1M cells. (**a**) radial distribution (**b**) stator vane distribution (**c**) rotor vane distribution.

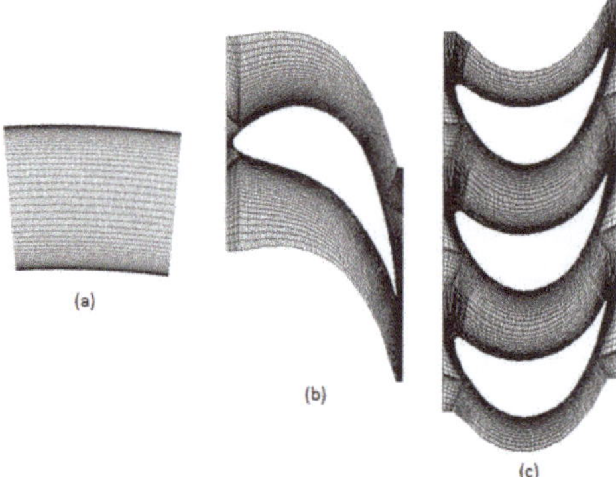



**Figure 6.** Mesh distribution for 8M cells. (**a**) radial distribution (**b**) stator vane distribution (**c**) rotor vane distribution.

The grid independence study highlights that 1M cells discretization is able to achieve the requested accuracy. The MPM approach showed unstable behaviour of the numerical solution hence it has been discarded.

For both the Moving Reference Frame Model and the Sliding Mesh Model, the first parameter that has been analysed is the pressure drop across the turbine. Particularly, for the MRF the pressure drop is equal to 0.091 bar, whilst for the SMM is equal to 0.095 bar, with an error, computed against the experimental data, of 9% and 5%, respectively.

Another comparison has been carried out considering the absolute velocity components (axial, radial and tangential) in a section located at 135% of the axial chord downstream of the rotor trailing edge, that is the distance corresponding to the location of the measuring probe at the experimental facility (Figures 7–9).

As reported in Figures 7–9 and from Table 2, MFR and SMM provide approximately the same results since the Mean Square Error Percentage (MSEP) shows minimal differences between the two models. Particularly, the tangential speed shows the best agreement between experimental data and the computational results. The value of the axial speed at the tip is underestimated by both the models, whilst at the hub the axial speed is underestimated or overestimated if the MRF or the SMM are used, respectively. The absolute speed differences can be found at the hub and at the tip. In detail, at the hub the simulation results match the experimental trend, whereas the models underestimate the axial velocity at the tip. Indeed at the hub it is possible to observe the same trend as the axial speed, whilst at the tip the simulation tends to underestimate the value of the absolute speed.

**Table 2.** Comparison of the Mean Square Error Percentage (MSEP) for the Multiple Reference Frame model (MRF) and Sliding Mesh Model (SMM) in the case of the three speed components.

|  | MRF | SMM |
| --- | --- | --- |
| Axial Speed | 0.9700 | 0.9710 |
| Absolute Tangential Speed | 0.8485 | 0.8695 |
| Absolute Speed | 0.9707 | 0.9719 |

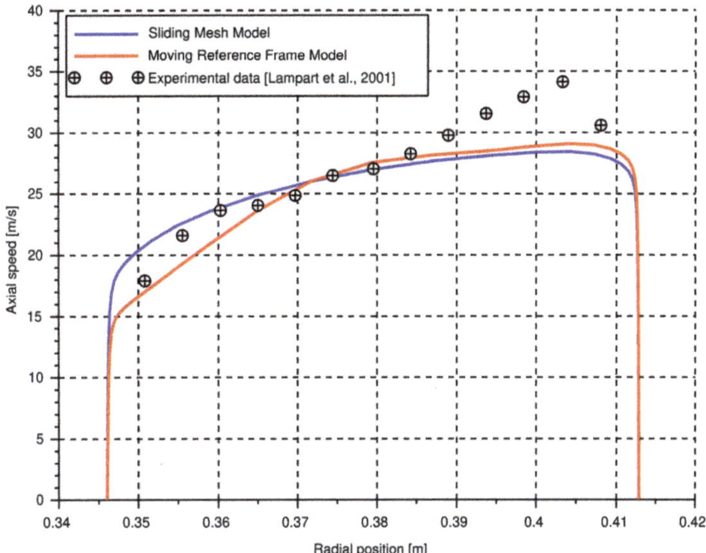



**Figure 7.** Comparison of computed and experimental radial distribution of axial speed at 135% of the axial chord downstream of the rotor trailing edge.

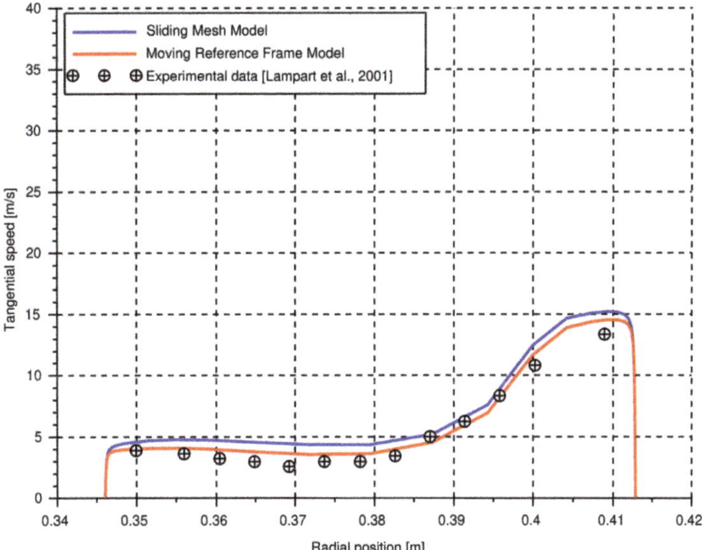



**Figure 8.** Experimental data Comparison of computed and experimental radial distribution of absolute tangential speed at 135% of the axial chord downstream of the rotor trailing edge.

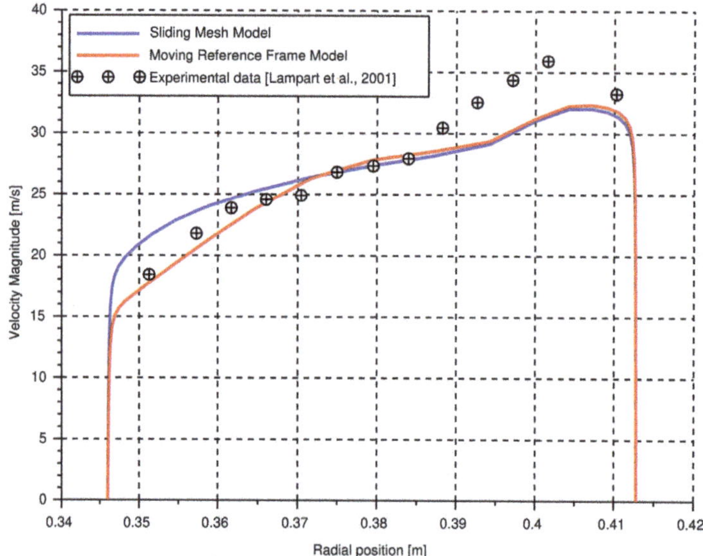

**Figure 9.** Comparison of computed and experimental radial distribution of absolute speed at 135% of the axial chord downstream of the rotor trailing edge.

Even though the SMM gives the best results in this assessment, it is extremely time consuming with respect to MRF. Hence, the MRF model represents the best compromise in terms of accuracy and computational effort.

## 4. Case Study

The proposed methodology has been applied in order to design a single stage axial turbine to be used in substitution of a pressure reduction valve, exploiting the pneumatic energy associated to its pressure drop. The application taken into account requires that the turbine operates with a loading factor, $\Psi$, at least equal to 0.8262, an expansion ratio, $\beta_e$, equal to 1.144 and a flow coefficient, $\phi$, equal to 0.31.

The turbine under investigation is composed by a nozzle, a single stage axial turbine, and a diffuser (Figure 10). In the 3D CFD calculation, the computational domain includes all the parts of the machine, considering air as an ideal gas. Furthermore, the design involves both the loss coefficients of the nozzle, $\varphi_C$, and of the diffuser, $\varphi_D$.

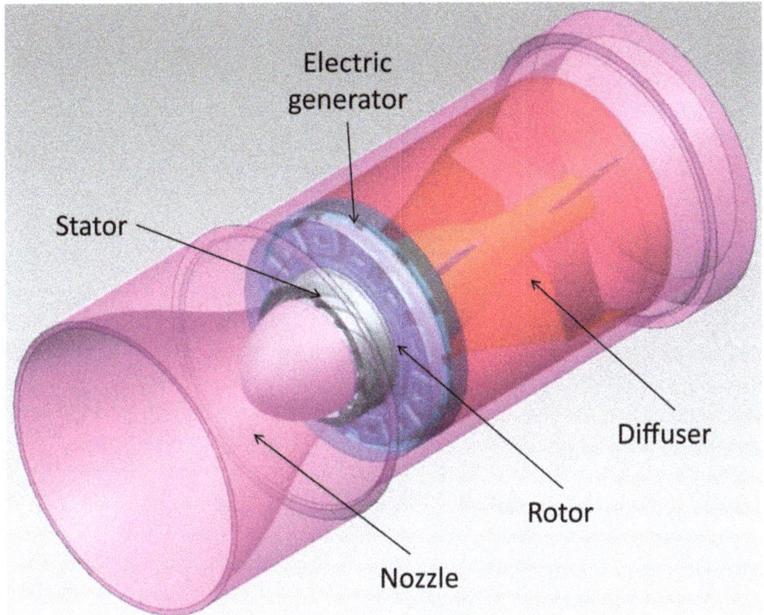


**Figure 10.** 3D model of the turbine under investigation.

Following the procedure, described in the previous chapter, the relative error becomes quickly lower than 1% already at the 3 rd iteration as shown in Table 3.

**Table 3.** Input Values (I.V.) of the first iteration and Supposed Values (S.V.) of the next iterations together with the Computed Values (C.V.) for loss coefficients off all the three iterations.

| Coefficients | | Iteration n.1 | | | Iteration n.2 | | | Iteration n.3 | | |
|---|---|---|---|---|---|---|---|---|---|---|
| | | I.V. | C.V. | Error [%] | S.V. | C.V. | Error [%] | S.V. | C.V. | Error [%] |
| $\beta_e$ | [−] | 1.144 | 1.191 | 4.15 | 1.144 | 1.168 | 2.10 | 1.144 | 1.154 | 0.91 |
| $\varphi_C$ | [−] | 1 | 0.9207 | 7.93 | 0.9207 | 0.9197 | 0.11 | 0.9197 | 0.9218 | 0.23 |
| $\varphi_N$ | [−] | 1 | 0.9354 | 6.46 | 0.9354 | 0.9335 | 0.20 | 0.9335 | 0.9306 | 0.30 |
| $\varphi_R$ | [−] | 1 | 0.8092 | 19.08 | 0.8092 | 0.8171 | 0.99 | 0.8171 | 0.81341 | 0.46 |
| $\varphi_D$ | [−] | 1 | 0.1805 | 81.95 | 0.1805 | 0.1909 | 5.80 | 0.1909 | 0.1890 | 0.99 |
| $\eta_v$ | [−] | 1 | 0.8973 | 10.17 | 0.8973 | 0.9042 | 0.77 | 0.9042 | 0.9040 | 0.02 |
| $k_w$ | [−] | 0 | 0.0005153 | [−] | 0.0005153 | 0.0005218 | 1.26 | 0.0005218 | 0.0005176 | 0.80 |

As described in the flow chart of the iterative procedure (see Figure 2), the process is initialized under the hypothesis of an ideal machine without losses.

The variation of the loss coefficients during the iterative procedure is reflected by the change of the machine geometry, and specifically of the heights (Table 4) and airfoil shapes of both stator and rotor blades (Figure 11). Looking at Figure 11, one can see that the substantial variations occur to the stator blade profiles going from the first to the third iterations. Moreover, the turbine rotational speed (Table 5) is also modified in order to preserve the optimum working conditions (namely, inlet flow velocities aligned with the inlet blade angles).

Indeed, as shown in Reference [54] and Reference [57], the loss coefficients affect the thermodynamic properties of the flow at each considered section of the turbine. According to the continuity equation, for increasing losses, the pressure drop increases, whereas the density decreases as well as the velocity.

**Table 4.** Blade heights at different iterations.

|  |  | 1st it. | 2nd it. | 3rd it. |
|---|---|---|---|---|
| $H_1/D_m$ | $[-]$ | 0.05304 | 0.05705 | 0.05718 |
| $H_2/D_m$ | $[-]$ | 0.05821 | 0.06279 | 0.06296 |
| $H_3/D_m$ | $[-]$ | 0.05821 | 0.06493 | 0.06562 |

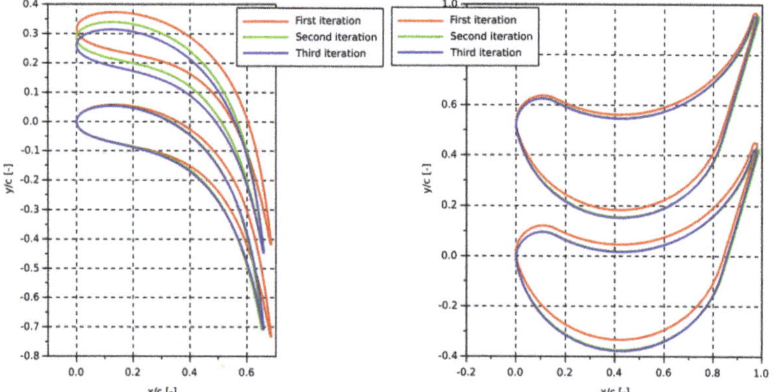


**Figure 11.** (**a**) Stator and (**b**) rotor blade shapes at different iterations.

**Table 5.** Non-dimensional rotational speed at different iterations.

|  |  | 1st it. | 2nd it. | 3rd it. |
|---|---|---|---|---|
| $\omega D_m/\sqrt{RT_0^0}$ | $[-]$ | 0.5057 | 0.4676 | 0.4673 |

Once the geometry of the machine has been defined, a simulation has been carried out with a finer mesh in order to study in detail the fluid dynamic behaviour of the machine. The mesh density has been selected by means of a mesh sensitivity study, which is not reported here for the sake of brevity.

The results of this last simulation (the one characterized by the finest used mesh) are summarized in Table 6 together with those obtained with the starting coarse mesh, for comparison. Actually, the mesh used during the convergence procedure underestimates the performance of the turbine. Furthermore, from Table 6 it is possible to observe that the turbine designed with the proposed tool allowed us to reach the desired results. Indeed, under the design working conditions in terms of flow coefficient ($\phi = 0.31$) and expansion ratio ($\beta_e = 1.144$), the turbine gives a loading factor, $\Psi = 1.800$, more than twice the minimum expected value of 0.8262 (Table 6).

**Table 6.** Mesh sensitivity analysis.

|  | Fine Mesh | Coarse Mesh | Difference [%] |
|---|---|---|---|
| $\beta_e$ | 1.144 | 1.181 | 3.13 |
| $\phi$ | 0.3100 | 0.3254 | 4.73 |
| $\Psi$ | 1.800 | 2.008 | 10.36 |
| $\eta_{tt}$ | 0.5800 | 0.6752 | 14.10 |
| $\phi_C$ | 0.9218 | 0.9585 | 3.83 |
| $\phi_N$ | 0.9306 | 0.9399 | 0.99 |
| $\phi_R$ | 0.8134 | 0.8419 | 3.39 |
| $\phi_D$ | 0.1890 | 0.2070 | 8.70 |
| $\eta_v$ | 0.9040 | 0.9169 | 1.41 |
| $k_w$ | $5.176 \times 10^{-4}$ | $1.201 \times 10^{-4}$ | 330.97 |

Figure 12, which represents the contours of the expansion ratio at the hub of the stator and rotor blade passages, confirms the impulse behaviour of the turbine: the entire pressure drop available is elaborated by the stator, whilst across the rotor the flow experiences only a minimum pressure drop essentially caused by distributed losses.

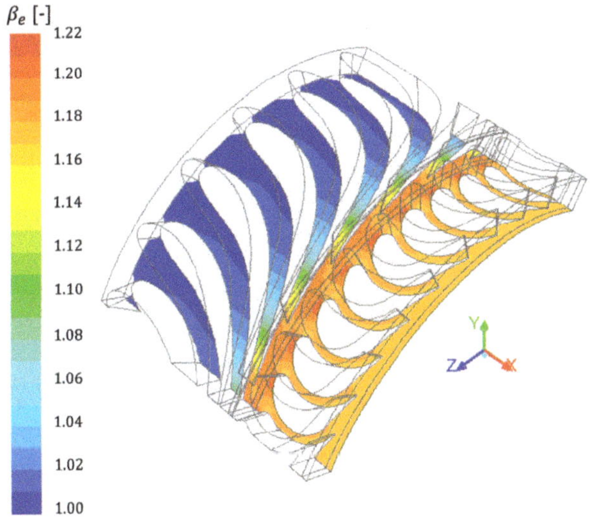


**Figure 12.** Contours of expansion ratio at the hub of stator and rotor blade passages.

This behaviour is also confirmed by Figure 13, which reports the absolute and the relative Mach number contours for the stator and the rotor, respectively. Again, the absolute Mach number in the stator evidences the flow acceleration in the stator whereas the relative Mach number shows only minimum variations especially after the leading edge. Due to the impulse behaviour of the rotor, actually the work extraction is mainly due to the change of the flow direction rather than its acceleration, which is negligible.

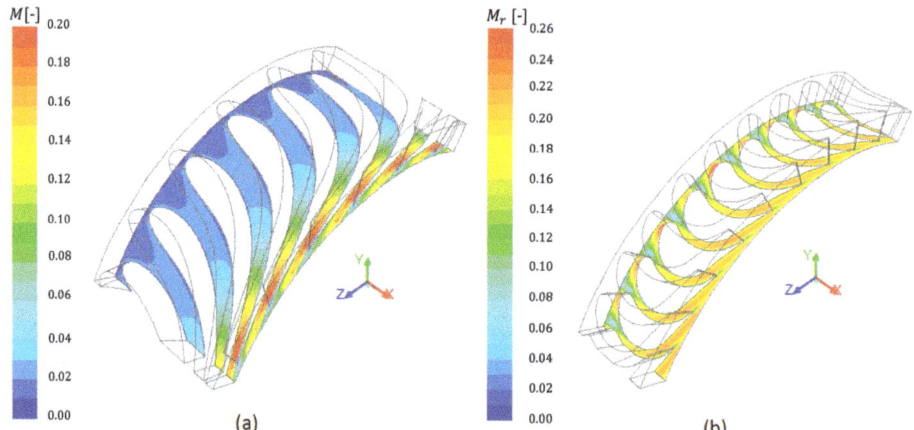

**Figure 13.** (**a**) Contours of Mach number at stator hub (**b**) Contours of relative Mach number at rotor hub.

The Mach number distribution is also analysed around the blade surfaces, in order to verify the presence of high pressure gradient zones which can determine flow separation (see Figure 14). The Mach number distribution is regular around the stator blades, whilst the relative Mach number distribution around the rotor blades shows a high gradient in the leading edge zone.

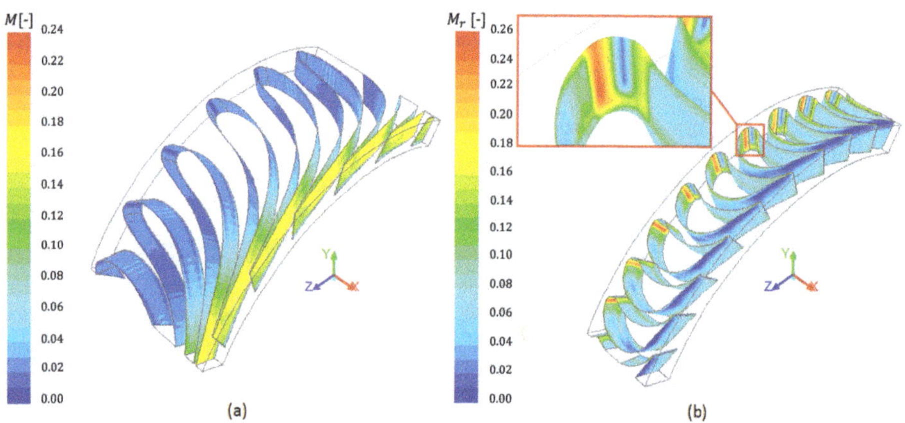

**Figure 14.** (**a**) Mach number distribution around stator blades (**b**) Relative Mach distribution around rotor blades.

The ratio between the axial and the tangential velocities at the mid and tip of the rotor is shown in Figure 15. The signature of the flow separation is the negative value of the axial velocity registered at the blade leading edge. This phenomenon takes place because of the particular design of the rotor blade. Indeed, in the connection of the leading edge arc with the pressure side and the suction side, only the continuity of the first derivative is preserved, whereas the continuity of the second derivative is not guaranteed.

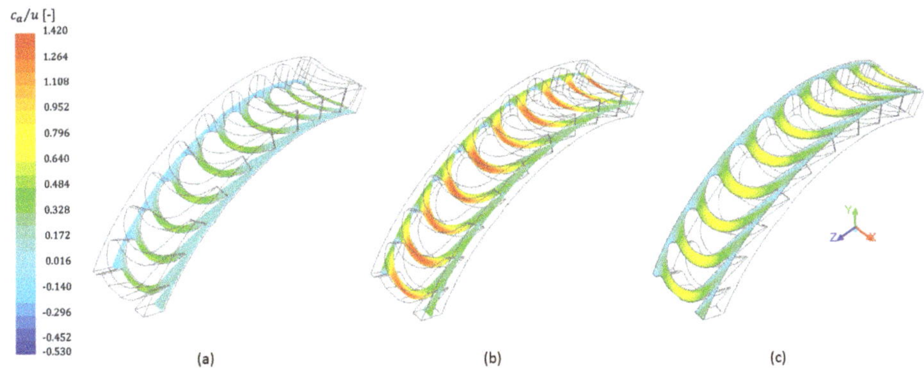

**Figure 15.** Contours of normalized axial velocity at (**a**) hub, (**b**) mid, and (**c**) tip of rotor blade passages.

Another issue which influences the global efficiency of the designed machine is represented by the leakage through the clearance between the rotor disk and the casing. In the preliminary steps, in order to minimize this problem, the inlet section for the leakage mass flow (Figure 16) was kept as lower as possible, compatibly with technological issues relative to manufacturing. Nevertheless, a volumetric efficiency exists and influences the performance of the machine. In Figure 17 it is shown how the streamlines just under the tip of the stator blades bypass the rotor blades flowing through the tip clearance between the rotor disk and the casing, determining a device work loss.

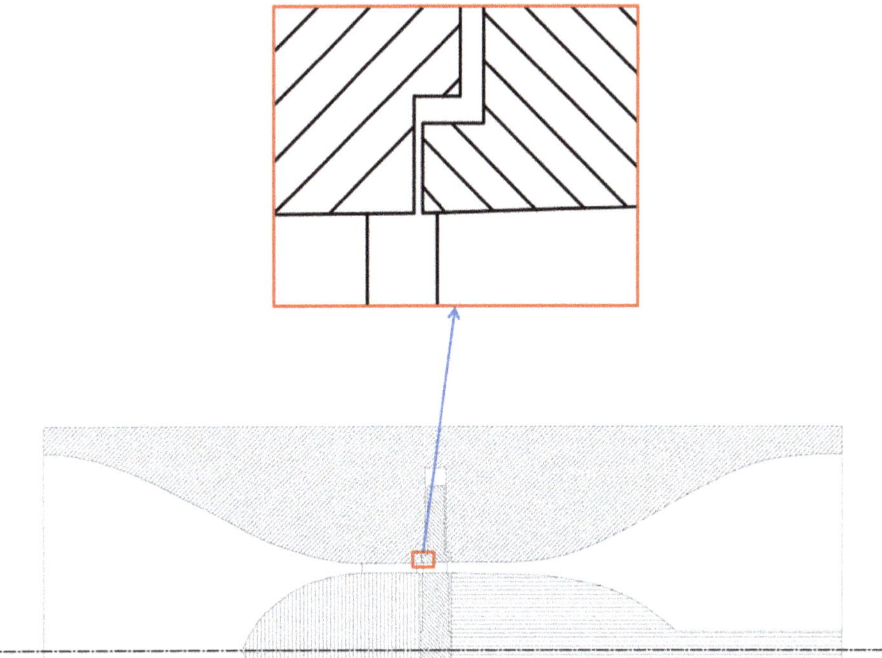

**Figure 16.** Cross section of the designed machine with a detail on the inlet section of the leakage mass flow.

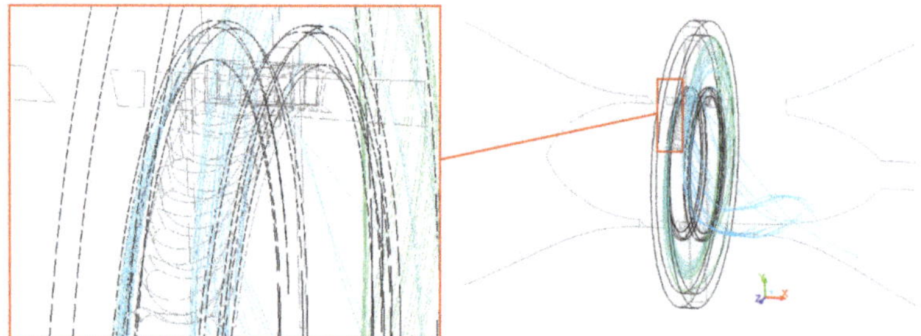

**Figure 17.** Visualization of streamlines through the clearance between rotor disk and casing.

Also the diffuser seems to have a bad influence on the global efficiency of the machine according to its loss coefficient, $\phi_D = 0.1890$. Actually, as shown in Figure 18, the streamlines are very irregular pointing out the kinetic energy losses. This is due to axysimmetric profile of the diffuser, especially when the available cross section area rapidly increases. Indeed, a smoother passage area increase is expected in order to improve the diffuser efficiency, but the available axial size constraint does not allow to modify the diffuser geometry. After this detailed insight in the fluid dynamic behaviour of the designed machines, it is evident that margins exist for a further improvement, but this will be subject of a next work.

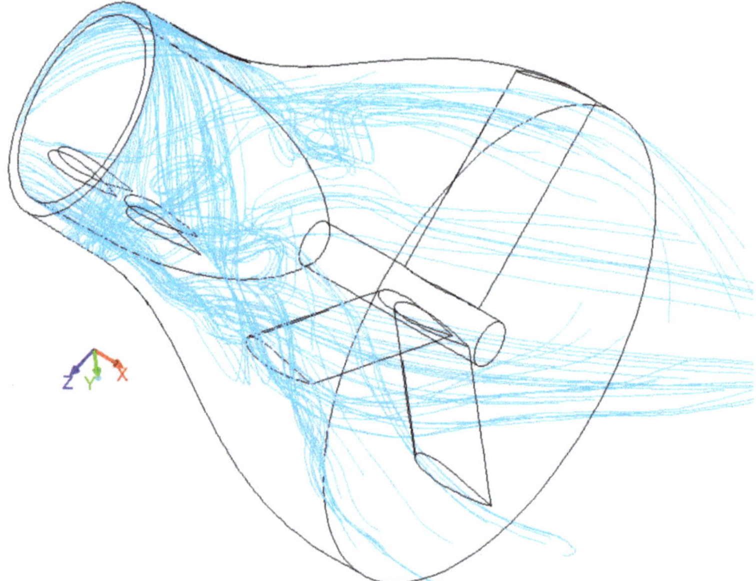

**Figure 18.** Visualization of the streamlines through the diffuser.

## 5. Conclusions

For a sustainable development of industrial processes, the energy recovery can be an effective strategy. Under this framework, a design procedure for turboexpanders has been developed, which is characterized by a high level of automatization and a fast convergence toward the optimal solution. In order to validate this procedure, this was applied to an innovative turboexpander characterized by a high specific speed (with an embedded permanent magnet direct drive generator) thought for substituting a Pressure Reduction Valve (PRV) in a pipeline. The design requirements were specified in terms of maximum allowable expansion ratio ($\beta_e$) and flow coefficient ($\phi$), while achieving the highest possible turbine loading factor ($\Psi$). The design procedure is iterative and based on the following four steps:

1.  a one-dimensional calculation aimed at the evaluation of the main geometric (blades angles and heights) and operating (output power, rotational speed and efficiency) parameters of the turbine;
2.  an optimization algorithm based on a two-dimensional viscous/inviscid interaction technique in order to evaluate both stator and rotor blade shapes;
3.  set up of a script file, based on the geometry resulting from the previous two steps, which is read by the mesh generator software and which allows the automatic generation of the mesh;
4.  a three-dimensional CFD simulation performed in order to verify the design in terms of performance and loss coefficients.

The need to adopt an iterative procedure lies in the fact that when the design process starts, the loss coefficients to be used in the one-dimensional calculation are not known, hence a hypothesis on their values has to be done. If the supposed values of the loss coefficients are equal to the ones resulting from the 3D CFD simulation, convergence is reached, otherwise the one-dimensional calculation is updated and the iterative procedure continues. By means of a test case, an assessment of the CFD model as been performed. In particular, the attention was focused on the mesh density to be used during the iterative procedure, which should be a right compromise between accuracy and computational costs. Furthermore, the possible models for dealing with rotor/stator interaction have been investigated and then, the Multiple Reference Frame model has been selected to be used in the design methodology. The iterative procedure allowed us to reach a satisfactory turbine design after only three iterations. When the design was completed, a fine mesh has been generated in order to obtain the final results.

Furthermore, the results show that the final turboexpander design is characterized by a loading factor ($\Psi = 1.800$), which is more than twice the minimum expected value (0.8262). Finally, an insight into the flow field has been carried out by investigating the contours of the Mach number and the streamlines across the turbine stage. The former points out small regions close to the rotor blade leading edges where flow separation occurs; the latter evidences that the diffuser profile can be improved to minimize the kinetic energy losses.

**Author Contributions:** G.M., M.T., P.O., F.F., M.S., and T.C. wrote the paper; G.M. performed CFD analysis, got the results and draft the paper; P.O., S.M.C., and M.T. organized the layout of the paper; T.C. and M.S. performed an analysis of the energetic and environmental impact of the proposed technology; G.M. and M.T. arranged the iterative procedure; S.M.C., and M.T. co-operated with G.M. on the theoretical aspects concerning this topic; M.T. and S.M.C. supervised the work. All authors have read and agreed to the published version of the manuscript.

**Funding:** This research received no external funding.

**Acknowledgments:** This work was financially supported by MIUR-"Ministero dell'Istruzione dell'Università e della Ricerca" under the Programme "Department of Excellence" Legge 232/2016 (Grant No. CUP-D94I18000260001).

**Conflicts of Interest:** The authors declare no conflict of interest.

## Abbreviations

| | |
|---|---|
| UN | United Nations |
| OpEx | Operating Expenditure |
| CapEx | Capital Expenditure |
| CFD | Computational Fluid Dynamics |
| SDG | Sustainable Development Goals |
| IEA | International Energy Agency |
| GRS | Gas Regulation Stations |
| DPP | Discounted Payback Period |
| CGS | City Gas Station |

| | |
|---|---|
| LNG | Liquefied Natural Gas |
| PaT | Pump as Turbine |
| RANS | Reynolds-averaged Navier-Stokes equations |
| PRV | Pressure Reduction Valve |
| MRF | Multiple Reference Frame model |
| MPM | Mixing Plane Model |
| SMM | Sliding Mesh Model |
| ITC | Institute of Thermal Engineering |
| MSEP | Mean Square Error Percentage |
| I.V. | Initial Value |
| C.V. | Computed Value |
| S.V. | Supposed Value |

## Nomenclature

| | | |
|---|---|---|
| $c_1$ | Nozzle outlet absolute speed in real case | $(m/s)$ |
| $c_{1,is}$ | Nozzle outlet absolute speed in ideal case | $(m/s)$ |
| $c_2$ | Stator outlet absolute speed in real case | $(m/s)$ |
| $c_{2,is}$ | Stator outlet absolute speed in ideal case | $(m/s)$ |
| $c_4$ | Diffuser outlet absolute speed in real case | $(m/s)$ |
| $c_{4,is}$ | Diffuser outlet absolute speed in ideal case | $(m/s)$ |
| $c_a$ | Axial speed | $(m/s)$ |
| $k_w = \dfrac{P_w}{\rho_2 u^3}$ | Windage power loss coefficient | $(-)$ |
| $p_0^0$ | Turbine inlet total pressure | $(Pa)$ |
| $p_4^0$ | Turbine outlet total pressure | $(Pa)$ |
| $u$ | Tangential speed | $(m/s)$ |
| $w_3$ | Rotor outlet relative speed in real case | $(m/s)$ |
| $w_{3,is}$ | Rotor outlet relative speed in ideal case | $(m/s)$ |
| $D_m$ | Mean diameter | $(m)$ |
| $D_S = \dfrac{D H_{ad}^{1/4}}{\sqrt{V_3}}$ | Specific diameter | $(ft)$ |
| $G$ | Mass flow rate | $(kg/s)$ |
| $G_{lkg}$ | Leakage mass flow rate | $(kg/s)$ |
| $H_1$ | Blade height at stator inlet | $(m)$ |
| $H_2$ | Blade height at stator outlet/rotor inlet | $(m)$ |
| $H_3$ | Blade height at rotor outlet | $(m)$ |
| $H_{ad}$ | Adiabatic Head | $(ft)$ |
| $M$ | Mach number | $(-)$ |
| $M_r$ | Relative Mach number | $(-)$ |
| $N$ | Rotational speed | $(rpm)$ |
| $N_S = \dfrac{N \sqrt{V_3}}{H_{ad}^{3/4}}$ | Specific speed | $(rpm)$ |
| $P_w$ | Windage power loss | $(W)$ |
| $R$ | Gas constant | $(J/(kg \cdot K))$ |
| $T_0^0$ | Turbine inlet total temperature | $(K)$ |
| $V_3$ | Volumetric flow rate at rotor outlet | $(ft^3/s)$ |
| $\beta_e = \dfrac{p_0^0}{p_4^0}$ | Pressure ratio | $(-)$ |
| $\zeta$ | Generic loss coefficient | $(-)$ |
| $\eta_v = 1 - \dfrac{G_{lkg}}{G}$ | Volumetric efficiency | $(-)$ |
| $\eta_{tt} = \dfrac{\Delta h^0}{\Delta h_{is}^0}$ | Total-to-total efficiency | $(-)$ |

| | | |
|---|---|---|
| $\rho_2$ | Density at rotor inlet | $(kg/m^3)$ |
| $\phi = \dfrac{c_a}{u}$ | Flow coefficient | $(-)$ |
| $\varphi_C = \dfrac{c_1}{c_{1,is}}$ | Nozzle loss coefficient | $(-)$ |
| $\varphi_D = \dfrac{c_4}{c_{4,is}}$ | Diffuser loss coefficient | $(-)$ |
| $\varphi_N = \dfrac{c_2}{c_{2,is}}$ | Stator loss coefficient | $(-)$ |
| $\varphi_R = \dfrac{w_3}{w_{3,is}}$ | Rotor loss coefficient | $(-)$ |
| $\omega$ | Angular speed | $(rad/s)$ |
| $\Gamma = \dfrac{G\sqrt{RT_0^0}}{p_0^0 D_m}$ | Non-dimensional mass flow rate | $(-)$ |
| $\Delta h^0$ | Actual work output | $(J/kg)$ |
| $\Delta h_{is}^0$ | Ideal work output operating with the same back pressure | $(J/kg)$ |
| $\Psi$ | $= \dfrac{\Delta h^0}{u^2}$ Loading factor | $(-)$ |

## References


1. International Energy Agency. Energy Efficiency 2019. Available online: https://www.iea.org/reports/energy-efficiency-2019 (accessed on 16 July 2020).
2. Kuczyński, S.; Laciak, M.; Olijnyk, A.; Szurlej, A.; Wlodek, T. Techno-Economic Assessment of Turboexpander Application at Natural Gas Regulation Stations. *Energies* **2019**, *12*, 755.
3. Borelli, D.; Devia, F.; Cascio, E. L.; Schenone, C. Energy recovery from natural gas pressure reduction stations: Integration with low temperature heat sources. *Energy Convers. Manag.* **2018**, *159*, 274 – 283.
4. Fornarelli, F.; Lippolis, A.; Oresta, P. Buoyancy Effect on the Flow Pattern and the Thermal Performance of an Array of Circular Cylinders. *J. Heat Transf.* **2016**, *139*, 022501. [CrossRef]
5. Oresta, P.; Prosperetti A. Effects of particle settling on Rayleigh-Bénard convection *Phys. Rev. E* **2013**, *87*, 063014. [CrossRef]
6. Schmidt, L.E.; Oresta, P.; Toschi, F.; Verzicco, R.; Lohse, D.; Prosperetti, A. Modification of turbulence in Rayleigh-Bénard convection by phase change. *New J. Phys.* **2011**, *13*, 025002.
7. Lo Cascio, E.; Borelli, D.; Devia, F.; Schenone, C. Key performance indicators for integrated natural gas pressure reduction stations with energy recovery. *Convers. Manag.* **2018**, *164*, 219–229.
8. Fornarelli, F.; Ceglie, V.; Fortunato, B.; Camporeale, S.M.; Torresi, M.; Oresta, P.; Miliozzi, A. Numerical simulation of a complete charging-discharging phase of a shell and tube thermal energy storage with phase change material. *Energy Procedia* **2017**, *126*, 501–508. [CrossRef]
9. Fornarelli, F.; Camporeale, S.M.; Fortunato, B. Simplified theoretical model to predict the melting time of a shell-and-tube LHTES. *Appl. Therm. Eng.* **2019**, *153*, 51–57. [CrossRef]
10. Lo Cascio, E.; Ma, Z.; Schenone, C. Performance assessment of a novel pressure reduction station equipped with parabolic though solar collectors. *Renew. Energy* **2018**, *164*, 177–187.
11. Zabihi, A.; Taghizadeh, M. Feasibility study on energy recovery at Sari-Akand city gate station using turboexpander. *J. Nat. Gas Sci. Eng.* **2016**, *35*, 152–159.
12. Neseli, M. A.; Ozgener, O.; Ozgener, L. Energy and exergy analysis of electricity generation from natural gas pressure reducing stations. *Energy Convers. Manag.* **2015**, *93*, 109–120.
13. Howard, C.; Oosthuizen, P.; Peppley, B. An investigation of the performance of a hybrid turboexpander-fuel cell system for power recovery at natural gas pressure reduction stations. *Appl. Therm. Eng.* **2011**, *31*, 2165–2170.
14. Poživil, J. Use of Expansion Turbines in Natural Gas Pressure Reduction Stations. *Acta Montan. Slovaca* **2004**, *9*, 258–260.
15. Golchoobian, H.; Taheri, M.H.; Saedodin, S. Thermodynamic analysis of turboexpander and gas turbine hybrid system for gas pressure reduction station of a power plant. *Case Stud. Therm. Eng.* **2019**, *14*, 100488.
16. Cao, L.; Liu, J.; Xu, X. Robustness analysis of the mixed refrigerant composition employed in the single mixed refrigerant (SMR) liquefied natural gas (LNG) process. *Appl. Therm. Eng.* **2016**, *93*, 1155–1163.

17. Ancona, M.; Bianchi, M.; Branchini, L.; Pascale, A.D.; Melino, F. Performance increase of a small-scale liquefied natural gas production process by means of turbo-expander. *Energy Procedia* **2017**, *105*, 4859–4865.
18. Taher, M.; Meher-Homji, C. Cryogenic turboexpanders in lng liquefaction applications. In Proceedings of the ASME Turbo Expo 2016: Turbomachinery Technical Conference and Exposition, Seoul, South Korea, 13–17 June 2016.
19. Shanju, Y.; Chen, S.; Chen, X.; Zhang, X.; Hou, Y. Study on the coupling performance of a turboexpander compressor applied in cryogenic reverse brayton air refrigerator. *Energy Convers. Manag.* **2016**, *122*, 386–399.
20. Zhang, Y.; Li, Q.; Wu, J.; Li, Q.; Lu, W.; Xiong, L.; Liu, L.; Xu, X.; Sun, L.; Sun, Y.; et al. Performance analysis of a large-scale helium brayton cryo-refrigerator with static gas bearing turboexpander. *Energy Convers. Manag.* **2015**, *90*, 207–217.
21. Hou, Y.; Yang, S.; Chen, X.; Chen, S.; Lai, T. Study on the matching performance of a low temperature reverse brayton air refrigerator. *Energy Convers. Manag.* **2015**, *89*, 339–348.
22. Qyyum, M.A.; Ali, W.; Long, N.V.D.; Khan, M.S.; Moonyong Lee, M.; Energy efficiency enhancement of a single mixed refrigerant LNG process using a novel hydraulic turbine. *Energy* **2018**, *144*, 968–976.
23. Stefanizzi, M.; Torresi, M.; Fornarelli, F.; Camporeale, S.M. Performance prediction model of multistage centrifugal Pumps used as Turbines with Two-Phase Flow. *Energy Procedia* **2018**, *148*, 408–415. [CrossRef]
24. Laux, C.H. Rückwärtslaufende mehrstufige Pumpen als Energierćkgewinnungs-Turbinens in Ölversorgunssyste. *Technische rundschau Sulzer* **1980**, *2*, 61–80.
25. Apfelbacher, R.; Etzold, F. Energy-Saving, Shock-Free Throttling with the Aid of a Reverse Running Centrifugal Pump. *KSB Technische Berichte* **1988**, *24e*, 33–41.
26. Hamkins, C.; Jeske, H.O.; Apfelbacher, R.; Schuster, O. Pumps as energy recovery turbines with two phase flow. In Proceedings of the ASME Pumping Machinery Symp, San Diego, CA, USA, 9–12 July 1989; pp. 73–81.
27. Gopalakrishnan, S. Power recovery turbines for the process industry. In Proceedings of the 3rd International Pump Symposium, Houston, TX, USA, 20–22 May 1986.
28. Bolliger, W. Pumpen und Turbinen in Anlagen für umgekehrte Osmos. *Technische Rundschau Sulzer* **1984**, *3*, 29–32.
29. Bolliger, W.; Menin, J.A. Pumpen als Turbinen erfolgreich. *Technische Rundschau Sulzer* **1997**, *2*, 27–29.
30. Slocum, A.H.; Haji, M.H.; Trimble A.Z.; Ferrara, M.; Ghaemsaidi, S.J. Integrated Pumped Hydro Reverse Osmosis systems. *Sustain. Energy Technol. Assess.* **2016**, *18*, 80–99.
31. Capurso, T.; Bergamini, L.; Camporeale, S.M.; Fortunato, B.; Torresi M. CFD analysis of the performance of a novel impeller for a double suction centrifugal pump working as a turbine. In Proceedings of the 13th European Turbomachinery Conference on Turbomachinery Fluid Dynamics and Thermodynamics, ETC, Lausanne, Switzerland 8–12 April 2019.
32. García-Gutiérrez, A.; Martínez-Estrella, J.I.; Ovando-Castelar, R.; Canchola-Félix, I.; Gutiérrez-Espericueta, S.; Vázquez-Sandoval, A.; Rosales-López, C Energy Recovery Using Turboexpanders in Production Wells and Gathering Systems with High Pressures. *Geotherm. Resour. Counc. Annu. Meet.* 2013, *37*, 703–707.
33. GE Oil & Gas. Turboexpander-Generators,2012. directindustry.com. Available online: https://pdf.directindustry.com/pdf/ge-steam-turbines/turboexpander-generators/116289-573764.html (accessed on 16 July 2020)
34. GE Oil & Gas.Turboexpander-Compressors—Increased Efficiency for Refrigeration Applications,2010. directindustry.com. Available online: https://pdf.directindustry.com/pdf/ge-steam-turbines/turboexpander-compressors/116289-573763.html (accessed on 16 July 2020)
35. Atlas Copco, 2017. Driving expaner technology.
36. Fiaschi, D.; Manfrida, G.; Maraschiello, F. Design and performance prediction of radial orc turboexpanders. *Appl. Energy* **2015**, *138*, 517–532.
37. Talluri, L.; Lombardi, G. Simulation and design tool for orc axial turbine stage. *Energy Procedia* **2017**, *129*, 277–284.
38. Ying, Q.; Zhuge, W.; Zhang, Y.; Zhang, L. Design optimization of a small scale high expansion ratio organic vapour turbo expander for automotive application. *Energy Procedia* **2017**, *129*, 1133–1140.
39. Sam, A.A.; Ghosh, P. Flow field analysis of high-speed helium turboexpander for cryogenic refrigeration and liquefaction cycles. *Cryogenics* **2017**, *82*, 1–14.

40. Alshammari, F.; Karvountzis-Kontakiotis, A.; Pesiridis, A.; Minton, T.; Radial expander design for an engine organic rankine cycle waste heat recovery system. *Energy Procedia* **2017**, *129*, 285–292.

41. Dong, B.; Xu, G.; Luo, X.; Zhuang, L.; Quan, Y. Analysis of the supercritical organic rankine cycle and the radial turbine design for high temperature applications. *Appl. Therm. Eng.* **2017**, *123*, 1523–1530.

42. Fiaschi, D.; Innocenti, G.; Manfrida, G.; Maraschiello, F.; Design of micro radial turboexpanders for orc power cycles: From 0d to 3d. *Appl. Therm. Eng.* **2016**, *99*, 402–410.

43. Sam, A.A.; Mondal, J.; Ghosh, P. Effect of rotation on the flow behaviour in a high-speed cryogenic microturbine used in helium applications. *Int. J. Refrig.* **2017**, *81*, 111–122.

44. Li, L.; Ge, Y.; Luo, X.; Tassou, S. Experimental analysis and comparison between co2 transcritical power cycles and r245fa organic rankine cycles for low-grade heat power generations. *Appl. Therm. Eng.* **2018**, *136*, 708–717.

45. Li, L.; Ge, Y.; Luo, X.; Tassou, S. An experimental investigation on a recuperative organic rankine cycle (orc) system for electric power generation with low-grade thermal energy. *Energy Procedia* **2017**, *142*, 1528–1533. Proceedings of the 9th International Conference on Applied Energy.

46. Li, L.; Ge, Y.; Luo, X.; Tassou, S.A. Experimental investigation on power generation with low grade waste heat and co2 transcritical power cycle. *Energy Procedia* **2017**, *123*, 297–304.

47. Morgese, G.; Torresi, M.; Fortunato, B.; Camporeale, S.M. Design of an Axial Impulse Turbine for Enthalpy Drop Recovery. In ASME Turbo Expo 2014: Turbine Technical Conference and Exposition, Düsseldorf, Germany, 16–20 June 2014. [CrossRef]

48. Morgese, G.; Torresi, M.; Fortunato, B.; Camporeale, S.M. Optimized aerodynamic design of axial turbines for waste energy recovery. *Energy Procedia.* **2015**, *82*, 194–200. [CrossRef]

49. Focchi, M.; Guglielmino, E.; Pane, G.; Cordasco, S.; Tacchino, C.; Caldwell, D. Device for Generating Electric Power From A Source Of Air Or Other Gas Or Fluid Under Pressure. WO 2012/004738, 12 January 2012.

50. Kenneth, E.; Nichols, P. *How to Select Turbomachinery for Your Application*; Barber-Nichols Inc.: Arvada, CO, USA, 2012; pp. 5–6.

51. Ainley, D.; Mathieson, G. *A Method of Performance Estimation for Axial-Flow Turbines*; Technical Report; Aeronautical Research Council, ARC: London, UK, 1957.

52. Dunham, J.; Came, P. *Improvements to the Ainley-Mathieson Method of Turbine Performance Prediction*; 1970; pp. 252–256.

53. Craig, H.R.M.; Cox, H.J.A. Performance estimation of axial flow turbines. *Proc. Inst. Mech. Eng.* **1970**, *185*, 407–424.

54. Kacker, S.; Okapuu, U. A mean line prediction method for axial flow turbine efficiency. *J. Eng. Power.* **1982**, *104*, 111—119.

55. Trans Austria Gasleitung GmbH. Available online: www.taggmbh.at (accessed on 16 July 2020).

56. ISPRA. Fattori di Emissione Atmosferica di $CO_2$ e Sviluppo Delle Fonti Rinnovabili Nel Settore Elettrico,2015. Istituto Superiore per la Protezione e la Ricerca Ambientale. Available online: https://www.isprambiente.gov.it/it/pubblicazioni/rapporti/fattori-di-emissione-atmosferica-di-co2-e-sviluppo-delle-fonti-rinnovabili-nel-settore-elettrico (accessed on 16 July 2020).

57. Dixon, S. *Fluid mechanics and thermodynamics of turbomachinery*, 4th ed.; Butterworth-Heinemann: Oxford, UK, 1998.

58. Coenen, E.G.M.; Veldman, A.E.P.; Patrianakos, G. .Viscous-inviscid interaction method for wing calculations. In Proceedings of the European Congress on Computational Methods in Applied Sciences and Engineering, ECCOMAS 2000, Barcelona, Spain, 11–14 September 2000.

59. Lewis, R. *Vortex Element Methods for Fluid Dynamic Analysis of Engineering Systems*; Cambridge Engine Technology Series; Cambridge University Press: Cambridge, UK, 2005.

60. Moran, J. *An Introduction to Theoretical and Computational Aerodynamics*; Dover Books on Aeronautical Engineering; Dover Publications: Mineola, NY, USA, 2013.

61. Luo, J.; Issa, R.; Gosman, A. Prediction of impeller-induced flows in mixing vessels using multiple frames of reference. *Inst. Chem. Eng. Symp. Ser.* **1994**, *136*, 549–556.

62. Lampart, P.; Swirydczuk, J.; Gardzilewicz, A. On the prediction of flow patterns and losses in hp axial turbine stages using 3d rans solver and two turbulence models. *Task Q.* **2001**, *5*, 191–206.

63. Wiechowski, S. *Results of Investigations of Annular Cascades TK8, TK9 and Models TK8-TW3, TK9-TW3 Report*; Institute of Thermal Engineering: Łódź, Poland, 1988.

*Energies* **2020**, *13*, 3669

64. Wilcox, D. *Turbulence Modeling for CFD*; DCW Industries, Incorporated: La Canada, CA, USA, 1993.

65. Wasserman, S. Choosing the Right Turbulence Model for Your Cfd Simulation. Engineering.com. 2016. Available online: https://www.engineering.com/DesignSoftware/DesignSoftwareArticles/ArticleID/13743/Choosing-the-Right-Turbulence-Model-for-Your-CFD-Simulation.aspx (accessed on 16 July 2020).

66. Li, Y. Compuctional fl'uid dyn'amics(cfd) stu'dy on f'ree surf'ace a'nti-Roll Tank and Experimental Validation. Master's Thesis, Elesund University College, Ålesund, Norway, 2015.

67. Chima, R. A k-Omega Turbulence Model for Quasi-Three-Dimensional Turbomachinery Flows. In 34th Aerospace Sciences Meeting, Reno, NV, USA, 15–18 January 1996.

68. Djouimaa, S.; Messaoudi, L.; Giel, P.W. Transonic turbine blade loading calculations using different turbulence models effects of reflecting and non-reflecting boundary conditions. *Appl. Therm. Eng.* **2007**, *27*, 779–787.

© 2020 by the authors. Licensee MDPI, Basel, Switzerland. This article is an open access article distributed under the terms and conditions of the Creative Commons Attribution (CC BY) license (http://creativecommons.org/licenses/by/4.0/).

MDPI
St. Alban-Anlage 66
4052 Basel
Switzerland
Tel. +41 61 683 77 34
Fax +41 61 302 89 18
www.mdpi.com

*Energies* Editorial Office
E-mail: energies@mdpi.com
www.mdpi.com/journal/energies

www.ingramcontent.com/pod-product-compliance
Lightning Source LLC
LaVergne TN
LVHW070124170726
843515LV00007B/1744